APPLIED INTEGER
PROGRAMMING

APPLIED INTEGER PROGRAMMING

Modeling and Solution

DER-SAN CHEN
The University of Alabama

ROBERT G. BATSON
The University of Alabama

YU DANG
Quickparts.com, Inc.

WILEY

A JOHN WILEY & SONS, INC., PUBLICATION

Published by John Wiley & Sons, Inc., Hoboken, New Jersey
Published simultaneously in Canada

For general information on our other products and services or for technical support, please contact our Customer Care Department within the United States at (800) 762-2974, outside the United States at (317) 572-3993 or fax (317) 572-4002.

Wiley also publishes its books in a variety of electronic formats. Some content that appears in print may not be available in electronic formats. For more information about Wiley products, visit our web site at www.wiley.com.

Library of Congress Cataloging-in-Publication Data:

Chen, Der-San, 1940–
    Applied integer programming : modeling and simulation / Der-San Chen, Robert G. Batson, Yu Dang.
        p. cm.
    Includes bibliographical references and index.
    ISBN 978-0-470-37306-4 (cloth)
  1. Integer programming. I. Batson, Robert G., 1950- II. Dang, Yu., 1977–
III. Title.

    T57.74.C454 2010
    519.7'7–dc22

                                    2009025987

10 9 8 7 6 5 4 3 2

Der-San dedicates this book to his blessed family,
Hannah, Suzy, and Benjamin

Bob dedicates this book to his wife Jane
and to his parents

Yu dedicates this book to her husband Qiu Fang
and to her parents

CONTENTS

PREFACE

Integer programming (IP) is a class of constrained optimization problems in which some or all variables are integers and all mathematical functions in the objective and constraints are conventionally linear. In the professional community, the acronym MIP (mixed integer programming) is more often used, because many real-world problems involve a mix of continuous and integer-valued decision variables.

PURPOSE, SCOPE, AND AUDIENCE

We set out to write an easy-to-read, applied textbook for students enrolled in multiple academic disciplines and for professionals. In academia, the textbook is intended for graduate- and senior-level students of industrial engineering, operations research, management science, computer science, and applied mathematics. Other disciplines (such as operations management, supply chain management, logistics management, transportation engineering) that need a course in applied optimization would find this text a relevant option. Because of its application emphasis, this textbook can also be used as a reference book by practitioners whose jobs require modeling and solving real-world optimization problems using commercial integer programming software, as well as MIP software developers and analysts.

Instructors who are preparing students for careers in the *practice* of operations research and management science will find this book appealing. However, because of its application emphasis rather than mathematical rigor, this book is not suitable for instructors who are looking for *theoretical* underpinnings, such as mathematicians who are selecting a text for a course in discrete or combinatorial optimization.

Instructors of operations research and management science will find this text a natural continuation of and complement to well-known introductory textbooks in operations research and management science. As the subtitle indicates, the major approach of this book is modeling and solution. Modeling is emphasized because the insertion of integer variables in a linear program (LP) enables much more rich and realistic representations of decision situations. Both in the examples and exercises, students develop advanced modeling skills. Integer and linear programming terminology commonly referenced in commercial MIP solution software is covered in the text. This text provides extensive coverage of modeling techniques and solution methods with algorithms that are implemented in today's commercial software.

TOPIC COVERAGE, LEVEL OF PRESENTATION, AND IMPORTANT FEATURES

This text is organized into three parts—Part I: Modeling, Part II: Review of Linear Programming and Network Flows, and Part III: Solutions. Part I (Chapters 1–6) includes areas of successful integer programming applications, systematic modeling procedure, types of integer programming models, transformation of non-IP models, automatic preprocessing for better formulation, and an introduction to combinatorial optimization. Part II (Chapters 7–10) reviews algebraic-geometric concepts and solution methods related to LP and network flows that are needed for understanding IP. Part III (Chapters 11–15) describes various solution approaches for large-scale IP and combinatorial optimization problems in addition to fundamentals of typical software systems. Solution approaches include classical, branch-and-cut, branch-and-price, primal heuristic, and Lagrangian relaxation. In Chapter 15, three popular modeling languages and one solver are introduced. Answers to selected problems from each chapter appear in an appendix. A more detailed preview of the text may be found in Section 1.5.

As an application-oriented text, we aim to teach students about the art and science of mathematical modeling for the collection of problems that fit the MIP framework and about the algorithms and associated practices that enable those models to be solved most efficiently. To make algorithms easier to comprehend, this book places unique emphasis not only on *how* the algorithms work but also on *why* they work. To achieve these goals, reasoning and interpretation are exercised more often than rigorous mathematical proofs of theorems, which may be located in referenced articles. The authors have been very thorough in searching out and synthesizing various modeling and solution approaches that have appeared in disparate publications over the past 40 years. We want the student, who we envision will become a practitioner, to have a well-organized and comprehensive reference that eases the learning hurdles in integer programming and provides suggestions/guidelines for practice, once on the job.

The book makes liberal use of examples and flowcharts. Each new concept or algorithm mentioned is illustrated by a numerical example. The book contains over

100 figures, either flowcharts or simple geometric drawings, to illustrate the concepts in the text. A unique feature is that where possible, we use graphics to draw together diverse problems or approaches into a well-structured whole. Chapters typically have between 10 and 20 exercises; some are simple applications similar to examples, and some are more comprehensive and challenging, such as choosing the appropriate methods from several presented, and applying them collectively to a problem. This again simulates the authors' experiences as practitioners. There are a few problems that require the reader to investigate a topic further or to attempt to prove an assertion or provide a counterexample. In summary, we attempted to write an applied integer programming text that emphasizes modeling and solution, with due attention to fundamentals of theory and algorithms. We believe it meets an unfulfilled need for an IP text that links together problem solving, theory, algorithms, and commercial software.

SUGGESTIONS FOR COURSE USE

This book is self-contained, requiring only a background in linear or matrix algebra. The book offers a great deal of flexibility to university course instructors. The entire book can be used for a two-semester sequence in linear and integer programming, at the level of seniors or masters students in engineering, computer science, or business schools. For students already completing a full course in linear programming, Parts I and III can be used as a masters-level course entitled Integer Programming or Integer and Combinatorial Programming. For students with a partial knowledge of linear programming obtained in an undergraduate survey of operations research, a compromise is to cover sections of Part I, II, and III. For instance, one coauthor taught a masters-level course Integer Programming using Chapters 1–4, 7–10, 11, 12, and 15.

ACKNOWLEDGMENTS

Der-San Chen would like to express his gratitude to Dr. Stanley Zionts, SUNY at Buffalo, for leading him to the field of integer programming through dissertation guidance. Similarly, Robert Batson recognizes Dr. Wei-Shen Hsia, The University of Alabama (UA), for introducing him to optimization theory while supervising his dissertation research. Drs Chen and Batson are affiliated with the College of Engineering at UA; Dr. Yu Dang received her Ph.D. in Operations Management from UA, and is affiliated with Quickparts.com, Inc. Former UA graduate student Sriram Venkataraman assisted Dr. Batson with preparation of the appendix. The authors would like to recognize Dr. Tao Huang of the SAS Institute who graciously contributed Section 15.3 to the text and suggested improvements to the organization and contents of the rest of the text. Also, many thanks to our Wiley Editor, Susanne Steitz-Filler, who offered encouragement when the going got tough.

PART I

MODELING

1

INTRODUCTION

1.1 INTEGER PROGRAMMING

A *linear programming problem* (*LP*) is a class of the *mathematical programming problem*, a *constrained optimization problem*, in which we seek to find a set of values for continuous variables $(x_1, x_2, \ldots, x_n)$ that maximizes or minimizes a linear objective function z, while satisfying a set of linear constraints (a system of simultaneous linear equations and/or inequalities). Mathematically, an LP is expressed as follows:

$$(\text{LP}) \quad \text{Maximize} \quad z = \sum_j c_j x_j$$

$$\text{subject to} \quad \sum_j a_{ij} x_j \le b_i \quad (i = 1, 2, \ldots, m)$$

$$x_j \ge 0 \qquad (j = 1, 2, \ldots, n)$$

An *integer* (*linear*) *programming problem* (*IP*) is a linear programming problem in which at least one of the variables is restricted to integer values. In the past two decades, there has been an increasing use of an alternate term—*mixed integer programming problem* (*MIP*)—for LPs with integer restrictions on some or all of the variables. In this text, the terms IP and MIP may be used interchangeably unless there is a chance of confusion. For clarity, we shall use the term *pure integer*

Applied Integer Programming: Modeling and Solution, By Der-San Chen, Robert G. Batson, and Yu Dang
Copyright © 2010 John Wiley & Sons, Inc.

programming problem (or pure IP) to emphasize an IP whose variables are all restricted to be integer valued.

The term "programming" in this context means *planning* activities that consume *resources* and/or meet requirements, as expressed in the m constraints, not the other meaning—*coding computer programs*. The resources may include raw materials, machines, equipments, facilities, workforce, money, management, information technology, and so on. In the real world, these resources are usually limited and must be shared with several competing activities. Requirements may be implicitly or explicitly imposed. The objective of the LP/IP is to allocate the shared resources, and responsibility to meet requirements, to all competing activities in an optimal (best possible) manner.

The term "programming problem" is sometimes replaced by *program*, for short. Thus, an integer programming problem is also called an *integer program*, and so are *mixed integer program*, *pure integer program*, and so on. Mathematically, an MIP is defined as

$$(\text{MIP}) \quad \text{Maximize} \quad z = \sum_j c_j x_j + \sum_k d_k y_k$$

$$\text{subject to} \quad \sum_j a_{ij} x_j + \sum_k g_{ik} y_k \le b_i \quad (i = 1, 2, \ldots, m)$$

$$x_j \ge 0 \qquad\qquad\qquad (j = 1, 2, \ldots, n)$$

$$y_k = 0, 1, 2, \ldots \qquad\quad (k = 1, 2, \ldots, p)$$

Note that all input parameters (c_j, d_k, a_{ij}, g_{ik}, b_i) may be positive, negative, or zero. Using matrix notation, a mixed integer program may be expressed as

$$(\text{MIP}) \quad \text{Maximize} \quad z = \mathbf{c}^T \mathbf{x} + \mathbf{d}^T \mathbf{y}$$

$$\text{subject to} \quad \mathbf{Ax} + \mathbf{Gy} \le \mathbf{b}$$

$$\mathbf{x} \ge \mathbf{0}$$

$$\mathbf{y} \ge \mathbf{0} \text{ integer}$$

where m = number of constraints

n = number of continuous variables

p = number of integer variables

$\mathbf{c}^T = (c_j)$ is a row vector of n elements

$\mathbf{d}^T = (d_k)$ is a row vector of p elements

$\mathbf{A} = (a_{ij})$ is an $m \times n$ matrix

$\mathbf{G} = (g_{ik})$ is an $m \times p$ matrix

$\mathbf{b} = (b_i)$ is a column vector of m constants (or right-hand-side column, rhs)

$\mathbf{x} = (x_j)$ is a column vector of n continuous variables

$\mathbf{y} = (y_k)$ is a column vector of p integer variables

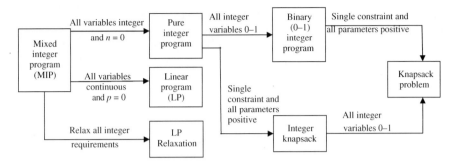

FIGURE 1.1 A simple classification of integer programs.

When $n = 0$, no continuous variables $\mathbf{x}$ are present and the MIP reduces to a *pure IP*. When $p = 0$, no integer-restricted variables $\mathbf{y}$ are present and the MIP reduces to a *linear program*. An LP is also obtained by relaxing (or ignoring) the integer requirements in a given MIP. Thus, the resulting LP is called the *LP relaxation* (of a given IP). Unlike the above-mentioned LP that contains only variables $\mathbf{x}$, the LP relaxation contains both $\mathbf{x}$ and $\mathbf{y}$ variables and treats $\mathbf{y}$ as a vector of continuous variables.

An integer program in which the integer variables are restricted to be 0 or 1 is called a 0–1 (binary) *integer program*, or *binary IP (BIP)*. A binary IP with a single $\leq$ linear constraint, whose objective function and constraint coefficients are all positive, is called a *knapsack (or backpack) problem*. An IP with a single constraint and all positive constraint coefficients is called an *integer knapsack program*, in which the values of an integer variable are not restricted to 0–1. In particular, an integer knapsack program is a knapsack program if all integer variables are restricted to be 0 or 1. Figure 1.1 depicts the relationships between various classes of MIPs under certain conditions. A box represents an IP class and an arrow represents the imposed condition(s) leading to a subclass from a class. There are many more subclasses than shown in this simple diagram, but the details of Figure 1.1 are adequate for this introductory chapter.

1.2 STANDARD VERSUS NONSTANDARD FORMS

Throughout this text, a mixed integer program will be said to be in *standard form* if (1) the objective function is maximized, (2) all the constraints are of $\leq$ form, (3) each integer variable is defined over consecutive integer numbers whose lower bound is 0 and upper bound infinity, and (4) each continuous variable is nonnegative with no finite upper bound.

Any MIP that does not conform to the conditions (1)–(4) is considered to be in *nonstandard form*, but may be converted to a standard one through simple mathematical manipulations. For ease of presentation, we shall use the standard form for the remainder of the text, except for special purposes. The following are various nonstandard forms that need to be converted:

- Minimization problem
- Inequality of $\geq$ form
- Equation (equality constraint)
- Unrestricted variable (continuous or integer)
- Variable with a positive or a negative lower bound
- Variable with a finite upper bound

If a given problem is a *minimization problem*, then it may be converted to an *"equivalent" maximization* problem. Two problems are considered equivalent if their optimal solutions are the same. Consider the given problem,

$$\text{Minimize} \quad z' = \sum_j c_j x_j + \sum_k d_k y_k$$

To convert to a standard form, we multiply the given objective function by -1 and change the minimization to the maximization as follows:

$$\text{Maximize} \quad -z' = -\sum_j c_j x_j - \sum_k d_k y_k$$

For example, we convert $\min z' = 3x_1 - 2x_2 + 4x_3$ to $\max z = -3x_1 + 2x_2 - 4x_3$, and the new objective value becomes $z = -z'$.

If a given inequality is in $\geq$ form, we then convert it to the standard $\leq$ form by multiplying the inequality by -1 and reversing the direction of the inequality sign. For example, the inequality $6x_1 - 5x_2 + 3x_3 \geq 10$ may be converted to $-6x_1 + 5x_2 - 3x_3 \leq -10$.

Converting an equation to the standard $\leq$ form requires two steps: (1) replace the equation by a pair of inequalities of opposite sense, and as before, (2) convert the inequality of $\geq$ form to the standard $\leq$ form. For example, we first convert $-2x_1 + 5x_2 - 3x_3 = 15$ to the following two inequalities: $-2x_1 + 5x_2 - 3x_3 \leq 15$ and $-2x_1 + 5x_2 - 3x_3 \geq 15$. We then convert the nonstandard inequality by multiplying it by -1 and reversing the sign of the inequality to get the second standard inequality: $2x_1 - 5x_2 + 3x_3 \leq -15$.

If a continuous or an integer variable is unrestricted in sign (i.e., it can be negative, positive, or zero), then we may replace an unrestricted variable by the difference of two new variables, x_j^+ and x_j^-, as follows:

$$x_j = x_j^+ - x_j^-, \quad x_j^+, x_j^- = 0$$

where $x_j^+ = x_j$ if $x_j > 0$
$\qquad\quad = 0$, otherwise
$\quad\ \ x_j^- = -x_j$ if $x_j < 0$
$\qquad\quad = 0$, otherwise

Note that the same variable t may be used for other unrestricted variables. Thus, only one variable is increased regardless of the number of unrestricted variables.

If a continuous or an integer variable, respectively, has a positive or negative lower bound, say, l_j or l_k, respectively, then it can be transformed to a new variable (say, x'_j or y'_k) by substituting

$$x'_j = x_j - l_j \quad \text{or} \quad y'_k = y_k - l_k$$

The transformed problem is equivalent to the original problem with a set of new variables. After solving the transformed problem, the optimum solution in terms of the original variables is recoverable from the above equations.

Recall that the upper bound of a continuous or an integer variable in the standard form of IP is infinite. Thus, a continuous or an integer variable having a *finite* (value of) upper bound needs to be transformed. However, the above substituting equation cannot be used to get a standard (an infinite) upper bound because the new transformed variable will still have a finite upper bound (why?). In this case, an upper bound constraint, $x_j \leq u_j$ or $y_k \leq u_k$, must be adjoined to the program. Basically, we treat a lower or an upper bound as a *simple* constraint consisting of a single variable.

1.3 COMBINATORIAL OPTIMIZATION PROBLEMS

A *combinatorial optimization problem* (COP) is a discrete optimization problem in which we seek to find a solution in a finite set of solutions that maximizes or minimizes an objective function. This type of problem usually arises in the selection of a finite set of mutually exclusive alternatives. These qualitative alternatives may be quantified by the use of discrete variables. Usually, the set of all possible solutions can be enumerated and their associated objective values can be evaluated to determine an optimum solution. But unfortunately, the number of solutions by complete enumeration is usually too huge even for a moderate-sized problem.

The COP is closely related to the IP in that most, if not all, COPs can be formulated as 0–1 integer programs. Well-known examples of COP include the classical *assignment problem* and *traveling salesman problem* (TSP). The assignment problem may be applied, for example, to assign n jobs to n workers in a most efficient manner so that each job is assigned to one and only one worker, and vice versa. The TSP originates from a salesman who starts from a home city to visit $n - 1$ cities so that each city is visited once and only once and then returns to the home city with a minimum travel distance. The assignment problem is "well solved" because any optimum solution to its LP relaxation is naturally integer. Moreover, there are special assignment algorithms such as Hungarian algorithm that are available to solve the problem much faster than the standard simplex method. This class of "well-solved (*easy*)" integer programs will be discussed in more detail in Chapter 10.

It is "*hard*" to find an exact optimum solution to a traveling salesman problem because of its combinatorial nature. Although there are many algorithms available for finding an approximate solution, the state of the art for finding an exact solution is to formulate and solve it as a 0–1 (binary) integer program. Unfortunately, the formulated model requires an enormous number of binary variables and constraints even for a moderate-sized problem. Modeling combinatorial optimization problems will be discussed in Chapters 5 and 6, and the solution methods to these problems will be a main theme of Chapters 11–13.

1.4 SUCCESSFUL INTEGER PROGRAMMING APPLICATIONS

The authors believe that integer programming plays a key role in operations research, an observation supported by analysis below. This textbook is grounded in theoretical developments in IP over the past five decades, but is written in hope of bridging the gap between academic developments in IP and modern OR practice.

Interfaces, a bimonthly journal publication of INFORMS, had published over 500 OR/MS application articles from 1979 to 2006, when we started writing this book. We reviewed all these articles and surprisingly found that about 23% of them used integer programming and that many of them were finalists of the annual Franz Edelman Award competitions over the years.

We further identified 44 IP application articles in *Interfaces* that claimed enormous savings in cost or increase in profit. Financial benefits cited were of a magnitude of tens or hundreds of million dollars per year. In Table 1.1, these 44 applications are classified by industry sector. They are transportation and distribution, manufacturing, communication, military and government, finance, energy, and others. In this count, the sectors of manufacturing and transportation and distribution tie for first place in terms of most IP application papers (13 each), followed by the communication, military and government, and finance sectors (4 articles each of three sectors). Within the sectors, the airline industry had the most application papers (9 articles).

These 44 articles also are classified in Table 1.1 by problem/model type: workforce/staff scheduling, transportation and distribution, supply chain management, production planning, government services, financial services, project management, and others. In this count, workforce/staff scheduling problem has the most papers (11 articles), followed by the transportation and distribution (10 articles), and the supply chain management (5 articles).

1.5 TEXT ORGANIZATION AND CHAPTER PREVIEW

This text is organized into three parts: Part I Modeling, Part II Review of Linear Programming and Network Flows, and Part III Solutions. Part I (Chapters 1–6) includes areas of successful integer programming applications, systematic modeling procedure, types of integer programming models, transformation of non-IP models, automatic preprocessing for better formulation, and an introduction to combinatorial optimization. Part II (Chapters 7–10) reviews algebraic–geometric concepts and solution methods relating to LP and network flows that are needed for understanding IP. Part III (Chapters 11–15) describes various solution approaches for large-scale IP

TABLE 1.1 Classification of IP Application Papers in *Interfaces* by Industry

Industry Category	Subcategory	Company Name (Year Published)	IP/LP	Nature of Primary Applications	Savings/Benefits (Projected/Actual)
Transportation and distribution	Airline	American Airlines (1981)	IP	Used an IP model to determine the least-cost crew schedule	$0.25 million
	Airline	American Airlines (1991[a])	IP	Crew pairing optimization	$20 million per year
	Airline	American Airlines (1991[b])	IP and LP	Implemented a network optimization-based system to help reduce delays caused by air traffic control	$5.2 million
	Airline	Air New Zealand (2001[a])	IP	Developed computer systems to solve the planning and rostering processes (IP problem)	$15.655 million per year
	Airline	American Airlines (1989)	IP	Used IP algorithm to build flight crew schedules	$18 million per year
	Airline	Continental Airlines (2004)	IP	Solved large-scale IP-formulated pilot staffing and training problems to save costs	$10 million per year
	Airline	Continental Airlines (2003[b])	IP	Developed IP-based system to generate optimal crew recovery solutions	$40 million
	Airline	Delta Airlines (2003[c])	IP	Developed an automated optimization system to minimize operating costs and maximize training assignments	$7.5 million
	Airline	Qantas Airways Limited (1979)	IP and LP	Used ILP model for planning annual manpower requirement for telephone reservation	$0.235 million
	Airline	United Airlines (1986[a])	IP and LP	Used IP/LP-based system to control the entire manpower scheduling process	$6 million per year

(continued)

TABLE 1.1 (*Continued*)

Industry Category	Subcategory	Company Name (Year Published)	IP/LP	Nature of Primary Applications	Savings/Benefits (Projected/Actual)
	Public transportation	The Société de transport de la communauté urbaine de Montréal (1990[a])	IP	Employs network flow methods (an IP formulation) to generate optimal vehicle schedules	$4 million per year
	Railway	The Canadian Pacific Railway (2004[b])	IP and LP	Used IP/network algorithms for planning locomotive use and distributing empty cars	CN$510 million
	Railway	NS Reizigers (Dutch Railway) (2005[c])	IP	Applied a set covering model to support the development of an alternative set of scheduling rules	$4.8 million per year
	Shipping	Menlo Worldwide Forwarding (2004[a])	IP	Developed a network routing optimization model to optimize its transportation network in North America	$80 million
	Shipping	UPS (2004[a])	IP	Created an IP-based system to optimize the design of package delivering networks	$87 million
	Container port	Hong Kong International Terminals (2005[a])	IP	Developed a decision support system to generate various decisions, including scheduling, storage, and so on	$100 million per year
Communication	Telephone	AT&T (1990[a])	IP	Developed an MIP-based system to minimize cost	$1 million
	Telephone	GTE (1992[a])	IP	Developed an IP-based optimization tool to improve productivity	$30 million per year
	Telephone	Bellcore (1995[a])	IP	Built an IP-based decision support software to design robust fiber-optic networks	$50–225 million

	Telephone	Motorola (2005[b])	IP	Used Emptoris's end-to-end Internet negotiations platform to identify the best procurement strategy	$600 million
	Television	NBC (2002[a])	IP	Used MIP-based sales systems to improve its revenues and productivity	$200 million
Manufacturing	Automobile	Ford Motor Company (2001[a])	IP	Developed an IP model to shorten the planning process and establish global procedures	$250 million
	Automobile	General Motors (1987[a])	IP/LP	Used network tools to reduce logistics cost	$2.9 million per year
	Automobile	General Motors (2004[d])	COP	Developed a heuristic-based decision support tool to schedule vehicle road tests	Millions of dollars of savings; 100% increase in throughput
	Automobile	Volkswagen of America (2000)	IP	Used a combination of simulation and MIP models to analyze supply chain	35% reduction in cost
	Chemical	Air Products and Chemicals (1983[b])	IP	Developed a decision support system for vehicle scheduling	$1.54–1.72 million
	Chemical	Proctor & Gamble (2006[a])	IP	Built a sourcing network that optimizes sourcing problem with suppliers	$294.8 million
	Chemical	Trumbull Asphalt (1985)	IP	Used MIP to assist planning of sourcing, distribution, blending, and facility configuration	$1 million per year
	Computer	Digital Equipment Corporation (1995[a])	IP	Used a large-scale MIP model to minimize supply chain cost	$100 million

(continued)

TABLE 1.1 (*Continued*)

Industry Category	Subcategory	Company Name (Year Published)	IP/LP	Nature of Primary Applications	Savings/Benefits (Projected/Actual)
	Food	Golden Vale Cooperative Creameries Ltd (1983)	IP and LP	Developed large-scale IP/LP program to analyze the problem of milk collecting and transporting	$4 million
	Food	Irish Milk Cooperative (1986)	IP and LP	Used large-scale network (graphic) method to solve the transshipment and lot sizing problem	IR £1.5 million per year
	Lumber	The Chilean Forest Sector (1999[a])	IP	Implemented MIP models to support decisions on truck scheduling, harvesting, and so on	$20 million per year
	Machinery	Schindler Elevator Corporation (2003[a])	IP	Provided an IP-based application to optimize preventive maintenance operations	$1 million per year
	Pharmacy	P&G (1997[a])	IP and LP	Developed MIP and network models to improve work processes	$200 million
	Photography	Kodak Australasia (1991[a])	IP	Developed a two-phase IP-based system for the problem of cutting photographic color papers	$2 million
	Steel	The Bethlehem Plant (1989[a])	IP	Developed a two-phase, IP-based procedure to determine new mold dimensions	$8 million per year
Energy	Electricity	Southern Company (1991[a])	IP	Installed an optimization software based on IP algorithm to reduce fuel cost	$140 million
	Gas	Exxon Corporation (1982[a])	IP	Developed an MIP model to evaluate projects and determine utility distribution	$100 million

Military and government	Water	Hidroeléctrica Española (1990[a])	IP	Developed and implemented a hierarchy of models, including IP and network models, to manage its system of reservoirs	$2 million per year
	Military	South African Defense Force (1997[b])	IP	Used MIP model to analyze the size and shape of defense force when no threat exists	$32–78 million
	Military	U.S. Army (1998[a])	IP	Used MIP model to allocate budget	$360 million
	Police	The San Francisco Police Department (1989[b])	IP	Implemented an IP-based support system for deploying patrol officers	$14 million per year
	Tax	Office of Tax Analysis, U.S. Treasury Department (1980)	IP	Used an IP model to minimize the loss of information by using a subset of the database instead of the whole file	3–13% improvement in accuracy
Finance	Bank	The Maryland National Bank (1983)	IP	Implemented a computerized IP model for transit check clearing	$0.1 million
	Bank	The World Bank, Chinese State Planning Commission (1995[a])	IP	Developed a coal transporting study system with MIP as an important element	$6.4 billion
	Insurance	PSI Insurance (1992[a])	LP and IP	Developed a series of optimization-based models, including LP/IP, to value and trade mortgage-backed securities	Over $10 billion increase in trading volume; rank increased from below No. 10 to No. 3
	Insurance	The Variable Annuity Life Insurance Company (1984)	IP	Used branch-and-bound method to solve an IP model to find out the best number of sales regions	$8.8 million

(continued)

TABLE 1.1 (*Continued*)

Industry Category	Subcategory	Company Name (Year Published)	IP/LP	Nature of Primary Applications	Savings/Benefits (Projected/Actual)
Others	Construction	Homart Development Company (1987[a])	IP	Designed an IP model to schedule the divestiture of shopping malls	$40 million
	Retail	Fingerhut Companies, Inc. (2001[a])	IP	Developed IP-based system to select the most profitable sequence of catalogs mailing stream	$3.5 million per year
	Retail, alcohol	Société des alcools du Québec (2005[c])	IP	Developed a solution engine that implements an IP model to reduce the costs of producing worker schedules	CN$1 million per year
	Restaurant	Taco Bell (1998[a])	IP	Used IP model to schedule and allocate crew members to minimize payroll	$53 million
	Waste collection	Waste Management (2005[a])	COP	Developed a comprehensive route management system to solve its vehicle routing problems	$18 million
	Education	Nanzan Gakuen (Nanzan Educational Complex (2006[a])	COP	Solved school bus problems, school time problems, and the problem of assigning supervisors for entrance examinations	$2 million

[a] Franz Edelman Award finalist of the previous year.
[b] Franz Edelman Award winner of the previous year.
[c] Daniel H. Wagner Prize finalist of the previous year.
[d] Daniel H. Wagner Prize winner of the previous year.

and combinatorial optimization problems in addition to fundamentals of typical software systems. Solution approaches include classical, branch-and-cut, branch-and-price, primal heuristics, and Lagrangian relaxation. In Chapter 15, three popular modeling languages and one solver are introduced. Answers to selected exercises from each chapter appear in an appendix.

This chapter (a) defines the IP model and associated notation to be used in the text, (b) classifies IP models and describes their relationships to linear and combinatorial optimization models, (c) previews the contents of each chapter, and (d) categorizes numerous successful IP applications arising in diverse industry/business sectors, based on survey data collected from the articles published in *Interfaces* (a bimonthly journal by INFORMS) 1979–2006, when we started writing this book.

Chapter 2 (a) explores the assumptions underlying the MIP mathematical model and explains their physical interpretations, (b) provides a step-by-step procedure for building a model from a given real-world problem, and (c) introduces fundamental formulations for the most utilized types of MIP models that are identified from the survey of successful applications described in this chapter. Seven assumptions underlying the MIP problem are fully uncovered through a careful examination of its mathematical anatomy. Some of these assumptions do not appear explicitly in other texts of operations research and integer programming.

In Chapter 3, beyond the simple use of 0–1 variables discussed in Chapter 2, the formulation power of 0–1 variables extends their ability to transform a variety of optimization models into integer programs. Transformable optimization models are identified and grouped together according to the types of decision variables, mathematical functions, and constraints. This chapter also describes the relation between logical (Boolean) expressions and 0–1 formulations, in addition to modeling the bundle pricing problem, which is a common business practice. These features appear for the first time in any integer programming text.

Chapter 4 (a) defines and explains what is meant by better formulation of an IP problem, (b) introduces several basic preprocessing techniques, for both general and special problems, that can automatically transform a user-supplied formulation into a better one, and (c) identifies primary preprocessing functions/areas that are covered by most preprocessors of current IP software.

Chapter 5 begins with defining the class of COPs and ends with a discussion of the computational complexity of a problem or an algorithm. Three classes of COPs are discussed: set covering, partitioning, and packing; matching problems; and cutting stock problems.

Chapter 6 is devoted to the best-known combinatorial optimization problem, the TSP, and its many variations. More details on TSP applications are given, expanding the discussion in this chapter. Solution approaches, which generally involve creating constraints that prevent inclusion of subtours in the IP search for the optimal tour, depend on whether the arcs connecting the nodes are one-way (asymmetric TSP) or bidirectional (symmetric TSP).

Chapter 7 reviews the fundamentals of linear programming theory and network flows that are essential to the understanding of the solution space and solution methods to be discussed in Chapters 11–13.

Chapter 8 reviews/introduces basic geometric concepts and terminology that are essential to the understanding of the properties of the solution spaces and cutting planes for both general and special IP problems. These concepts are prerequisites for full understanding of the branch-and-cut method to be discussed in Chapter 12.

The modern methods for solving a large-scale integer program require the *optimization* and *reoptimization* of a usually long sequence of LP relaxation problems that in turn are often solved by a variety of simplex-based methods (and/or an interior point method). Chapter 9 reviews four simplex-based methods that serve as building blocks for solving integer programs. The *simplex method* provides the foundation for optimizing a long sequence of LP relaxations. The *simplex method for upper-bounded variables* is used for reducing the problem size by implicitly handling the upper and lower bounds on variables (equivalent to *single-variable constraints*). The *dual simplex method* is most effective for reoptimizing the current optimum, after addition of constraints, without resolving the augmented LP problem from scratch. The *revised simplex method* produces the same sequence of bases as the simplex method, but depends on updating the basis inverse (m columns) rather than the entire simplex tableau (n columns) in each iteration.

Chapter 10 (a) identifies a class of *easy* network optimization problems whose IP formulations are solvable as LPs by simply ignoring the integer requirements, (b) describes the sufficient conditions (or model structure) that characterize this class of problems, and (c) introduces a more efficient algorithm than the ordinary simplex for solving this class of network optimization problem.

Chapter 11 introduces three classical approaches for solving integer programs: branch-and-bound, cutting plane, and group theoretic. Currently, these approaches are not implemented in practice as stand-alone solvers. However, they are integrated parts of a modern solution approach such as the branch-and-cut to be described in Chapter 12.

The recent advances in solving large-scale integer programs have been made possible by great improvements in modeling, preprocessing, solution algorithms, LP software, and computer hardware. We have already discussed modeling and pre-processing. Chapter 12 addresses a modern solution approach known as the *branch-and-cut*, in which a substantial portion of the discussion centers on the generation of cuts that are useful for solving general and special integer programs.

In the previous chapter, branch-and-bound is generalized to include generation of cuts or rows, hence the name *branch-and-cut*. In Chapter 13, branch-and-bound is first generalized to include generation of *columns* by solving pricing problems, hence the name *branch-and-price*, and then generalized to include columns and rows, hence the name *branch-and-price-and-cut*. Basically, all these generalizations solve a sequence of LP relaxations of a given IP. Branch-and-cut tightens the LP relaxations (or polyhedra) by adding cuts or constraints (rows). Branch-and-price tightens the LP relaxations by generating a subset of profitable columns associated with variables to join the current basis. These columns are generated iteratively by solving subproblems or *pricing problems.*

Chapter 14 introduces a variety of primal heuristic algorithms that can be used to obtain a good solution or an approximate solution for an integer program or a combinatorial optimization problem. Both classical and artificial intelligence (AI) heuristic algorithms are provided. The traveling salesman problem is used for the purpose of illustration. This chapter also (a) describes various relaxation methods for solving IP problems, (b) lists examples of IP model types to which the Lagrangian relaxation approach is applied, (c) derives the associated Lagrangian dual problems for both linear and integer programs, (d) provides efficient methods for solving the Lagrangian dual, and (e) develops Benders' decomposition algorithm for integer programming.

Chapter 15 (a) provides some practical considerations when algorithms are implemented in software, (b) describes the key components and features of a typical software system, (c) introduces three commonly used modeling languages (AMPL®, LINGO®, and MPL®) in more depth than earlier chapters, and (d) briefly describes other modeling languages and systems.

1.6 NOTES

Section 1.1

General IP textbooks that are referenced in this text include Hu (1969), Garfinkel and Nemhauser (1972), Zionts (1974), Taha (1975), Nemhauser and Wolsey (1988), Parker and Rardin (1988), Salkin and Mathur (1989), and Wolsey (1998).

Introductory OR/MS textbooks that are referenced in this text include Wagner (1975), Winston (1994), Hillier and Lieberman (2005), and Taha (2007).

Journals that are referenced include *Interfaces, Operations Research, Management Science, European Journal of Operational Research, IIE Transactions, Transportation Science, Naval Research Logistics Quarterly, Journal of the Association for Computing Machinery, Mathematical Programming, Discrete Applied Mathematics*, and *SIAM Journal on Algebraic and Discrete Methods*.

Many textbooks, like this one, use a maximization problem as a standard MIP, while others use a minimization problem. In a minimization MIP, the standard inequality constraint is of $\geq$ form.

Section 1.2

Conversion from a nonstandard MIP to standard form is similar to that for linear programs. For references of conversion techniques, see any introductory OR/MS textbooks such as Winston (1994) and Hillier and Lieberman (2005).

Section 1.3

Some authors, for example, Parker and Rardin (1988), view discrete optimization problems as a combination of integer programming and combinatorial optimization problems. Literally speaking, a *discrete optimization problem* is an optimization problem defined over discrete variables. However, a *discrete variable* is different from an integer variable in that an integer variable may take on any consecutive integral values, while a discrete variable may take on specified discrete values, consecutive or not, integer number or not—essentially what mathematicians call a countable set. Thus, an integer variable is a discrete variable, but a discrete variable may or may not be an integer variable. For example, both solution sets of y_5 and y_3, defined by $\mathbf{Z}_5 = \{3, 4, 5, 6, 7\}$ and $\mathbf{Z}_3 = \{4, 6, 7, 10\}$, respectively, are discrete variables. But y_3 is not an integer variable, while variable y_5 is both an integer and a discrete variable. In Chapter 2, we shall show how a discrete variable can be converted to a set of binary (0–1) variables.

Section 1.4

INFORMS is a professional society that was founded through the merger of two older societies: the former Operations Research Society of America (ORSA) and The Institute of Management Science (TIMS).

Interfaces, a bimonthly journal publication of INFORMS, has published over 500 OR/MS application articles since 1971. All articles are available in both electronic form and hard copy.

The Franz Edelman Award was founded in 1972 (initially under the name of "the Annual International Management Science Achievement Award"). From 1975 to 1984 (the year in which the award name was changed to Franz Edelman Award), the papers of the finalist and the winners were published in *Interfaces* in the last issue of that year. From 1985 up to today, the first issue each year is dedicated to the finalist and the winner(s) of the previous year. "The Edelman Award recognizes outstanding implemented operations research that has had a significant, positive impact on the performance of a client organization. The top finalist receives a $10,000 first prize" (*OR/MS Today*).

The Daniel H. Wagner prize was founded in 1998. It "emphasizes the quality and coherence of the analysis used in practice. Dr. Wagner strove for strong mathematics applied to practical problems, supported by clear and intelligible writing. This prize recognizes those principles by emphasizing good writing, strong analytical content, and verifiable practice successes. The competition is held each year in the fall at the INFORMS Annual Meeting" (see http://www2.informs.org/Prizes/WagnerPrize. html for details). Papers of each year's finalists are published in the fifth issue of *Interfaces* of the following year.

1.7 EXERCISES

1.1 Read one of the successful application articles from the category of transportation and distribution published by INFORMS in *Interfaces* as shown in Table 1.1. Do the following:

    (a) Verify the entries described in the row associated with the company.

    (b) Use your own words to describe the objective, sets of constraints, decision variables, and types of variables (continuous or integer, binary or general).

1.2–1.7 Do the same for each of the remaining categories: (1.2) communications, (1.3) manufacturing, (1.4) energy, (1.5) military and government, (1.6) finance, and (1.7) others.

1.8–1.14 Read one of the application articles from each of the following problem types published in *Interfaces*, as given in Table 1.1: (1.8) project management, (1.9) production planning, (1.10) workforce scheduling, (1.11) transportation and distribution, (1.12) supply chain management, (1.13) cutting stock, and (1.14) machine scheduling and sequencing. Do the following:

(a) Verify the entries described in the row associated with the company.

(b) Use your own words to describe the objective, sets of constraints, decision variables, and types of variables (continuous or integer, binary or general).

1.15–1.19 Transform each of the following nonstandard integer programs into a standard form of IP defined in this text.

1.15

$$\text{Minimize} \quad 3x_1 - 11x_2 + 5x_3 + x_4$$
$$\text{subject to} \quad x_1 + 5x_2 - 3x_3 + 6x_4 \leq 7$$
$$-x_1 + x_2 + x_3 - 2x_4 \geq 3$$
$$x_1, \ x_2, \ x_3, \ x_4 \geq 0$$

1.16

$$\text{Maximize} \quad -x_1 + 5x_2 + 2x_3 - 7x_4 - x_5$$
$$\text{subject to} \quad x_2 + x_3 + x_4 \geq 13$$
$$x_1 - x_2 + 2x_4 + 2x_5 \leq 4$$
$$x_1 \text{ unrestricted in sign}$$
$$x_2, x_4, x_5 \geq 0$$
$$x_3 \geq -2$$

1.17

$$\text{Maximize} \quad 7x_1 + 2x_2 + x_3 - 4x_4$$
$$\text{subject to} \quad 2x_1 - x_2 + x_3 \leq 10$$
$$x_1 + x_4 = 12$$
$$x_1, x_2, x_4 \geq 0$$
$$x_3 \geq 0$$

1.18   Minimize $-11x_1 + 13x_2 - 15x_3$

                      subject to $x_2 + x_3 = 7$

                                    $x_1 - x_3 \leq 3$

                                    x_1 unrestricted in sign

                                    $x_2 \geq 5$

                                    $x_3 \geq 0$

1.19   Maximize $x_1 + x_2 + x_3$

                      subject to $-x_1 + x_2 \geq 8$

                                    $x_1 - x_2 + x_3 \leq 2$

                                    $x_1, x_3 \geq 0$

                                    $x_2 \leq 15$

2

MODELING AND MODELS

In Chapter 1, we mathematically defined a mixed integer program (MIP). A mathematical definition in general has the advantages of being precise, concise, and capable of data manipulation. But to most managers and even some practitioners, it may be too abstract to comprehend and difficult to relate to reality. To alleviate this difficulty, we begin this chapter with an explanation of the real-world meanings of the MIP assumptions (or conditions). Section 2.1 describes these assumptions and their physical implications. Having this background, we then introduce a three-step procedure for modeling real-world problems in Section 2.2. This procedure systematically leads the practitioner toward an MIP model. In case the constructed model is not an MIP, the transformation techniques introduced in the next chapter may be used to obtain an equivalent MIP.

Recall in Chapter 1 we tabulated many successful IP application papers published in *Interfaces* and classified them by problem type. Each problem type will be given one to three examples in Sections 2.3–2.9. These examples may appear to be simpler than the real-world problems described in the application articles. Nevertheless, they do provide primary characteristics of the model types.

Applied Integer Programming: Modeling and Solution, By Der-San Chen, Robert G. Batson, and Yu Dang
Copyright © 2010 John Wiley & Sons, Inc.

2.1 ASSUMPTIONS ON MIXED INTEGER PROGRAMS

Recall the following mixed integer program defined in Chapter 1:

$$(\text{MIP}) \quad \text{Maximize} \quad z = \sum_j c_j x_j + \sum_k d_k y_k$$

$$\text{subject to} \quad \sum_j a_{ij} x_j + \sum_k g_{ik} y_k \le b_i \quad (i = 1, 2, \ldots, m)$$

$$x_j \ge 0 \qquad\qquad\qquad (j = 1, 2, \ldots, n)$$

$$y_k = 0, 1, 2, \ldots \qquad\quad (k = 1, 2, \ldots, p)$$

The mixed IP comprises two fundamental building blocks: variables (including continuous x_j and integer y_k) and input parameters (including c_j, d_k, a_{ij}, g_{ik}, b_i, m, n, p). The objective function is a summation of several functions, each containing a single variable. Likewise, each constraint function (the left-hand side of an inequality) is also a summation of several functions of single variables. Both the objective and the constraint functions in the mixed IP are separable and linear. Figure 2.1 gives an anatomy of all assumptions imposed on a mixed integer program.

The above mathematical definition of an MIP implies the following assumptions:

- *Divisibility assumption* for each continuous variable ($x_j \ge 0$)
- *Integrality assumption* for each integer variable ($y_k = 0, 1, 2, \ldots$)
- *Certainty (constant) assumption* for each input parameter (c_j, d_k, a_{ij}, g_{ik}, b_i)
- *Proportionality assumption* for each term in the constraint and objective function ($c_j x_j$, $d_k y_k$, $a_{ij} x_j$, $g_{ik} y_k$)
- *Additivity and separability assumption* for each combined function in the objective and constraints ($\sum_j c_j x_j$, $\sum_k d_k y_k$, $\sum_j a_{ij} x_j$, $\sum_k g_{ik} y_k$)
- *Single-objective assumption* (max or min $z = \sum_j c_j x_j + \sum_k d_k y_k$)
- *Simultaneousness (conjunction) assumption* for the system of all constraint equations and inequalities

Now we interpret each of the above assumptions in detail.

The *divisibility assumption* implies that each continuous variable in a solution is allowed to be any real value, which may carry an arbitrary number of decimal places. For example, a production level of 2534.397 cars per week is an acceptable computed solution because in practice it may be rounded up to 2535 or rounded down to 2534 without making any difference in a practical sense. Continuous commodities such as the quantity of water flowing through a segment of a pipeline obviously satisfy divisibility.

The *integrality assumption* implies that each integer variable is restricted to be one of the integral values {0, 1, 2, ...} or binary values {0 or 1}. A solution carrying a fractional value is unacceptable under this assumption. For example, we are to determine whether plant A or plant B should be built and a computed solution of linear

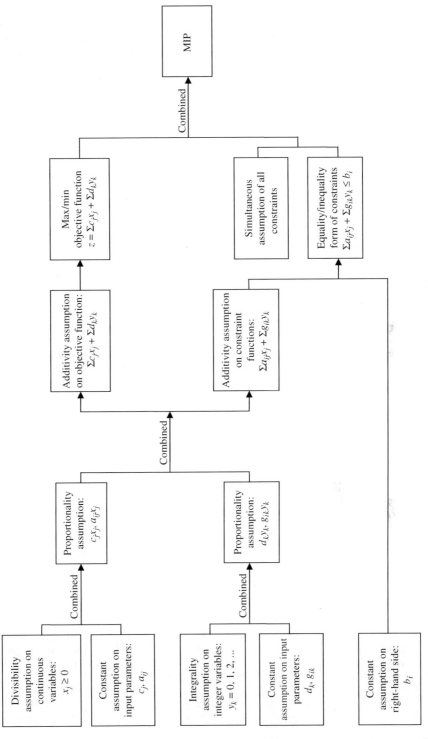

FIGURE 2.1 Anatomy of MIP assumptions.

23

program results in building 0.57 plant A and 0.43 plant B. Obviously, this fractional solution does not make sense in decision making. Even if sometimes a sensible solution were obtained after rounding, chances are this solution might not be optimal.

The *certainty assumption* implies that the values of all input parameters can be estimated or predicted with almost certainty, if not certainty. In other words, under this assumption, each input parameter (data point) is constant or fixed, and any variation about this fixed value is negligible. Consider a counterexample. Suppose a profit of a certain product per unit is $1.2 if the economy is good, a profit of $0.3 if the economy is mediocre, and a loss of $1.1 if the economy is bad. There are three possible values regarding the unit profit or loss depending on the economic conditions.

There is another class of mathematical program in which the unit profit is a random variable following a certain probability distribution. The integer program with random parameter(s) is called a *stochastic integer program*. Another class of mathematical program in which the unit price is a mathematical function of a certain parameter is called a *parametric integer program*.

The *proportionality (linearity) assumption* implies that the total contribution to a function value is proportional to the values of a variable. In other words, the marginal contribution to the function value by each unit of a variable is constant. Figures 2.2 and 2.3 depict, respectively, the proportionality assumption of a continuous variable x and an integer variable y. Note that both increasing and decreasing functions are linear, and each linear function has a constant slope over the domain defined by the variable(s).

Recall that the slope of a continuous function at any continuous point x is defined as

$$\frac{\mathrm{d}f(x)}{\mathrm{d}x} = \lim_{\Delta \to 0} \frac{f(x+\Delta)-f(x)}{\Delta}$$

where Δ is arbitrarily small and approaches to 0. For the function of continuous variable x_1 defined in Figure 2.2, the slope $= \lim_{\Delta \to 0}(2(x_1+\Delta)-x_1)/\Delta = 2$, a constant for every value of x_1. For a function of continuous variable x_2, the slope $= \lim_{\Delta \to 0}(-3(x_2+\Delta)-(-3x_2))/\Delta = -3$, a constant for every value of x_2.

The slope of a discrete function at any discrete point y is defined as

$$\frac{\Delta f(y)}{\Delta y} = \frac{f(y+\Delta)-f(y)}{\Delta}$$

where Δ is a positive increment (equal to 1 in this case). Applying this slope definition to the two functions given in Figure 2.3, we obtain the following constant slopes, respectively:

$$\frac{f(y_1+1)-f(y_1)}{(y_1+1)-y_1} = \frac{2}{1} = 2$$

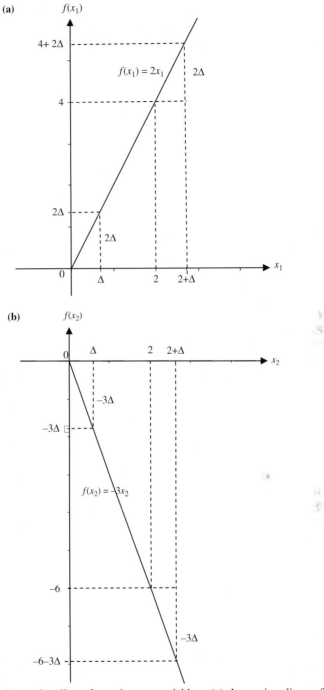

FIGURE 2.2 Proportionality of continuous variables: (a) Increasing linear function; (b) decreasing linear function.

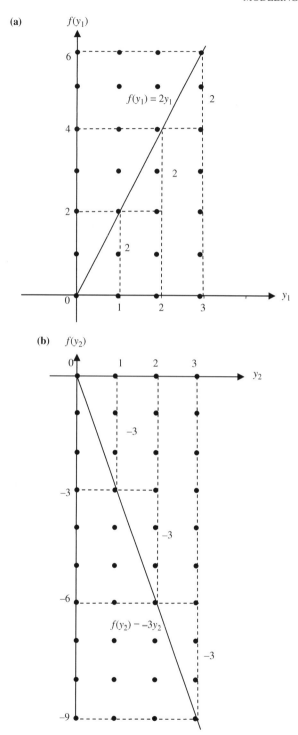

FIGURE 2.3 Proportionality of integer variables: (a) Increasing linear function; (b) decreasing linear function.

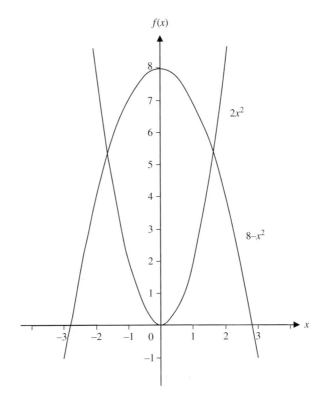

FIGURE 2.4   Counterexamples for the proportionality assumption.

and

$$\frac{f(y_2 + 1) - f(y_2)}{(y_2 + 1) - y_2} = \frac{-3}{1} = -3$$

Counterexamples to the proportionality assumption of continuous variables are $2x^2$ and $8 - x^2$. They are *nonlinear functions*, as shown in Figure 2.4. Any math programming problem containing any nonlinear function in the objective or a constraint function is called a *nonlinear programming problem (nonlinear program)*.

The *additivity/separability assumption* implies that every function (in the objective or in each of the constraints) can be expressed as a sum of several functions, each containing a single variable. Note that the function $3x_1 - 5x_2$ is equivalent to the sum of two single-variable functions: $3x_1 + (-5x_2)$, with a negative coefficient in the second function. Also note that a function is separable if it is an algebraic sum of functions of single variables.

Mathematically, a *separable function* is defined as $f(x_1, x_2, \ldots, x_n) = f_1(x_1) + f_2(x_2) + \cdots + f_n(x_n)$. Counterexamples of additivity/separability include functions that contain product terms such as $x_1 x_2$ and $x_1^2 - 2x_1 x_2 + x_2^2$. These functions are nonseparable and nonlinear.

Precaution: When formulating an objective function or a constraint equation/ inequality, make sure that the units or dimensions of all terms in the same function are identical.

The *single-objective assumption* implies that an optimization problem satisfying the above assumptions, but with multiple objectives, is not a mixed integer program. However, there are cases where a multiobjective problem may be converted into a single-objective problem. The multiobjective problem is beyond the scope of this book. For further information, read the references given in Section 2.10.

The *simultaneousness assumption* implies that a feasible solution must simultaneously satisfy all the constraint equations and inequalities. That is, any feasible solution must not violate any constraint in a given mixed IP. If a problem requires only a subset of constraints to be satisfied, then it must be transformed into an equivalent problem in which all constraints must be satisfied simultaneously. Chapter 3 will discuss how to perform this transformation.

2.2   MODELING PROCESS

Many definitions of operations research (OR) have been published over the five-decade history of ORSA/INFORMS. INFORMS recently defined OR to be "the discipline of applying advanced analytical methods to help make better decisions." One way to understand how such methods apply to a decision situation (a real system to be optimized or problem to be solved) is to consider the three phases of an OR study in Figure 2.5:

   i. Construction of the model
  ii. Solution of the model
 iii. Validation of the model results and interpretation back to the decision situation

Two other phases in the OR approach to problem solving are important, but are not shown in Figure 2.5. There is a premodel phase "Definition of the problem." This

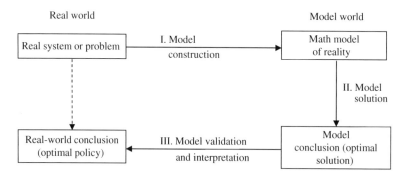

FIGURE 2.5   Three phases of an OR study.

phase establishes objectives and scope of the model and is carried out by the OR analyst in conjunction with the client and his staff, or by the appointed "OR team." There is also a postmodel phase "Implementation of the optimal policy" in the organizational environment. The policy is translated into action by managers and workers under authority of the client.

Modeling is therefore central to any application of OR, and the construction of an OR model is in part both art and science. There are many cases where practitioners who follow the following three-step "model construction process" naturally arrive at a model formulation:

Step 1. Verbally identify and define decision variables, input data or parameters, constraints, state variables (if any), and the objective from the given problem description. Then assign appropriate symbols to decision variables, input parameters (or data), and state variables (if any).

Step 2. Translate the verbal description of the objective and constraints into functions, equations, and inequalities. Check whether each of the seven MIP assumptions is satisfied. If all are satisfied, then an MIP is obtained; otherwise, go to Step 3.

Step 3. Check whether the non-MIP factors such as a discrete (but not integer) variable, a nonlinear function, or nonsimultaneous constraints can be transformed into equivalent mathematical expressions that satisfy all MIP assumptions. If yes, we obtain an MIP; otherwise, the problem is not an MIP.

Decision variables are variables under the control of the decision authority. Appropriate symbols for the decision variables are selected, and data needed to express objective and constraint functions are organized into tables. In large-scale applications, these tables are more appropriately called "decision databases."

In Step 1, the objective to be achieved by the decision should be expressed verbally. Constraints that often relate to resources, requirements, and regulations should also be verbally described. Sometimes, these symbols, data, and verbal descriptions may be augmented by graphical (or iconic) or analog models, for example, an input–output diagram or a network flow diagram with appropriate labels.

Step 2 translates the verbal and/or graphical description into a mathematical model using the selected symbols for the decision variables, and using functions of these variables to represent objectives (to be maximized or minimized), and other functions of the variables combined with a constant to create equation or inequality constraints. These constraints express the nature of resource limitations or requirements, and how the values of the variables are converted into resource demands (performance versus requirement). It is desirable, of course, if the functions created in the mathematical expression of objectives and constraints are linear. In that case, the tables from Step 1 become matrices in MIP model.

Check the formulated math model to see if it satisfies each of the seven assumptions pertaining to an integer program. If any assumption is violated, the math program is not an MIP, but it may be possible to transform it into an MIP. Step 3 performs the transformation using techniques to be introduced in Chapter 3.

Now we are ready, in Sections 2.3–2.9, to apply this modeling process to the following problem types selected from the IP applications from *Interfaces*:

- Project selection
- Production planning
- Workforce/staff scheduling
- Fixed-charge transportation and distribution
- Multicommodity network flow
- Side-constrained network optimization
- Supply chain planning

2.3   PROJECT SELECTION PROBLEMS

Two types of project selection (or capital budgeting) problems will be discussed here. One type covers a single time period and the other multiple time periods. In fact, the single-period project problem may be viewed as a knapsack problem. We begin this section with the knapsack problem and then proceed to more complicated, and more realistic, problems.

2.3.1   Knapsack Problem

The simplest form of integer program is the knapsack problem (or 0–1 knapsack problem) that contains a single constraint with 0–1 variables. The name is taken from a decision problem faced by a hiker who is to select items of a given set to be included in his backpack (or knapsack) within the limit of a specified weight. Each item selected contributes a (relative) value to the hiking trip and the objective of this decision problem is to maximize the total value of all the items selected. Following the modeling procedure described above, we now formulate this problem in two steps.

Step 1

Input parameters:	number of items (n), weight of each item (a_j), value of each item (c_j), total weight limit (b)
Decision variables:	whether or not to select item j ($y_j = 1$ or 0)
Constraint:	total weight of selected items cannot exceed weight limit (b)
State variables:	none
Objective:	maximize total value of selected items

Step 2. The knapsack problem can be formulated as follows: Find values of y_j ($j = 1, 2, \ldots, n$) so as to

$$\text{Maximize} \quad z = \sum_j c_j y_j$$

$$\text{subject to} \quad \sum_j a_j y_j \leq b$$

$$y_j = 0 \text{ or } 1 \quad j = 1, 2, \ldots, n$$

where a_j, c_j, and b are assumed nonnegative.

Sometimes, a_j and b are further assumed integer, while other times they are assumed *rational* (integer or fractional). The integrality assumption does not affect the generality of the problem definition because any constraint containing fractional coefficients can be made integer by multiplying through by an appropriate number. For example, the fractional constraint, $2.4x_1 + x_2 \leq 5.6$, can be converted to an integer constraint by multiplying it by 5 on both sides.

Depending on its application area, the knapsack problem carries many different names. In project management, for example, a project manager is faced with the problem of selecting a subset of n projects to be undertaken because of budget limitation that prohibits funding them all. Each project j will cost a_j dollars if selected, and benefits to the firm in the future have a present value of c_j dollars. The manager has a budget of b dollars to be allocated to the selected projects. Thus, the knapsack problem can be viewed as a single-period *project selection problem* (based on its decision variables) or a single-period *capital budgeting problem* (based on its constraint).

Furthermore, the knapsack problem sometimes is also referred to as the *cargo loading problem* when cargos of various weights are being selected for loading onto a vessel having a limited weight capacity. Similarly, the knapsack problem is sometimes called the *flyaway kit problem* when a number of valuable items are being considered for loading on an airplane.

Obviously, volume can be the deciding factor and can replace weight as the criterion of the constraint. Volume can also be "another" constraint criterion if volume of each item and total volume capacity limit are also specified. This two-constraint problem is known as the two-dimensional knapsack problem. There are obvious extensions to multiple criteria (multidimensional knapsack problem).

In reality, there may be other conditions or requirements about the selection of projects. Examples are the following: (1) the number of projects selected in each period may not exceed a certain number, (2) project 3 may not be selected unless both projects 1 and 2 have been undertaken in the previous periods, (3) either project 4 or project 6, but not both, may be selected in the same time period, and others. These additional conditions may be formulated as linear constraints and will be discussed in detail in Chapter 3.

2.3.2 Capital Budgeting Problem

The capital budgeting problem often arises over a planning horizon of multiple time periods. The time period may be quarterly, semiannually, or annually. The multiperiod problem may be described as follows. A project manager has n projects that he would like to undertake but not all can be selected because of budget limitation in each time period over a prescribed planning horizon. Assume project j has a present value of c_j dollars and requires an investment of a_{tj} dollars in time period t ($t = 1, \ldots, T$). The capital available in time period t is b_t dollars. The objective of this problem is to maximize the total present value subject to the budgetary constraint in each time period over a prescribed planning horizon T. The problem may be mathematically modeled as follows: Find a set of values for y_j so as to

$$\text{Maximize} \quad z = \sum_j c_j y_j$$

$$\text{subject to} \quad \sum_j a_{tj} y_j \leq b_t \quad t = 1, \ldots, T$$

$$y_j = 0 \text{ or } 1 \quad j = 1, 2, \ldots, n$$

where input parameters n and b_t are positive, a_{tj} nonnegative, and c_j unrestricted in sign. Again, the negative coefficient can be made positive by changing the associated variable to its complement. In a real application, a model may have additional constraints such as requiring contingency and/or mutual exclusion among projects. Chapter 3 will discuss how to handle these types of constraints.

2.4 PRODUCTION PLANNING PROBLEMS

Production planning problems often arise in multiple periods. As shown in Figure 2.6, there is a demand in each time period. The demand can be met by two sources: production in the same time period and the inventory carried over from the previous period (assuming no backorder is allowed). A production run incurs a fixed setup cost (per run) and a variable production cost (function of production quantity). The inventory carried over from the previous period incurs a variable "carrying" or holding cost (function of carryover quantity). The planning objective is to minimize

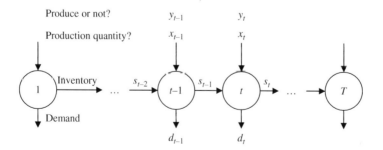

FIGURE 2.6 Uncapacitated lot sizing.

the sum of these three costs. In what follows, we shall discuss three examples of production planning.

2.4.1 Uncapacitated Lot Sizing

A lot sizing problem in production planning is to find an optimal lot size (or quantity of a production run) for each time period, so that the total cost of production and inventory is minimized while the demand in each period is satisfied. The uncapacitated lot sizing problem assumes unlimited production capacity (lot size) in each period. This implies one lot size (production run) per time period. For the following problem formulation, we further assume that (1) the production cost is proportional to the production quantity and (2) the carrying cost is proportional to the ending inventory level of the previous period.

Step 1

Input parameters:	number of periods (T), demand in each period (d_t), setup cost for each period (f_t), unit production cost (c_t), unit holding cost (h_t)
Decision variables:	whether or not to produce in each time period $(y_t = 1$ or $0)$ and how much if the decision is to produce (x_t)
Constraints:	satisfy the demand in each period t
State variables:	inventory level at the end of each period (s_t), assuming the beginning inventory level $s_0 = 0$
Objective:	minimize the total production and inventory costs

Step 2. Let M be a "sufficiently" large number (say, $M = \sum_t d_t$). Note that $y_t = 1$ if and only if $x_t > 0$. The problem can be formulated as follows: Find values of x_t and y_t $(t = 1, 2, \ldots, T)$ so as to

$$\text{Minimize} \quad \sum_t (c_t x_t + f_t y_t + h_t s_t)$$

subject to	$s_{t-1} + x_t - s_t = d_t$	for all t
	$x_t \leq M y_t$	for all t
	$x_t \geq 0$	for all t
	$s_t \geq 0$	$t = 0, 1, \ldots, T$
	$y_t = 0$ or 1	for all t

Note that if backorder is allowed, we simply change the constraint from $s_{t-1} + x_t - s_t = d_t$ to $s_{t-1} + x_t - s_t - b_{t-1} + b_t = d_t$, to include a backorder amount in the inventory balancing equation, where b_t is the backorder amount cumulated at the end of time period t.

2.4.2 Capacitated Lot Sizing

In the event that the production quantity in a given time period cannot exceed a certain amount, for instance, due to plant capacity, then the problem becomes a capacitated lot sizing problem. When the given capacity is constant over periods, we simply replace the big "M" in the uncapacitated lot sizing model with a capacity upper limit u (i.e., replace $x_t \le M y_t$ with $x_t \le u y_t$).

The capacity may also vary from period to period with an upper limit u_t in period t, which is reflected in the model by using $x_t \le u_t y_t$ to replace $x_t \le M y_t$; the complete model becomes

$$\text{Minimize} \quad \sum_t (c_t x_t + f_t y_t + h_t s_t)$$

$$\begin{aligned}
\text{subject to} \quad & s_{t-1} + x_t + s_t = d_t && \text{for all } t \\
& x_t \le u_t y_t && \text{for all } t \\
& x_t \ge 0 && \text{for all } t \\
& s_t \ge 0 && \text{for all } t \\
& y_t = 0 \text{ or } 1 && \text{for all } t
\end{aligned}$$

2.4.3 Just-in-Time Production Planning

Now we present a multiproduct, multiperiod production planning problem under the just-in-time environment. This type of production planning seeks to determine a production level for each product in each time period with the right quantity at the right time. The ideal for just-in-time manufacturing is to maintain a zero inventory level (i.e., to prevent any surplus or shortage of inventory for each product at each time). However, in practice, there may occur a small surplus of inventory that can be temporarily stored on the plant floor in buffer area(s) or there may occur a temporary shortage of inventory. In either case, a penalty is imposed on each unit of excess or shortage of inventory. If no amount of shortage is allowed, a very large penalty should be imposed. Note that any excessive inventory implies production "too soon" and any shortage of inventory implies tardy production.

Thus, the primary objective of the just-in-time production problem may be modeled as to minimize the total penalties caused by the earliness/tardiness for all products over the planning horizon. The unit penalty of earliness and of tardiness, which may or may not be the same, may be assessed by the management. The model formulation follows.

Step 1

Input parameters:	number of product types (n), number of periods (T), demand of product j in each period (d_{jt}), prescribed production lot size for each product (l_{jt}), unit penalty of earliness (p_j), unit penalty of lateness (q_j)
Decision variables:	production level of each product in each period $(x_{jt} \geq 0)$, number of production runs in each period t for each product (y_{jt})
Constraints:	satisfy demand of each product j in each period and constraints relating to prescribed lot size, number of production runs per period, and production level
State variables:	surplus and shortage inventory levels for each product in each time period $(d_{jt}^{+}$ and $d_{jt}^{-})$, ending inventory level of each product (s_{jt})
Objective:	minimize total penalty cost of all products due to earliness and lateness over all periods

Step 2. Recall the inventory balancing equation that relates the beginning inventory level, production level, demand level, and the ending level given below:

$$s_{j,t-1} + x_{jt} - d_{jt} = s_{jt} \quad \text{for all} \quad j, t$$
$$\text{or} \quad s_{j,t-1} + x_{jt} - s_{jt} = d_{jt}$$

Let d_{jt}^{+} and d_{jt}^{-}, respectively, be a nonnegative amount of surplus and shortage for each period t and each product j. Let $s_{jt} = d_{jt}^{+} - d_{jt}^{-}$. Note that variable s_{jt} may be positive, negative, or zero; all x_{jt}, d_{jt}^{+}, and d_{jt}^{-} are nonnegative variables.

To model the relationships between the production level (a continuous variable), prescribed production lot size (an integer constant), and number of production runs (an integer variable), caution must be exercised because the production level may not be divisible by the prescribed lot size, which may result in a fractional number of production runs. To overcome this modeling difficulty, we introduce the following pair of inequality constraints. For example, assume the prescribed lot size in period t $(l_{jt}) = 150$ units and the production level in period t $(x_{jt}) = 700$ units. Then the number of lots in period t is $700/150 = 4.67$ or 5 after rounding up. Thus, the fifth (the last) lot contains only 100 units instead of 150. To resolve this problem, we introduce the following pair of inequality constraints:

$$x_{jt} = l_{jt} y_{jt} \quad \text{for all} \quad j \text{ and } t$$
$$\text{and} \quad x_{jt} = l_{jt}(y_{jt} - 1) \quad \text{for all} \quad j \text{ and } t$$

where $y_{jt} \geq 0$ and integer for all j and t.

Thus, the objective is to

$$\text{Minimize } z = \sum_j \left(p_j \sum_t d_t^+ + q_j \sum_t d_t^- \right)$$

2.5 WORKFORCE/STAFF SCHEDULING PROBLEMS

2.5.1 Scheduling Full-Time Workers

Many companies or institutions, especially those operating 24 h daily, usually divide the daily schedule into discrete (say, T) time windows. Examples include hospitals, restaurants, call centers, and police departments. The number of staff required typically varies among time windows. Staff members are scheduled to n different (work) shifts, each covering $m(m < T)$ consecutive time windows. Staff members assigned to different shifts may be paid differently, depending on which shift they work. For example, those working overnight are usually paid at a higher rate. The scheduling problem is to determine the number of workers to be assigned to each shift so that the company meets the demand in each time window.

Following is an example of a 24 h fast food restaurant. The daily operation is divided into eight consecutive time windows, each of 3 h duration. A shift covers three consecutive time windows (i.e., 9 h), as shown in Table 2.1. Information about the number of workers required within each time window as well as the wage level for each shift is also listed in the table.

Step 1

Input parameters:	number of shifts (n), number of time windows (T), number of workers required per time window (d_t), wage rate per shift (w_j)
Decision variables:	number of workers needed per work shift (y_j)
Constraints:	demand within each time window t must be satisfied
State variables:	none
Objective:	minimize the total wages paid to all workers

TABLE 2.1 Time Windows for Shift Workers

Time Window	Shift 1	2	3	4	Workers Required
6 a.m.–9 a.m.	X			X	55
9 a.m.–12 noon	X				46
12 noon–3 p.m.	X	X			59
3 p.m.–6 p.m.		X			23
6 p.m.–9 p.m.		X	X		60
9 p.m.–12 a.m.			X		38
12 a.m.–3 a.m.			X	X	20
3 a.m.–6 a.m.				X	30
Wage rate per 9 h shift	$135	$140	$190	$188	

Step 2. Let $a_{jt} = 1$ if shift j covers time window t ($j = 1, \ldots, n; t = 1, \ldots, T$) and $a_{jt} = 0$ otherwise. Then the model formulation is

$$\text{Minimize} \quad \sum_j w_j y_j$$

$$\text{subject to} \quad \sum_j a_{jt} y_j \geq d_t \qquad t = 1, \ldots, T$$

$$y_j \geq 0 \text{ and integer} \quad t = 1, \ldots, T$$

where the matrix (a_{jt}) is of the following form:

$$\begin{bmatrix} 1 & 0 & 0 & 1 \\ 1 & 0 & 0 & 0 \\ 1 & 1 & 0 & 0 \\ 0 & 1 & 0 & 0 \\ 0 & 1 & 1 & 0 \\ 0 & 0 & 1 & 0 \\ 0 & 0 & 1 & 1 \\ 0 & 0 & 0 & 1 \end{bmatrix}$$

Note that in the previous model if the integer requirement is relaxed, the problem might generate fractional solutions. In reality, a fractional staff member can be interpreted as a part-time worker. For example, a solution with 4.2 workers in shift 2 means that we have 4 full-time shift-2 workers, and a part-time worker who works 20% of the time and is paid 20% of a full-time workers. Hence, if it is allowed for some shifts to have part-time staff, then the problem becomes a mixed integer program. However, this is not the only way to handle the part-time situation. In Section 2.5.2, we will discuss another way to formulate the personnel scheduling problem when both full-time and part-time staffs are necessary in the model.

2.5.2   Scheduling Full-Time and Part-Time Workers

We still consider the problem described in Section 2.5.1. Now assume that part-time workers may be hired per time window. That is, during time window t, if a part-time worker is used, then he/she is paid c_t. However, at least one full-time worker has to be present when part-time workers are hired. The problem is to determine how many full-time and part-time workers need to be hired to minimize the total workforce cost.

Step 1

Input parameters:	number of shifts (n), number of time windows (T), number of workers required during each time window (d_t, $t = 1, 2, \ldots, T$), wage rate per 12 h shift for a full-time worker (w_j), wage rate per 6 h time window per part-time worker (c_t)
Decision variables:	number of full-time workers needed for each work shift (y_j), number of part-time workers needed for each time window (x_t)
Constraints:	demand within each time window t must be satisfied, restriction on using part-time workers (can be used only if one or more full-time workers are available in the same time window)
State variables:	none
Objective:	minimize the total wages paid to all workers

Step 2. Let $a_{jt} = 1$ if shift j covers time window t, 0 otherwise. Let M be a "sufficiently" large number (say, $M = \sum_t d_t$). Then the model formulation is

$$\text{Minimize} \quad \sum_j w_j y_j + \sum_t c_t x_t$$

$$\text{subject to} \quad \sum_j a_{jt} y_j + x_t \geq d_t \qquad t = 1, \ldots, T$$

$$M \sum_j a_{jt} y_j - x_t \geq 0 \qquad t = 1, \ldots, T$$

$$x_t, y_j \geq 0 \text{ and integer} \quad j = 1, \ldots, n; \ t = 1, \ldots, T$$

2.6 FIXED-CHARGE TRANSPORTATION AND DISTRIBUTION PROBLEMS

2.6.1 Fixed-Charge Transportation

Units of a product (single commodity) are to be shipped from m source nodes to supply the demands at n destinations (as shown in Figure 2.7). Shipping cost from source i to destination j includes a unit shipping charge c_{ij} in addition to a fixed charge f_{ij} if arc (i, j) is used in the solution, regardless of the shipping quantity (as long as a positive amount, of course). Find a minimum cost shipping plan so that the demand at each destination is met. Assume that each source node can supply all the demands at destinations.

Source   Destination

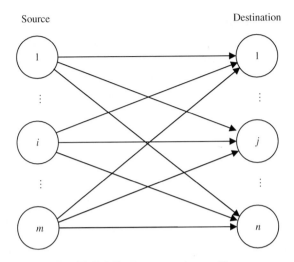

FIGURE 2.7   Transportation problem.

Step 1

Decision variables:	whether or not source i will supply destination j ($y_{ij}=1$ or 0). If yes, how much (x_{ij})
Input parameters:	unit shipping cost (c_{ij}), fixed cost (f_{ij}) from source i to destination j, demand at destination j (d_j)
Constraints:	demand at each destination must be satisfied (assuming unlimited product availability at each source node)
State variables:	none
Objective:	minimize sum of fixed and variable costs

Step 2. Let M be a "sufficiently" large number (we can let $M = \sum_j d_j$). Note that $y_{ij}=1$ if and only if $x_{ij}>0$. The transportation model can be formulated as

$$\text{Minimize} \quad \sum_i \sum_j (c_{ij}x_{ij} + f_{ij}y_{ij})$$

$$\text{subject to} \quad \sum_i x_{ij} = d_j \qquad j = 1,\ldots,n$$

$$x_{ij} \le My_{ij} \qquad i = 1,\ldots,m; \, j = 1,\ldots,n$$

$$x_{ij} \ge 0 \qquad i = 1,\ldots,m; \, j = 1,\ldots,n$$

$$y_{ij} = 0 \text{ or } 1 \qquad i = 1,\ldots,m; \, j = 1,\ldots,n$$

2.6.2   Uncapacitated Facility Location

A company needs to build several distribution centers to supply its retail stores located at n different cities, each with different demand. There are m candidate locations for the distribution centers. There is a unit transportation cost for shipping from distribution center i to retail store j, and a fixed cost for opening distribution center i. Decide on which distribution centers to open so that total cost (including opening cost and transportation cost) is minimized, while the demand at each retail store is satisfied.

Step 1

Decision variables:	whether or not distribution center i should be opened ($y_i = 1$ or 0). If opened, how much should be shipped from distribution center to retail store (x_{ij})
Input parameters:	unit shipping cost from center i to retail j (c_{ij}), fixed cost for opening distribution center (f_i)
Constraints:	all demands are to be met at all retail stores
State variables:	none
Objective:	minimize total cost of opening and transportation cost

Step 2. Let M be a "sufficiently" large number (we can let $M = \sum_j d_j$). Note that $y_i = 1$ if and only if $\sum_j x_{ij} > 0$. The uncapacitated facility location problem can be formulated as

$$\text{Minimize} \quad \sum_i \sum_j c_{ij} x_{ij} + \sum_i f_i y_i$$

$$\text{subject to} \quad \sum_i x_{ij} = d_j \qquad j = 1, \ldots, n$$

$$\sum_j x_{ij} \le M y_i \qquad i = 1, \ldots, m$$

$$x_{ij} \ge 0 \text{ and integer} \qquad i = 1, \ldots, m; \ j = 1, \ldots, n$$

Substituting $x'_{ij} = x_{ij}/d_j$ or $x_{ij} = d_j x'_{ij}$ into the above model, an alternate formulation is obtained:

$$\text{Minimize} \quad \sum_i \sum_j c'_{ij} x'_{ij} + \sum_i f_i y_i$$

$$\text{subject to} \quad \sum_i x'_{ij} = 1 \qquad j = 1, \ldots, n$$

$$\sum_j x'_{ij} \le n y_i \qquad i = 1, \ldots, m$$

$$x'_{ij} \ge 0 \qquad i = 1, \ldots, m; \ j = 1, \ldots, n$$

$$y_i = 0 \text{ or } 1 \qquad i = 1, \ldots, m$$

where $c'_{ij} = c_{ij}/d_j$. Note that x'_{ij} can be interpreted as the fraction (between 0 and 1 inclusive) of demand, rather than the quantity supplied, at store j satisfied by distribution center i. Also note that the "big M" is replaced by "n", which is the total number of retail stores. This replacement is valid because the demand at each store location, after the transformation, is equal to 1. Thus, the total demand at n store locations is n.

A third formulation can also be obtained by using a set of m constraints, $x'_{ij} \leq y_i$ ($j = 1, 2, \ldots n$), to replace each i of $\sum_j x'_{ij} \leq ny_i$. Although this replacement multiplies the number of constraints, this alternative does give a better formulation. We will justify this claim later in Chapter 4.

2.6.3 Capacitated Facility Location

When each distribution center has limited supply u_i, the uncapacitated facility location problem becomes a capacitated facility location problem.

Let u_i be the supply amount at distribution center i, then the first model formulation is

$$\text{Minimize} \quad \sum_i \sum_j c_{ij} x_{ij} + \sum_i f_i y_i$$

$$\text{subject to} \quad \sum_i x_{ij} = d_j \qquad\qquad j = 1, \ldots, n$$

$$\sum_j x_{ij} \leq u_i y_i \qquad\qquad i = 1, \ldots, m$$

$$x_{ij} \geq 0 \text{ and integer} \qquad i = 1, \ldots, m; \ j = 1, \ldots, n$$

$$y_i = 0 \text{ or } 1 \qquad\qquad i = 1, \ldots, m$$

2.7 MULTICOMMODITY NETWORK FLOW PROBLEM

A set of p commodities is to be shipped from m sources to n sinks. A source i can supply up to s_i^k units of commodity k. A sink j has demand d_j^k on commodity k. A transshipment node t is used as a connecting point between sources and sinks, but does not have its own supply or demand. The shipping amount between any pair of nodes is subject to a capacity limit, and for each unit of commodity k shipped from node i to t, or t to j, a cost is incurred. The problem requires finding a shipping plan that minimizes the total shipping cost as well as meets the demand for each commodity at each sink. Assume the nodes have been numbered consecutively and grouped into the three classes of source, transshipment, and sink nodes with indices i, t, and j, respectively. Also, assume no "backflow" is permitted from sink to transshipment or source, nor from transshipment to source.

Step 1

Decision variables:	units of commodity k to be shipped from source i to sink $j(x_{ij}^k)$, from source i to transshipment $t(x_{it}^k)$, and from transshipment t to sink $j(x_{tj}^k)$
Input parameters:	supply s_i^k of each commodity k at each source i, demand d_j^k for each commodity k at each sink j; maximum combined shipping capacity for all commodities from source i to sink $j(u_{ij})$, from source i to transshipment node $t(u_{it})$, from transshipment t to sink $j(u_{tj})$; unit transportation cost for commodity k that can be transported from source i to sink $j(c_{ij}^k)$, from source i to transshipment $t(c_{it}^k)$, from transshipment t to sink $j(c_{tj}^k)$
Constraints:	supply constraints for all sources, demand constraints for all sinks, flow conservation constraints for each transshipment node (total outflow equals total inflow for each commodity), maximum combined flow capacity for all commodities between any two nodes
State variables:	none
Objective:	minimize total transportation cost

Step 2. The problem can be mathematically modeled as

$$\text{Minimize} \quad z = \sum_k \sum_{(i,j)} c_{ij}^k x_{ij}^k + \sum_k \sum_{(i,t)} c_{it}^k x_{it}^k + \sum_k \sum_{(t,j)} c_{tj}^k x_{tj}^k$$

subject to $\quad \displaystyle\sum_{t,j} (x_{ij}^k + x_{it}^k) = s_i^k \quad$ for each i, k (node i supplies commodity k)

$$\sum_i x_{it}^k - \sum_j x_{tj}^k = 0 \quad \text{for each } t,\ k \text{ (node } t \text{ is a transshipment node)}$$

$$\sum_{i,t} (x_{ij}^k + x_{tj}^k) = d_j^k \quad \text{for each } j,\ k \text{ (sink } j \text{ demands commodity } k)$$

$$\sum_k x_{ij}^k \le u_{ij}$$

$$\sum_k x_{it}^k \le u_{it}$$

$$\sum_k x_{tj}^k \le u_{tj}$$

$$x_{ij}^k, x_{it}^k, x_{tj}^k \ge 0 \text{ and integer}$$

2.8 NETWORK OPTIMIZATION PROBLEMS WITH SIDE CONSTRAINTS

All the network problems we have discussed so far have common constraints: (1) each arc has some capacity limit, (2) flow on an arc is subject to unit cost, and (3) each node satisfies a flow conservation constraint. Sometimes, additional or side constraints are required. One of the most frequently seen side constraints are *proportional constraints* and *blending constraints*.

Proportional constraints are usually seen in production where raw materials are refined into different semiproducts, in which the amount of each semiproduct is specified as a proportion of raw materials. Figure 2.8 shows an example of such a production scenario. Requirements like this can be expressed by a set of side constraints in the following way.

Suppose the nodes are labeled with numbers, that is, the nodes associated with raw materials 1 and 2 are labeled nodes 1 and 2, respectively. The processor is node 3, semiproducts made from raw material 1 are nodes 4–6, and semiproducts made from raw material 2 are nodes 7–9. Let x_{ij} be the flow to be determined from node i to node j. Then the mathematical expressions for the proportional constraints are

$$x_{34} = 0.3x_{13}, \quad x_{35} = 0.5x_{13}, \quad x_{36} = 0.2x_{13}$$
$$x_{37} = 0.4x_{23}, \quad x_{38} = 0.1x_{23}, \quad x_{39} = 0.5x_{23}$$

Blending constraints are used to reflect a mixing or blending process. Several ingredients are mixed according to different ratios to get different products. Figure 2.9 shows such a blending process. With the raw material nodes labeled 1 and 2, as before, the processor node 3, and semiproduct nodes 4–6, such requirements can then be reflected by the following equality constraints: $x_{34} = 0.4x_{13} + 0.3x_{23}$, $x_{35} = 0.2x_{13} + 0.6x_{23}$, $x_{36} = 0.4x_{13} + 0.1x_{23}$.

Side constraints can take other forms too, either of special structure or of some arbitrary structure. Side constraints can be adjoined to many network optimization problems, such as multicommodity flow, facility location, and production lot sizing.

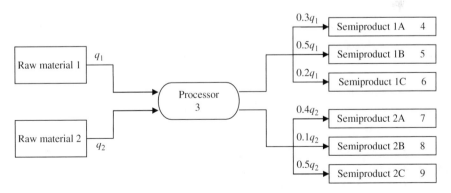

FIGURE 2.8 Proportional constraints.

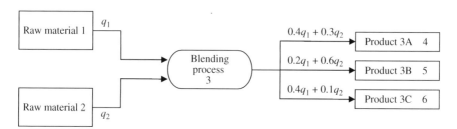

FIGURE 2.9   Blending constraints.

When embedded in a problem, side constraints might dramatically increase the difficulty in solving the problem, and new efficient algorithms must be developed to handle these constraints.

2.9   SUPPLY CHAIN PLANNING PROBLEMS

Two broad classes of operations research models are used to support supply chain management: normative models in the form of MIPs, which provide insight into pending decisions about supply chain structure, and descriptive models in the form of simulation models, which capture the dynamics of a proposed or existing supply chain after the structure is decided. Shapiro (2001) states that "optimization models provide templates for integration of concepts and constructs from multiple disciplines," which make up supply chain planning (SCP). According to Shapiro, "a company's supply chain is comprised of geographically dispersed facilities where products are acquired, transformed, stored, or sold, and transportation links connecting facilities along which products flow."

If product demand is assumed fixed, the SCP optimization problem is to minimize the total supply chain cost of satisfying demand, which may involve a simple transportation model (which distribution centers supply which products) or a complex, sequential decision involving multiple suppliers, multiple plants, and multiple distribution centers—and the transportation links among them. Furthermore, the time frame may vary from an operational planning model run weekly (for production or logistics planning) to strategic network models run once per year, with a planning horizon of 1–5 years. See Table 2.2 for time frames and horizons of typical MIP modeling situations in SCP.

Obviously, many supply chain problems have one of the network structures previously discussed, hence may be modeled as MIPs:

- Transportation model
- Assignment model
- Transshipment model
- Multicommodity flow model
- Single- and multicommodity capacitated flow model
- Multiple choices of mode of transport on the same arcs, each with costs and capacities

TABLE 2.2   Typical MIP Modeling Situations in Supply Chain Planning

Model Type and Objective Function	Planning Horizon	Model Structure	Use Frequency
Strategic network optimization (maximize net revenues or if demand is fixed, minimize the cost)	1–5 years	Yearly, or multiple linked years	Once/year
Tactical optimization model (minimize total cost of meeting forecasted demands)	12 months	Next 3 months and 3 quarters beyond	Once/month
Production planning optimization model (minimize avoidable production and inventory costs)	13 weeks	Next 4 weeks and 2 months beyond	Once/week
Logistics optimization model (minimize avoidable logistics costs)	13 weeks	Next 4 weeks and 2 months beyond	Once/week

The last item listed hints only at the broad applicability of 0–1 variables in supply chain modeling. Other well-recognized applications are to capture

- Fixed and investment costs
- Economies and diseconomies of scale
- Sole sourcing of markets
- A wide range of logical (if–then) conditions

Although it is recognized that integer variables should be used sparingly (only when necessary) in SCP models, their use in conjunction with MIP provides the company with powerful insights into decision situations that can literally convert a marginally profitable product line or supply chain into a profit maker. For a simple introduction to the use of MIP modeling constructs in SCP, see Shapiro (2001), Chapter 4.

As an example of a supply chain model, consider the following strategic distribution network model (Karabakal et al., 2000) implemented at Volkswagen of America.[1] Sources and markets were fixed, as was the variety of vehicle types and which sources would provide which vehicle type. The processing centers (which receive vehicles

---

[1] Reprinted with permission of authors (see Bibliography). Copyright 2000, the Institute for Operations Research and Management Sciences, 7240 Parkway Drive, Suite 300, Hanover, MD 21076, USA.

from sources) and the distribution centers (which receive vehicles from processing centers and provide them to markets) had to be located, the type of facility at the distribution centers had to be decided, and shipping quantities on each node of the network had to be determined. The objective was to minimize the total combined cost of transportation and fixed-facility installation. Therefore, the following model is a multicommodity, transshipment model with fixed and variable transportation costs between nodes and investment costs for the centers that are included in the network.

Step 1

Decision variables:	annual shipment of type k vehicles from source i to processing center $p(x_{ip}^k)$, annual shipment of type k vehicles from processing center p to distribution center $j(x_{pj}^k)$, annual shipment of type k vehicles from distribution center j to market $t(x_{jt}^k)$, yes–no variable on whether to install type $f(f = 1, 2)$ facility at distribution center $j(y_{jf} = 1$ or $0)$, yes–no variable on whether to install processing center $p(z_p = 1$ or $0)$
Input parameters:	annual demand for type k vehicles in market $t(d_t^k)$, mileage between distribution center j and market $t(m_{jt})$, cost of shipping one vehicle from source i to processing center $p(c_{ip})$, cost of shipping one vehicle from processing center p to distribution center $j(s_{pj})$, number of vehicles shipped to market t each load (L_t), fixed shipment cost per load of truck (C), shipment cost per mile traveled by each truck (v), fixed cost for installing a type f facility in distribution center $j(g_{jf})$, fixed cost for operating processing center $p(h_p)$, annual shipment capacity of a type 1 facility at distribution center $j(u_j)$
Constraints:	demand at each market area for each vehicle type must be met; vehicle flows from sources to processing centers and from processing centers to distribution centers must be balanced; total vehicle flow to each distribution center must satisfy the capacity limitation of the facility installed; no shipment to a distribution center (processing center) is possible if no facility installed there; facility type 2 (large) is installed only if facility type 1 (total capacity u_j at DC$_j$) does not have enough capacity to meet the shipment requirement to distribution center j, for each j
Objective:	minimize total cost, including shipping cost, facility installation cost, and processing centers operation cost

Step 2. Let M be a "sufficiently" large number, say, $M = \sum_t \sum_k d_t^k$, then the SCP problem can be mathematically formulated as

$$\text{Minimize} \quad \sum_j \sum_t \sum_k \left(\frac{x_{jt}^k}{L_t} \right) (C + vm_{jt}) + \sum_p \sum_j s_{pj} \sum_k x_{pj}^k + \sum_i \sum_p c_{ip} \sum_k x_{ip}^k$$

$$+ \sum_j \sum_f g_{jf} y_{jf} + \sum_p h_p z_p$$

subject to

$$\sum_j x_{jt}^k = d_t^k \qquad\qquad t = 1, \ldots, T; \, k = 1, \ldots, K$$

$$\sum_p x_{pj}^k = \sum_t x_{jt}^k \qquad\qquad j = 1, \ldots, J; \, k = 1, \ldots, K$$

$$\sum_i x_{ip}^k = \sum_j x_{pj}^k \qquad\qquad i = 1, \ldots, n; \, k = 1, \ldots, K$$

$$u_j y_{j2} \leq \sum_p \sum_k x_{pj}^k \leq u_j y_{j1} + M y_{j2} \qquad j = 1, \ldots, J$$

$$\sum_j \sum_k x_{pj}^k \leq M z_p \qquad\qquad p = 1, \ldots, P$$

$$x_{ip}^k, x_{pj}^k, x_{jt}^k \geq 0 \qquad\qquad i = 1, \ldots, n; \, j = 1, \ldots, J; \, p = 1, \ldots, P$$
$$t = 1, \ldots, T; \, k = 1, \ldots, K$$

$$y_{jf} = 1 \text{ or } 0 \qquad\qquad j = 1, \ldots, J; \, f = 1, 2$$

$$z_p = 1 \text{ or } 0 \qquad\qquad p = 1, \ldots, P$$

2.10   NOTES

Section 2.1

The seven MIP assumptions described in this section are extended from the four well-known LP assumptions described in introductory OR/MS textbooks such as Hillier and Lieberman (2005) and Winston (1994). The three additional assumptions are simultaneousness (conjunction), single objective, and integrality. Figure 2.1 is our original contribution intended to help practitioners understand and exploit these assumptions.

Section 2.2

The three-phase process of an OR study discussed in this section is similar to the three-phase modeling process given in Ravindran et al. (1987) and the five-phase process given in Taha (2007).

Section 2.3

Traditionally, "the knapsack problem" refers to the problem involving only one item of each type, each represented by a 0–1 variable. A problem that allows multiple items of each type is called *integer, general,* or *multi-item knapsack problem.* The knapsack problem received considerable attention in the literature during the early development of OR algorithms (1950–1970) mainly because it can be used as a subproblem in developing a decomposition algorithm for the well-known cutting stock (or trim loss) problem and because a general integer linear problem can be converted to a knapsack problem. Dozens of specialized algorithms for knapsack problems have been developed, encompassing dynamic programming, enumeration, Lagrangian multiplier, and network approaches.

Section 2.4

The just-in-time production planning model discussed in this section is based on a recent article by Li et al. (2006).

Section 2.8

SAS/OR User's Guide: Mathematical Programming (retrieved online at http://www. csc.fi/cschelp/sovellukset/stat/sas/sasdoc/sashtml/ormp/chap4/sect4.htm).

Section 2.9

For those interested in using MIPs in modeling supply chain planning problems at all levels (strategic, tactical, and operational), Shapiro (2001) is recommended.

2.11   EXERCISES

2.1   Consider the case of a quantity discount to a buyer, that is, the unit cost is lower when quantity purchased reaches a certain level. How would you express the quantity discount in the objective function of the lot sizing problem? Does the revised model still satisfy the assumptions of integer programming? (Assumption)

2.2   Give a situation (with side constraints) in which a project selection problem cannot be modeled as an MIP. (*Hint:* Include some special structure in the specification of the objective function or the constraint function.) (Assumption)

2.3   (A Diet Problem) Mrs. Bradley is on diet according to the instruction from her family doctor. Every day she can eat only several specific types of food and drink several specific beverages. There is even a limitation on how many ounces of each type of food she can eat at maximum. And she cannot take more than two types of beverages each day. Suppose if every day she eats *W* ounces of food and drinks *L* ounces of beverages, then she feels full. Given that each

TABLE 2.3   Stock Selection Options

Stock	Expected Annual Return in Present Value	Budget Requirement		
		Year 1	Year 2	Year 3
A	90	10	20	15
B	120	15	15	20
C	100	12	25	20
D	80	9	15	15
E	130	13	10	10
Money available		45	60	50

type of food or drink has a different unit price, how should she plan her diet to minimize the total daily cost? (Modeling Process)

(a) Follow the modeling process strictly and try to formulate the problem.

(b) Does the problem belong to any of the model types discussed in this chapter?

(c) What feature (variables or constraints) is unique about this problem?

2.4 Jimmy plans to invest in several stocks in the coming 3 years, each with a different expected return for each dollar invested and a specific amount of investment, as shown in Table 2.3 (all in thousands of dollars). Given that the amount Jimmy can invest in stock purchases is limited each year, help Jimmy to decide which stocks to invest in each year so as to maximize the total returns. (Project Selection)

2.5 (The Cutting Stock Problem) A standard fabric is usually L yards long. Based on customer need, it will be cut into small pieces of different lengths, say, $l_1, l_2, \ldots, l_n$. Any cutting combination will typically result in some unusable "leftover" material, of length less than $\min\{l_i\}$. Suppose the daily demand for the respective pieces is $d_1, d_2, \ldots, d_n$. Find a cutting pattern so that the leftover is minimized. (Modeling Process)

(a) Identify the parameters provided in this problem.

(b) Identify decision variables, objective, and constraints.

(c) What information is important for formulating this model but is not included in the problem description?

(d) If the information needed in part (c) is given, construct a model to solve the problem.

2.6 Nurses in large hospitals usually work 3 days a week. Daily demand for nurses is summarized in Table 2.4. Determine the number of nurses required per schedule type so that the total wage cost is minimized.

(a) What is the coefficient matrix $\mathbf{A} = (a_{jt})$?

(b) Use the numbers (not symbols) in the table to model this problem instance.

TABLE 2.4   Weekly Scheduling of Nurses

Day	1	2	3	4	5	Nurses Required
			Schedule Type			
Monday	X		X			20
Tuesday	X				X	25
Wednesday			X	X		26
Thursday		X			X	26
Friday		X		X	X	30
Saturday		X		X		30
Sunday	X		X			35
Weekly wage	525	470	550	500	425	

2.7   In Exercise 2.6, if part-time nurses are hired at the rate of $150/day, formulate the problem to minimize the total cost. If part-time nurses must be accompanied by at least four full-time nurses, how would you formulate this constraint?

2.8   XYZ University is planning on the construction of parking lots to solve the parking problem. There are m possible locations for parking lots, each with a specific amount of maintenance cost f_i, and a projected number of parking positions s_i. Students go to classes located at n different blocks. Distance from parking lot i to block j is a_{ij}. Forecast shows that the number of students attending classes at block j each day is around d_j. Assuming that one unit distance of walking costs $1, help the university to decide which parking lots to construct, and the most ideal parking situation, so that the total cost including walking and maintenance is minimized. (Facility Location)

2.9   Is the problem in Exercise 2.8 capacitated or uncapacitated? Under what situation(s) will it convert to the other? Do you believe such situation is realistic? Why? What if the maintenance cost of a lot is comprised of a fixed cost plus a variable cost that is proportional to the number of parking positions it contains? How does the model change?

2.10   (A Modified Caterer Problem) A caterer to "The Ritz" motel collects the dirty napkins and sends them to laundry every day. Due to different room occupation levels during a week, the number of dirty napkins on day i is d_i ($i = 1, \ldots, 7$). The caterer can wash and dry at most u napkins every day. If a dirty napkin is not cleaned on the same day, a new one is purchased at the price of c. If the laundry room is used on day i, a fixed cost of f_i is incurred. Assume that at the beginning of a week (Sunday), there are no dirty napkins left. That is, all dirty napkins are discarded at the end of the week, and Sunday's napkins are all new or clean. Find the best laundry plan for the caterer so that the entire week's cost is minimized. (Lot Sizing)

2.11   Cool Summer is a beverage company. It has 20 distribution centers located in different states to supply its 500 retail partners. Each retail partner j has a weekly demand level of d_j. Shipping cost per bottle of beverage from distribution center i to retail store j is c_{ij}. Once it is decided to ship from distribution center i to retail store j, a labor cost of f_i is incurred. Find the minimum cost shipping plan so that demand from each retail partner is satisfied. (Fixed Charge Transportation)

2.12 In Exercise 2.11, what changes will happen to the problem if distribution center i can only supply u_i bottles of beverages?

2.13 Consider Exercise 2.11 again. Now assume that Cool Summer produces four types of beverages. Each retail partner has different demand for each type of beverage. Shipping cost per bottle is the same for all four types and labor cost remains the same. The fixed cost is incurred once, if any quantity of any type of beverage is shipped from i to j. Formulate the problem to minimize the total cost. (Multicommodity Flow)

2.14 Formulate the following multicommodity flow problem as an IP: The Farmer's Orchard is a large fruit supplier in Georgia. It has three branch stores supplying five types of fruits to the distributors in six different cities. Due to the long distance and the fruit freshness requirements, some cities cannot be directly reached. Instead, the trucks have to stop at some other connecting cities, repack the fruit, and deliver from that city to the destination. The shipping network is shown in Figure 2.10. Demand for fruit type t in city i is shown in Table 2.5. Supply of fruit type $t(t = 1, \ldots, 5)$ from each branch store is shown in Table 2.6. Shipping cost for fruit t from city i to j is labeled below the arc (i, j) as a vector. Shipping capacity (regardless of fruit type) from city i to j is labeled above the associated arc. Develop a shipping policy to minimize the total shipping cost, while satisfying the demand from each retailer. (Multicommodity Flow)

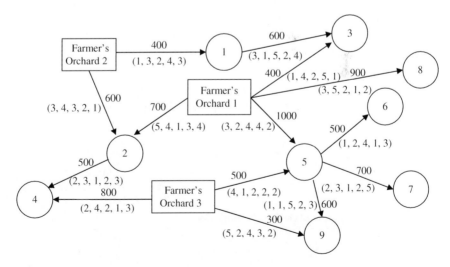

FIGURE 2.10 Multicommodity flow network.

TABLE 2.5 Multicommodity Demand

| | | | Fruit Type | | |
City	A	B	C	D	E
1	0	0	0	0	0
2	0	0	0	0	0
3	250	150	300	100	0
4	100	300	0	250	400
5	0	0	0	0	0
6	50	50	100	50	100
7	200	100	200	0	100
8	0	300	300	200	50
9	150	100	250	300	100

TABLE 2.6 Multicommodity Supply

| | | | Fruit Type | | |
Branch Store	A	B	C	D	E
1	400	350	350	0	150
2	100	500	300	700	0
3	250	150	500	200	600

2.15 Consider the following lot sizing problem with side constraints: The production plan for some product A is to be determined for the next T time periods. At the end of each period, 60% of the products unsold will go back to the assembly line and be renewed (assuming that this does not take up the capacity of the assembly line). The other 40% will be carried to the next time period as inventory. Demand at period t is d_t. Production and reassembling cost per unit is c_t for time period t. Inventory holding cost per unit is I_t. No backorders are allowed. Formulate the problem of finding the minimum cost production plan as an IP model. (Network with Side Constraints)

2.16 Consider the following multicommodity production–distribution problem with side constraints: Happy Bakery is a company making breads, cakes, muffins, and so on. It supplies 10 retailers in the city, including supermarkets, gas stations, and bakery thrift stores. Happy Bakery receives raw materials (flour, sugar, and butter) from two suppliers. Supplier A can provide up to 300 lb of flour, 500 lb of sugar, and 100 lb of butter. Supplier B can provide up to 700 lb of flour, 200 lb of sugar, and 150 lb of butter. Shipping cost for each raw material from each supplier is listed in Table 2.7. The ratios of the raw

TABLE 2.7 Raw Material Shipping Costs

Supplier	Flour	Sugar	Butter
A	0.2	0.05	0.8
B	0.3	0.04	0.7

TABLE 2.8 Travel Times from Depots to Neighborhoods

Neighborhoods	Depots						Population
	1	2	3	4	5	6	
1	15	17	27	5	25	22	12
2	10	12	24	4	22	20	8
3	5	6	17	9	21	17	11
4	7	6	8	15	13	10	14
5	14	12	6	23	6	8	22
6	18	17	10	28	9	5	18
7	11	10	5	21	10	9	16
8	24	22	22	33	6	16	20

materials in bread, cake, and muffin are 5:2:1, 4:4:1, and 3:2:1, respectively (assuming that the weight of water can be omitted). If the demand for bread is 400 lb, for cake is 300 lb, and for muffin is 200 lb, develop the minimum cost shipping plan. (Network with Side Constraints)

2.17 Read Chapter 4 of Shapiro (2001). Consider the example of "strategic planning at Ajax" in Section 4.3. Study the strategic model carefully. The model is a combination of which models discussed in this chapter (regardless of the objective function)? Identify them. Try to list the complete mathematical formulation of the problem using your own symbols.

2.18 (Shapiro, 2001, Exercise 4.3, p. 165[2]) Home Grocery is a new company that makes same-day deliveries of groceries to people's homes. The company is launching its business in Metropolis, a large urban area. The marketing department has identified eight neighborhoods in Metropolis where the company should concentrate its business. The logistics manager has identified six locations where the company may locate grocery depots. Table 2.8 shows the average time (in minutes) required to travel from each of the six potential depot locations to the center of each of the eight neighborhoods. It also shows the target population (in thousands) for the company's service in each neighborhood.

The company wishes to locate two depots so that they maximize the population served within 12 min of average travel time. Formulate the problem as an IP model.

[2] From Shapiro, *Modeling the Supply Chain*, 1st edition. Copyright 2001, South-Western, a part of Centage Learning, Inc. Reproduced with permission, www.centage.com/permissions.

3

TRANSFORMATION USING 0–1 VARIABLES

The ability to use 0–1 *(binary) variables* to formulate a wide variety of optimization problems expands the applicability of MIP and adds precision to modeling the real-world concerns of managers. In Chapter 2, many types of optimization problems are formulated by linear functions containing 0–1 variables. For example, in project selection, a 0–1 variable is used to represent whether or not a certain project is included in the project portfolio. In production planning, a 0–1 variable is used to represent whether or not a certain lot size is produced. In fixed-charge transportation and distribution problems, a 0–1 variable is used to represent whether or not an existing facility is utilized (or a new facility is built). All these examples share a common feature—any yes–no decision can be naturally formulated by using a 0–1 variable.

In this chapter, we move beyond the simple use of 0–1 variables to represent yes–no or on–off decisions. Though not obvious, 0–1 variables can also be used to transform a variety of optimization models into integer programs that model real-world considerations as follows:

1. Logical (Boolean) expressions
2. Nonbinary variables (discrete, integer)
3. Piecewise linear functions (arbitrary, concave—economic of scale)
4. Functions with products of 0–1 variables
5. Functions with products of binary and continuous variables (the bundle pricing problem)
6. Nonsimultaneous constraints (either/or, if/then, *p* out of *m*, negation)

Applied Integer Programming: Modeling and Solution, By Der-San Chen, Robert G. Batson, and Yu Dang
Copyright © 2010 John Wiley & Sons, Inc.

Clearly, a logical expression does not conform to the MIP format, but its true/false outputs correspond naturally to the values of a binary variable (1 for true, 0 for false). But, complications arise in almost any real-world MIP modeling effort. For example, the presence of a discrete variable (with nonconsecutive integer values) does not conform to the MIP format, the presence of a piecewise linear function violates the assumption of linear function, the presence of product terms of variables violates the MIP linearity assumption, and the presence of nonsimultaneous constraints violates the assumption of simultaneousness imposed on an integer program. Each of these violations must be resolved before the problem can be modeled and solved as an MIP. Such problem features will be addressed, one by one, in each of the following sections.

3.1 TRANSFORM LOGICAL (BOOLEAN) EXPRESSIONS

In some applications, using logical expressions may be easier, and even a more natural way, to describe problem requirements than mathematical expressions. That is, during the modeling process, the first model constructed may be in the form of logical expressions rather than mathematical expressions that conform to the MIP assumptions. The purpose of this section is to use 0–1 variables to transform logical relations into linear equations/inequalities that conform to the MIP assumptions.

3.1.1 Truth Table of Boolean Operations

Binary variables can be used to represent a variety of go/no–go or on/off decisions in the analysis of networks, such as transportation, electrical, and others. Binary variables are sometimes called *Boolean variables* in honor of the logician George Boole, who developed the rules of Boolean algebra for manipulating variables that can take on only two values. Originally these values were "true" and "false." However, a natural extension represents "true" by the value 1 and "false" by the value 0.

A basic logical relation deals with putting two *statements A* and *B* together to form a new combined (compound) statement, or to form a complement of a statement. In the context of Boolean algebra, a statement may represent a single Boolean variable or a *Boolean expression. In the context of MIP, a statement may represent a binary variable, a linear constraint, or even a set of linear constraints.* In this section, we shall focus on the operations of Boolean variables, leaving logical operations on linear constraints to the section of nonsimultaneous constraints. The following are the basic logical relations/operations of statements:

- Conjunction (A and B, $A \cap B$)
- Disjunction (A or B, $A \cup B$)
- Simple implication (If A then B, $A \rightarrow B$)
- Double implication (A if and only if B, $A = B$)
- Negation (not A, $\sim A$)

TABLE 3.1   Truth Table

(1)	(2)	(3)	(4)	(5)	(6)	(7)
Statement	Statement	A and B	A or B	If A then B	A if and only B	Negation
A	B	$A \cap B$	$A \cup B$	$A \rightarrow B$	if $A \Leftrightarrow B$	Not $A \sim A$
$T/1$	$T/1$	$T/1$	$T/1$	$T/1$	$T/1$	$F/0$
$T/1$	$F/0$	$F/0$	$T/1$	$F/0$	$F/0$	$F/0$
$F/0$	$T/1$	$F/0$	$T/1$	$T/1$	$F/0$	$T/1$
$F/0$	$F/0$	$F/0$	$F/0$	$T/1$	$T/1$	$T/1$

Each statement has two possible values (*true* or *false*), and four possible values for combining any two statements, as shown in columns 1 and 2 of Table 3.1. Five logical relations are listed as columns 3–7. Note that the negation of any statement has only two possible values.

3.1.2   Basic Logical (Boolean) Operations on Variables

We shall use the project selection problem described in Section 2.3 to illustrate various logical operations on 0–1 variables. Recall that the problem is to select a subset of n projects in a manner that maximizes the total present value while satisfying the budget limitation. To formulate it, we let $y_j = 1$ if project j is selected, and 0 otherwise. Translating to the logical expression, we have

Statement A: project A is selected ($y_A = 1$) or not selected ($y_A = 0$)
Statement B: project B is selected ($y_B = 1$) or not selected ($y_B = 0$)

To obtain a correct MIP model, keep in mind that (1) only linear equation(s)/inequalities are allowed, (2) if more than one linear constraint is required, these must be satisfied simultaneously, and (3) only the *true* value is of interest in the final logical output (i.e., the final value must be 1).

3.1.2.1   Conjunction (A and B, A ∩ B)   The conjunction of two statements, A and B, implies that both projects A and B are selected, or symbolically

$$y_A = 1 \text{ and } y_B = 1$$

Note that these two equations already satisfy the "simultaneousness" assumption. We can also use column 3 of Table 3.1 to verify the result. Note that the only true (T) value of four possible cases is when statement A is true ($y_A = 1$) and B is true ($y_B = 1$). An alternate formulation is $y_A + y_B = 2$.

3.1.2.2   Disjunction (A or B, A∪B)   The disjunction relation of two statements, A or B, implies that either A or B or both are true. In other words, at least one of the projects A or B must be selected. Clearly, the corresponding linear constraint is $y_A + y_B \geq 1$.

This constraint can be verified by column 4 of Table 3.1, where three possible cases are shown to be "true":

1. Project A is selected but not B ($y_A = 1$ and $y_B = 0$)
2. Project B is selected but not A ($y_A = 0$ and $y_B = 1$)
3. Both projects A and B are selected ($y_A = 1$ and $y_B = 1$)

Observe that in cases 1 and 2, $y_A + y_B = 1$, but $y_A + y_B = 2$ in case 3. To satisfy all the three cases, we have $y_A + y_B \geq 1$. Instead, if the problem requires that exactly one of the projects A and B can be selected, then the constraint must be an equality, $y_A + y_B = 1$.

3.1.2.3   Simple Implication (If A Then B, A $\rightarrow$ B)   "If statement A then statement B" (or "statement A implies statement B") means that if statement A is true, then statement B must be true; and if statement A is false, then statement B can be either true or false. Substituting statement A for project A and statement B for project B, we have the following three possible true cases:

1. Project A is selected ($y_A = 1$) and project B is selected ($y_B = 1$)
2. Project A is not selected ($y_A = 0$) and project B is selected ($y_B = 1$)
3. Project A is not selected ($y_A = 0$) and project B is not selected ($y_B = 0$)

Comparing the values of y_A and y_B in each case, we find that the following inequality captures all the three cases:

$$y_A \leq y_B$$

3.1.2.4   Negation (Not A, $\sim$A)   The negation of statement A is called "not A" or "$\sim A$" symbolically. The statement simply reverses "true to false," or "false to true," as shown in Column 7 of Table 3.1. That is, if $y_A = 1$, then $\sim y_A = 0$; or if $y_A = 0$, then $\sim y_A = 1$.

3.1.2.5   Relation Between Either/Or and If/Then Statements   There is an important relation between "either/or," "if/then," and "not A." That is, the following two statements are equivalent:

If A then B

$\sim A \cup B$

We may verify this by examining the values of all four possible cases in columns 3 and 5 of Table 3.2.

TABLE 3.2   Simple Implication and Negation

A	B	If A then B	$\sim A$	$\sim A \cap B$
1	1	1	0	1
1	0	0	0	0
0	1	1	1	1
0	0	1	1	1

TABLE 3.3   Linear Expressions for Boolean Relations

Logical Relation	Linear Inequality/Equation
$y_C = y_A \cap y_B$	$y_C \leq y_A$
	$y_C \leq y_B$
	$y_C \geq y_A + y_B - 1$
$y_C = y_A \cup y_B$	$y_C \geq y_A$
	$y_C \geq y_B$
	$y_C \leq y_A + y_B$
$y_A \rightarrow y_C$	$y_A \leq y_C$
$y_C = \sim y_A$	$y_C = 1 - y_A$

3.1.2.6   Double Implication or Biconditional (A If and Only If B)   Double implication means **A** implies **B**, and **B** also implies **A**. Applying this relation to the project selection problem, it means that project A is selected if and only if project B is selected. There are only two true cases for this compound statement, as shown in column 6 of Table 3.1:

(1) Project A is selected ($y_A = 1$) and project B is selected ($y_B = 1$)
(2) Project A is not selected ($y_A = 0$) and project B is not selected ($y_B = 0$)

To satisfy both cases, we must have $y_A = y_B$.

3.1.3   Multiple Boolean Operations on Variables

If there are two Boolean operations performed on three binary variables, for instance, $y_A \cap y_B \cup y_C$, then two steps are required: (1) perform $y_A \cap y_B$ and output a new variable (say, y_D), and (2) perform $y_D \cup y_C$ and evaluate the true value. Table 3.3 summarizes the logical operations and their corresponding 0–1 linear equations or inequalities. Table 3.3 illustrates the use of appropriate constraint combinations to represent compound operations.

3.2   TRANSFORM NONBINARY TO 0–1 VARIABLE

There are two types of nonbinary variables to be considered. In this context, we define a general integer variable that can take on *consecutive* integer values between 0 and infinity. If the smallest value of the variable is nonzero, then we add a simple lower bound constraint to the model. If the largest value is bounded, then we add a simple upper bound. All these cases, except for binary, are considered general integer variables. In this section, we refer to a discrete variable as one that takes on nonconsecutive integer values. For example, $z \in Z = \{2, 5, 9, 21\}$. Both general integer variables and discrete variables are called nonbinary integer variables.

3.2.1   Transform Integer Variable

Some algorithms apply only to problems with pure 0–1 variables. Conceptually, this places no limitation on their solution ability as any general integer variable $x \geq 0$ with a finite upper bound can be converted to a set of 0–1 variables. To illustrate how this is done, consider the integers restricted to 21 values $x = 0, 1, \ldots, 20$, which may be represented by a string of binary digits (bits):

$$x = 2^0 y_0 + 2^1 y_1 + 2^2 y_2 + 2^3 y_3 + 2^4 y_4$$
$$= 1 y_0 + 2 y_1 + 4 y_2 + 8 y_3 + 16 y_4$$

where $y_j = 0$ or 1 for $j = 0, 1, 2, 3, 4$. Note that possible combinations of binary variables y_j yield a range of 0–31, which can cover the integer values of x ranging from 0 to 20. Also note that the sum of coefficients that associate with x in all preceding terms is always 1 less than the coefficient of any term. In this example, $1 = 2 - 1$, $1 + 2 = 4 - 1$, $1 + 2 + 4 = 8 - 1$, and $1 + 2 + 4 + 8 = 16 - 1$. Based on these observations, we are able to calculate (or predetermine) the number of binary variables that are required for representing a given general integer variable $x \geq 0$.

Assume the upper bound of x is u. Then the required minimum number of binary variables, $k + 1$, must satisfy

$$2^k \leq u < 2^{k+1}$$

Taking $\log_2$ on the formula, we have

$$k \leq \log_2 u < k + 1$$

where k and $k + 1$, respectively, are integers obtained by rounding down and rounding up the value of $\log_2 u$. In this example, $u = 20$ giving $\log_2 (20) = 4.34$ or $k + 1 = 5$. For this problem, the proper binary representation of x can be obtained by substituting $k = 4$ into

$$x = 2^0 y_0 + 2^1 y_1 + 2^2 y_2 + \cdots + 2^k y_k$$

Note that the required number of binary variables is $k + 1$ or 5 because the set of binary variables begins with y_0.

Transformation of a bounded (but not necessarily positive) integer variable x, where $b \leq x \leq u$, is analogous to that described above for nonnegative integer variables. This is because $0 \leq x - b \leq u - b$, or $0 \leq x' \leq u'$, where $x' = x - b$ and $u' = u - b$ may be substituted into the expressions above for nonnegative integer variables bounded above.

Substituting this binary representation for each integer variable in the given IP problem will reduce the problem to a binary integer program but will increase the number of variables in the model. The increase may be large if the upper bound u of x is large. But the increase is not as fast as one might think. See Table 3.4 to get a feeling about the magnitude of the increase on the number of binary variables as u increases.

Therefore, the conditions that could make the conversion to binary variables useful are (1) a small number of general integer variables, each having a low upper bound, and (2) the proposed 0–1 algorithm is much more efficient than the existing general integer algorithms. The choice of using 0–1 transformation is more or less problem dependent. The practitioner should weigh the trade-off before using the transformation.

TABLE 3.4   Representing Integer Variable Using Binary Variables

u	10	100	1000	10,000
$\log_2 u$	3.32	6.64	9.96	13.29
$k + 1$	4	7	10	14

3.2.2   Transform Discrete Variable

When a discrete variable is limited to take only one value in a given list, then the discrete variable can be expressed by a set of binary variables. For example, the discrete variable z may take on only one value in the set $\mathbf{Z} = \{1, 5, 7, 9, 23\}$. To do this, we let a new 0–1 variable $y_i = 1$ $(i = 1, \ldots, 5)$ to represent the choice of the ith element in the set and then add the following set of constraints to the problem:

$$z = 1y_1 + 5y_2 + 7y_3 + 9y_4 + 23y_5$$

$$y_1 + y_2 + y_3 + y_4 + y_5 = 1$$

$$y_i = 0 \text{ or } 1 \text{ for all } i$$

The above constraint equation, which is the sum of all of the binary variables equal to 1, is called a *multiple choice constraint*.

3.3   TRANSFORM PIECEWISE LINEAR FUNCTIONS

3.3.1   Arbitrary Piecewise Linear Functions

Consider the following price structure offered by a seller for a certain commodity. The price is $10 per unit for the first 100 units, $9 per unit for the next 200 units, and $6 per unit for the next 200 units. Suppose at most 500 units may be purchased. Let x denote the number of units purchased and let $f(x)$ represent the total cost associated with the purchase of x units.

To represent the cost function $f(x)$ for this example, we first write a mathematical expression for $f(x)$. For the interval $0 \leq x \leq 100$, clearly $f(x) = 10x$. For the next interval $100 \leq x \leq 300$, then $f(x) = 10(100) + 9(x - 100)$. For the interval $300 \leq x$ 500, then $f(x) = 10(100) + 9(200) + 6(x - 100 - 200)$. Simplifying, we obtain the following mathematical function:

$$f(x) - 10x \qquad \text{if } 0 \leq x \leq 100 \tag{3.1}$$

$$f(x) = 100 + 9x \qquad \text{if } 100 \leq x \leq 300 \tag{3.2}$$

$$f(x) = 1000 + 6x \qquad \text{if } 300 \leq x \leq 500 \tag{3.3}$$

Note that the function $f(x)$ within each interval of x represents a line segment, bounded by two end points. Plotting this function graphically, we obtain Figure 3.1. Extending each line segment to reach the vertical axis, we obtain the *intercept* of that line. This

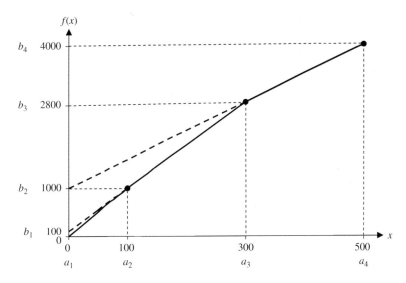

FIGURE 3.1 Representing a piecewise linear function.

intercept corresponds to the constant term of $f(x)$, and the slope of a line segment corresponds to the coefficient of x in $f(x)$. Specifically, the intercept of line segment 1 is 0 and the slope is 10, the intercept of line 2 is 100 and the slope is 9, and the intercept of line 3 is 1000 and the slope is 6. The collection of these three individual line segments form a "piecewise linear function" with four break-points located at $a_1 = 0$, $a_2 = 100$, $a_3 = 300$, and $a_4 = 500$. Note that the entire function is still considered nonlinear even though all individual segments are linear. Therefore, the piecewise linear function must be converted to an equivalent formulation involving only linear functions so that the resulting model can be solved by an MIP algorithm.

The piecewise linear cost function in this example is a *concave* function with a special property of having decreasing slopes of line segments ($10 > 9 > 6$). By taking advantage of this special property, there is a better formulation than the one for an arbitrary piecewise linear function. In this section, we first present an MIP formulation for the arbitrary piecewise linear function. Using the same example, we then introduce a better formulation for the "concave" piecewise linear cost function in Section 3.3.2.

Toward this end, consider two consecutive breakpoints, a_k and a_{k+1}, and the line segment between them. If x is any point lying on the line segment with end points, a_k and a_{k+1}, then x can be expressed by

$$x = \lambda_k a_k + (1-\lambda_k)a_{k+1}$$

where $0 \leq \lambda_k \leq 1$. Since $f(x)$ is also a line segment between $f(a_k)$ and $f(a_{k+1})$, it also follows that

$$f(x) = \lambda_k f(a_k) + (1-\lambda_k)f(a_{k+1})$$

Generalizing this idea to include all breakpoints, we can write

$$x = \lambda_1 a_1 + \lambda_2 a_2 + \cdots + \lambda_{r+1} a_{r+1}$$

$$f(x) = \lambda_1 f(a_1) + l_2 f(a_2) + \cdots + \lambda_{r+1} f(a_{r+1})$$

where $\lambda_1 + \lambda_2 + \cdots + \lambda_{r+1} + 1 = 1$, $\lambda_k \geq 0$ for all k, and at most two adjacent λ_k can be positive.

The condition that "at most two adjacent λ_k can be positive" is a nonmathematical expression, which must be replaced by a mathematical expression. To do this, we introduce a binary variable y_k for each line segment of the piecewise linear function and add the following set of linear constraints:

$$\lambda_1 \leq y_1$$

$$\lambda_2 \leq y_1 + y_2$$

$$\lambda_3 \leq y_2 + y_3$$

$$\vdots$$

$$\lambda_r \leq y_{r-1} + y_r$$

$$\lambda_{r+1} \leq y_r$$

$$\sum_{k=1}^{r} y_k = 1$$

$$\lambda_k \geq 0 \text{ for all } k$$

$$y_k = 0 \text{ or } 1 \text{ for all } k$$

Note that each y_k controls the value of λ_k and λ_{k+1}. That is, if $y_k = 0$, the above constraints force λ_k and λ_{k+1} to be 0. Likewise, if $y_k = 1$, then λ_k and λ_{k+1} are in the range $[0, 1]$. Since the constraint $\sum_{k=1}^{r} y_k = 1$ restricts all y_k so that exactly one y_k will have the value 1, exactly two adjacent λ_k are allowed to be nonzero in any solution. In fact, the y_k that assumes the value of 1 corresponds precisely to the line segment that is being used.

Applying this conversion to the example problem, the piecewise linear function of Figure 3.1 can be written as

$$x = 0\lambda_1 + 100\lambda_2 + 300\lambda_3 + 500\lambda_4$$

$$f(x) = f(0)\lambda_1 + f(100)\lambda_2 + f(300)\lambda_3 + f(500)\lambda_4$$

$$= 0\lambda_1 + 1000\lambda_2 + 2800\lambda_3 + 4000\lambda_4$$

$$\lambda_1 \leq y_1$$

$$\lambda_2 \leq y_1 + y_2$$

$$\lambda_3 \leq y_2 + y_3$$

$$\lambda_4 \leq y_3$$

$$y_1 + y_2 + y_3 = 1$$

$$\lambda_1 + \lambda_2 + \lambda_3 + \lambda_4 = 1$$

$$\lambda_k \geq 0 \text{ for all } k$$

$$y_k = 0 \text{ or } 1 \text{ for all } k$$

To utilize this technique in the context of integer programming, all occurrences of x and function $f(x)$ in the original objective and constraints should be replaced by continuous variables λ_k and binary variables y_k defined by the first two equations. The resulting new problem contains only λ_k and y_k variables. After adding the above remaining constraints to this problem, the augmented problem is equivalent to the original problem. After solving this equivalent problem, the solution in original variables x could be recovered by using the first equation.

3.3.2 Concave Piecewise Linear Cost Functions: Economy of Scale

"Economy of scale" is a common business practice. For example, suppliers offer various discounted unit prices for various scales of purchase quantities. Likewise, shippers offer various unit freight charge "breaks" for various scales of weights, and so on. The common property of these two is that the unit costs decrease as the quantity scales increase.

Consider the piecewise linear function defined in (3.1)–(3.3). There are three line segments with slopes $s_1 = 10$, $s_2 = 9$, and $s_3 = 6$, respectively. Because these slopes are in the decreasing order ($s_1 > s_2 > s_3$), the piecewise linear function is *concave*. The line segment can be expressed by

$$f(x) = b_i + s_i x \quad \text{for } i = 1, 2, 3$$

The intercepts of the three line segments are determined by

$$t_1 = 0 \text{ at } a_1 = 0$$
$$t_2 = t_1 + s_1 a_2 - s_2 a_2 = 0 + 10(100) - 9(100) = 100$$
$$t_3 = t_2 + s_2 a_3 - s_3 a_3 = 100 + 9(300) - 6(300) = 1000$$

To formulate the objective function, we define

$$y_i = 1 \quad \text{if } a_{i-1} < x \le a_i, \text{ 0 otherwise}$$
$$x_i = x \quad \text{if } a_{i-1} < x \le a_i, \text{ 0 otherwise}$$

In the objective function to be minimized, the formulation should include the following terms:

$$t_1 y_1 + s_1 x_1 + t_2 y_2 + s_2 x_2 + t_3 y_3 + s_3 x_3$$

and replace everywhere x appears by $x_1 + x_2 + x_3$. The required constraints similar to that of the fixed-charge problem are

$$x_i \le a_{i+1} y_i \quad \text{and} \quad a_i y_i \le x_i \quad \text{for } i = 1, 2, 3$$
$$y_1 + y_2 + y_3 \le 1$$
$$y_i = 0 \text{ or } 1 \qquad\qquad\qquad \text{for } i = 1, 2, 3$$

This formulation, by taking advantage of special concave property, is simpler than the one in Section 3.3.1.

Note that the above formulation considers a piecewise linear function for just a certain variable x. The extension to formulating a problem with multiple piecewise linear functions is straightforward as long as these functions are separable.

3.4 TRANSFORM 0–1 POLYNOMIAL FUNCTIONS

Consider a simple quadratic function in which the variables must be 0 or 1,

$$f(y_1, y_2, \ldots, y_n) = \sum_j y_j^2 + \sum_{i \ne k} y_i y_k$$

Obviously, each y_j^2 can be replaced by y_j without affecting the value of the function. Also, a new variable y_{jk} is needed to replace a product of $y_j y_k$ such that its values correspond to the values of y_j and y_k in Table 3.5.

Two linear constraints are added to give lower and upper bounds of $y_j + y_k$ for every pair j, k:

$$2y_{jk} \le y_j + y_k \le y_{jk} + 1$$

$$y_j, y_k, y_{jk} = 0 \text{ or } 1$$

TABLE 3.5 Linearization of a Quadratic Function in Two Binary Variables

Combination					
y_j	y_k	$y_{jk} = y_j y_k$	$2y_{jk}$	$y_j + y_k$	$y_{jk} + 1$
0	0	0	0	0	1
0	1	0	0	1	1
1	0	0	0	1	1
1	1	1	2	2	2

Functions with product terms of three binary variables can be transformed in a similar manner. A new binary variable y_{ijk} is introduced to replace a product of y_i, y_j, and y_k such that its values correspond to the values of y_i, y_j, and y_k in Table 3.6.

Two linear constraints are required to give lower and upper bounds of $y_i + y_j + y_k$, for every triple i, j, and k:

$$3y_{ijk} \leq y_i + y_j + y_k \leq y_{ijk} + 2$$

$$y_i, y_j, y_k, y_{ijk} = 0 \text{ or } 1$$

Higher degree functions can be generalized in a similar manner. In general, given a set, Q, composed of q 0–1 variables, the product $\prod_{j \in Q} y_j^p$, for any positive integer value of p, can be replaced by a single variable y_Q and imposing the additional constraints

$$\sum_{j \in Q} y_j \leq y_Q + (q-1) \tag{3.4}$$

$$\sum_{j \in Q} y_j \geq q y_Q \tag{3.5}$$

$$y_j, y_Q = 0 \text{ or } 1$$

Note that if any $y_j = 0$, then constraint (3.4) is nonrestrictive, constraint (3.5) becomes $y_Q < 0$, and therefore $y_Q = 0$. If all $y_j = 1$, then constraint (3.4) becomes $y_Q \geq 1$, the equality holds, and the desired relation is obtained.

TABLE 3.6 Linearization of a Cubic Function in Three Binary Variables

Combination						
y_i	y_j	y_k	y_{ijk}	$3y_{ijk}$	$y_i + y_j + y_k$	$y_{ijk} + 2$
0	0	0	0	0	0	2
0	0	1	0	0	1	2
0	1	0	0	0	1	2
0	1	1	0	0	2	2
1	0	0	0	0	1	2
1	0	1	0	0	2	2
1	1	0	0	0	2	2
1	1	1	1	3	3	3

Example 3.1   Consider the following 0-1 polynomial programming problem:

$$\text{Maximize} \quad 2y_1 y_2 y_3^2 + y_1^2 y_2$$

$$\text{subject to} \quad 12y_1 + 7y_2^2 y_3 - 3y_1 y_3 \le 16$$

$$y_1, y_2, y_3 - 0 \text{ or } 1$$

The conversion procedure is as follows:

1. Drop all positive exponents from the problem.
   Since $y^n = y$ for any binary y and $n > 0$, we can drop all positive exponents.
2. Replace each product term with a new binary variable.

Let $y_{123} = y_1 y_2 y_3$, $y_{12} = y_1 y_2$, $y_{23} = y_2 y_3$, and $y_{13} = y_1 y_3$. To ensure that the new variables correctly relate to the original variables, we must add a pair of linear inequalities for each new variable.

$$y_1 + y_2 + y_3 \ge 3y_{123} \tag{3.6}$$

$$y_1 + y_2 + y_3 \le y_{123} + 2 \tag{3.7}$$

$$y_1 + y_2 \ge 2y_{12} \tag{3.8}$$

$$y_1 + y_2 \le y_{12} + 1 \tag{3.9}$$

$$y_2 + y_3 \ge 2y_{23} \tag{3.10}$$

$$y_2 + y_3 \le y_{23} + 1 \tag{3.11}$$

$$y_1 + y_3 \ge 2y_{13} \tag{3.12}$$

$$y_1 + y_3 \le y_{13} + 1 \tag{3.13}$$

The new formulation becomes

$$\text{Maximize} \quad 2y_{123} + y_{12}$$

$$\text{subject to} \quad 12y_1 + 7y_{23} - 3y_{13} \le 16$$

$$\text{and } (3.6) - (3.13)$$

3.5   TRANSFORM FUNCTIONS WITH PRODUCTS OF BINARY AND CONTINUOUS VARIABLES: BUNDLE PRICING PROBLEM

Bundling products or services is a widespread marketing strategy. This strategy arises naturally when the products or services being offered are comprised of components. A firm must decide on prices for individual components and for bundled components so that its total revenue or profit is maximized. Examples of bundling products are ample across various industries:

1. A software firm has a product line composed of multiple software modules. Each module provides a unique set of features, ranging from statistics to graphics to database management to optimization.
2. A computer distributor has components such as basic computer, monitor, printer, hard disk, and memory board.
3. An insurance company has components such as auto, home, and life insurance policies; each component may also have several options.
4. A fast food restaurant has components of a hamburger, fries, and soft drink.
5. A travel agency offers products of airfare, rental car, and hotel.

All of the above components may be purchased individually or by bundling two or more components. If all n individual components can be bundled in any combination (including individually), then there are totally $2^n - 1$ possible component bundles. In practice, only a small subset of these will be considered for bundling. The main concern here is setting the prices for all product options, both individual and bundled components, so that the seller's total revenue is maximized. Toward this end, construction of a constrained optimization model has been attempted (Schrage, 2000), which turns out to be a nonlinear program in which the objective function contains products of binary and continuous variables. To fit into MIP format, the product term must be transformed to a set of linear functions. To describe the model, we give the following example.

Office Barn sells computers and accessories to its customers using the following strategy. Customers may choose from buying a computer only, a monitor only, or a bundle of computer and monitor. Its potential customers are categorized into four segments: home users, government and educational institutions, small firms, and medium and large firms. Assume that the size of each customer segment can be forecast accurately as can the maximum price each customer segment is willing to pay for each purchase option. Table 3.7 tabulates the data for a certain type of desktop computer and LCD monitor. If each customer buys exactly one option or buys nothing, how should Office Barn set the price for each option to maximize its total revenue?

TABLE 3.7   Office Barn Bundle Pricing Problem

Customer Segment	Expected Size (in 10,000)	Maximum Price Customer is Willing to Pay		
		Computer Only	LCD Monitor Only	Both
Home	5	600	350	850
Government and educational	15	700	350	1000
Small firms	8	650	300	900
Medium/large firms	12	700	300	900

Stigler's (1963) economic model for consumer behavior suggests that the relevant customer demand information is captured by a vector of *reservation prices* for the products. The reservation price is defined as the maximum price a customer is willing to pay. A customer will choose the product that maximizes *consumer surplus*, which is defined as the difference between the reservation price and the product price. For example, if the product prices for the three options, namely, computer only, monitor only, and both together, were set respectively at 550, 320, and 750, then for the home users the consumer surplus is respectively 50 (=600 − 550), 30 (=350 − 320), and 100 (=850 − 750). Then the home segment will buy both because the consumer surplus is the largest.

To develop a general model of this problem, we define

Input parameters: n_i = size of customer segments (number of individual customers in segment i), r_{ij} = reservation price of customer i for bundle j

Decision variable: x_j = price of bundle j to be determined

State variables: y_{ij} = 1 if customer segment i purchase bundle j, 0 otherwise; s_i = consumer surplus (=reservation price − selling price) achieved by customer segment i

The seller would like to determine x_j, the selling price of bundle j, to

$$\text{Maximize} \sum_i \left(n_i \sum_j y_{ij} x_j \right)$$

subject to constraint sets:

(1) Every customer buys exactly one bundle:

$$\sum_j y_{ij} = 1 \quad \text{for each } i$$

(2) Customer i will buy bundle j only when consumer surplus is maximum:

$$s_i \geq r_{ij} - x_j$$
$$\text{where } s_i = \sum_j (r_{ij} - x_j) y_{ij}$$

(3) Restrictions on variables:

$$y_{ij} = 0 \quad \text{for all} \quad i, j$$
$$x_j \geq 0$$

Note that there is a product term of $y_{ij} x_j$ in the objective function and in constraint set (2).

To linearize it, we replace $y_{ij}x_j$ by z_{ij} everywhere it appears in the model, giving the following modifications on the objective function and constraint set (2):

$$\text{Maximize} \sum_i \left(n_i \sum_j z_{ij} \right)$$

$$(2)' \quad \sum_j (r_{ij}y_{ij} - z_{ij}) + x_j \geq r_{ij}$$

in addition to $z_{ij} \geq 0$. Also add the following constraint sets to enforce the correct representation:

$$(4) \quad z_{ij} \leq x_j$$
$$(5) \quad z_{ij} \leq r_{ij}y_{ij}$$
$$(6) \quad z_{ij} \geq x_j - (1 - y_{ij})M_j$$

where M_j is an upper bound on x_j. Constraint set (4) ensures that z_{ij} ($=y_{ij}x_j$) cannot exceed the market price of bundle j. Constraint set (5) ensures that z_{ij} cannot exceed the maximum price that customer i is willing to pay for bundle j when it is purchased, and z_{ij} is 0 when bundle j is not purchased. If $y_{ij} = 1$, then $z_{ij} = x_j$ due to constraint (4). If $y_{ij} = 0$, then constraint (6) is satisfied due to redundancy.

3.6 TRANSFORM NONSIMULTANEOUS CONSTRAINTS

Recall that MIP assumes that all constraints must be satisfied simultaneously. In this section, we examine various types of nonsimultaneous constraints and show how to convert them to simultaneous constraints. Many of them are related to the following logical operations on constraints as they perform on binary variables:

- Either/or constraints
- Negation of a constraint
- If/then constraints

3.6.1 Either/Or Constraints

A decision variable may be defined by two disjunctive regions. For instance, variable x is defined outside the interval (3, 10). That is, either $x \leq 3$ or $x \geq 10$. To satisfy the simultaneousness assumption of MIP, they must be transformed.

To do this, rewrite the pair to $x - 3 \leq 0$ and $-x + 10 \leq 0$, and let M be a very big number such that $M \geq \max\{x - 3, -x + 10\}$. Let y be a binary variable. Then the disjunctive constraints can be replaced by two simultaneous constraints:

$$x - 3 \leq My$$
$$\text{and} \quad -x + 10 \leq M(1 - y)$$

Note that if $y = 1$, the constraint $x - 3 \leq M$ is redundant (and evidently satisfied) and one of the given constraints, $10 - x \leq 0$, is also satisfied. If $y = 0$, the other given constraint, $x - 3 \leq 0$, is satisfied, and the constraint $-x + 10 \leq M$ is also satisfied because it is redundant.

Consider the problem of scheduling jobs on a single machine. Let x_i and x_j respectively denote the start time of job i and job j to be scheduled. The start times may be continuous variables. Also, let t_i and t_j respectively represent the known machine time of job i and job j. Then the completion times of job i and job j are $x_i + t_i$ and $x_i + t_j$, respectively. Assume that a job once commenced must be processed until it completes. Since only one machine is available, it is impossible for two jobs to be scheduled during the same time interval. Thus, for any two jobs i and j, it must be true that

$$\text{Either } x_i + t_i \leq x_j$$

$$\text{or } x_j + t_j \leq x_i$$

The either-constraint enforces that job j cannot start before the completion of job i, and the or-constraint ensures that job i cannot start before the completion of job j. Obviously, both constraints cannot be satisfied simultaneously. Rewriting the constraints, we have

$$\text{Either } x_i - x_j + t_i \leq 0$$

$$\text{or } x_j - x_i + t_j \leq 0$$

Again introducing a binary variable y and a big M value, the either/or constraints can be converted to two simultaneous inequalities:

$$x_i - x_j + t_i \leq My$$

$$\text{and} \quad x_j - x_i + t_j \leq M(1-y)$$

Rewriting in MIP standard form, we have

$$x_i - x_j \leq t_i + My$$

$$\text{and} \quad x_j - x_i \leq t_j + M(1-y)$$

3.6.2 *p* Out of *m* Constraints Must Hold

Consider the case where the model has a set of m constraints but in addition requires only some p out of m (assuming $p < m$) constraints to hold. The problem allows selection of any combination of p constraints, and wants to select which p constraints so as to optimize the specified objective function. The $m - p$ constraints that are not selected are in effect dropped from the problem, although feasible solutions might coincidently satisfy some of them. This case is a direct generalization of the either/or case in which $m = 2$ and $p = 1$.

The formulation is similar to that of the either/or case. We let $y_i = 1$ if constraint i is selected, and 0 otherwise. In effect, p such constraints must have the following form of the constraint enforced for feasible $\mathbf{x}$, that is, $f_i(\mathbf{x}) - b_i \leq 0$. In addition, the remaining $p - m$ constraints are dropped from consideration, which can be accomplished by imposing the redundant constraints, $f_i(\mathbf{x}) - b_i \leq M$. To satisfy these two conditions, we thus use the constraints below:

$$f_i(\mathbf{x}) - b_i \leq My_i \quad \text{for } i = 1, 2, \ldots, m$$

$$\sum_i y_i = m - p$$

and y_i is binary for all i.

3.6.3 Disjunctive Constraint Sets

Now we consider a more generalized case where either one subset of constraints or another subset of constraints must be satisfied, but not both. These two subsets are disjunctive. We can convert them to form a set of simultaneous constraints. Let the two subsets be defined as follows:

Either subset 1: $\{\mathbf{a}_i^T\mathbf{x} - b_i \leq 0, \quad i = 1, 2, \ldots, m_1\}$
or subset 2: $\{\mathbf{c}_i^T\mathbf{x} - d_i \leq 0, \quad i = 1, 2, \ldots, m_2\}$

Again, let y be a binary variable and M be a big number such that it is greater than or equal to all constraints involved. Then the corresponding simultaneous constraints can be expressed as

$$\mathbf{a}_i^T\mathbf{x} - b_i \leq My \qquad i = 1, 2, \ldots, m_1$$

$$\mathbf{c}_i^T\mathbf{x} - d_i \leq M(1-y) \quad i = 1, 2, \ldots, m_2$$

3.6.4 Negation of a Constraint

Suppose the given constraint is $f(x_1, x_2, \ldots, x_n) - b \leq 0$ or $f(\mathbf{x}) - b \leq 0$, where b is the right-hand side constant. Then the negation of this constraint must be $f(\mathbf{x}) - b > 0$ or $-f(\mathbf{x}) + b < 0$.

3.6.5 If/Then Constraints

Previously we have shown that the logical statement

$$\text{If } A \text{ then } B$$

is equivalent to the logical statement

$$\sim A \cup B$$

In the context of MIP, we view a constraint as a statement. Specifically, constraints $f_1(\mathbf{x}) - b_1 \leq 0$ and $f_2(\mathbf{x}) - b_2 \leq 0$ are viewed as statements A and B, respectively. Then the negation of $f_1(\mathbf{x}) - b_1 \leq 0$ must be $f_1(\mathbf{x}) - b_1 > 0$ or $-f_1(\mathbf{x}) + b_1 \leq 0$. Therefore, the simple implication constraint:

$$\text{If } f_1(\mathbf{x}) - b_1 \leq 0 \quad \text{then} \quad f_2(\mathbf{x}) - b_2 \leq 0$$

is equivalent to

$$\text{Either } -f_1(\mathbf{x}) + b_1 \leq 0 \quad \text{or} \quad f_2(\mathbf{x}) - b_2 \leq 0$$

Applying the same transformation rule to either/or constraints, we obtain two simultaneous constraints:

$$-f_1(\mathbf{x}) + b_1 \leq My$$
$$f_2(\mathbf{x}) - b_2 \leq M(1-y)$$

where M is a big number such that $M \geq \max\{-f_1(\mathbf{x}) + b_1, f_2(\mathbf{x}) - b_2\}$ and y is a binary variable.

3.7 NOTES

Except for Sections 3.1 and 3.4, the materials of the remaining sections are based on the following OR, IP, and LP textbooks: Dantzig (1963), Garfinkel and Nemhauser (1972), Hillier and Lieberman (2005), LINDO Systems Inc. (2004), Nemhauser and Wolsey (1988), Salkin and Mathur (1989), Schrage (2003), Schriver (1986), Taha (1975, 2007), Wagner (1975), and Winston (1994).

Section 3.1

The truth table of logical relations can be found in an introductory textbook on finite or discrete mathematics, for example, Kemeny et al. (1959). This section describes basics about how Boolean operations (or computational problems of logic) can be formulated as 0–1 integer programs. For details about the connections between the methods of computational logic (or constraint logic programming) and integer programming, see Williams (1993, 1999), Williams and Wilson (1998), and Williams and Brailsford (1999).

Section 3.3

A function $f(\mathbf{x})$ is a convex function if for any two points, $\mathbf{x}_1$ and $\mathbf{x}_2$, $f[\lambda \mathbf{x}_1 + (1 - \lambda)\mathbf{x}_2] \geq \lambda f(\mathbf{x}_1) + (1 - \lambda) f(\mathbf{x}_2)$ holds for all λ, $0 \leq \lambda \leq 1$.

Section 3.4

The transformation of a 0–1 polynomial function to a pair of 0–1 linear functions was introduced by Watters (1967). The transformation has been applied to a high-risk investment problem in which a quadratic 0–1 function is maximized. Glover and Woolsey (1974) developed an improved transformation in which product terms are replaced by a continuous variable rather than integer variable and three new constraints, instead of two, are added. Other references include Glover and Woolsey (1973) and Hansen et al. (1993).

Section 3.5

Hanson and Martin (1990) first formulated the optimal bundle pricing problem as an MIP model. The model in the text is a simplified version, as given in Schrage (2003).

3.8 EXERCISES

3.1 E-Shop is an e-commerce company. For each transaction (can include multiple items) you make with it, a $9.50 transaction fee is charged. For every item you sell, E-Shop takes a commission that depends on the selling price. The commission rate is 5% for $(0, 45], 8% for $(45, 80], 12% for $(80, 100], 15% for $(100, 120], and 20% for sales above $120. Lee has three books to sell. If the acceptable price ranges for the three books are [40, 65], [75, 90], and [90, 110], respectively, formulate an IP problem to help Lee decide on the exact price for each book?

3.2 Give the final result of the following logical expressions, and verify your result using a truth table:

A, B, C, and D are four different events, where $A = T, B = F, C = F, D = T$

(a) $C \cap [(A \cup B) \rightarrow D] \cup \sim A$

(b) $(A \cap D) \cup [C \Leftrightarrow (B \cup D) \cup F]$

(c) $D \cup \{A \rightarrow [(C \cap A) \cup B] \cup \sim D\} \cap (C \cap B)$

3.3 Express the following statements using a linear integer formulation:

(a) A nurse can choose from a shift starting either before 11:00 a.m. or after 5:00 p.m.

(b) A task must be finished no earlier than 8:00 p.m., but no later than 6:00 p.m.

(c) If the completion time of a job is larger than its due date, it is counted as a late job.

3.4 Change the following functions to a linear integer formulation:

(a) $x_1 x_2 = 0, x_1, x_2 \geq 0$

(b) $y = \begin{cases} 2x + 3, & 0 < x \leq a, \quad a > 0 \\ 3x - 5, & a < x \leq b, \quad b > a \end{cases}$

3.5 A cell phone carrier provides several different plans for customers to choose. Plan A charges $0.10 per minute of usage, and has no monthly fee. Plan B has a monthly fee of $30, and an extra charge of $0.40 per minute if the usage exceeds 400 min. Plan C has a monthly fee of $40, and an extra charge of $0.60 per minute if the usage exceeds 600 min. Find the minimum cost plan for Donna if her monthly usage is at least 410 min. Formulate as an IP.

3.6 John is buying stocks. His broker suggests six different stocks, namely, 1, ..., 6. Formulate the stock selection problem subject to the following constraints, using 0–1 variables as needed:

(a) To lower the risk of losing money, John should buy at least two stocks.

(b) Due to John's budget limit, he cannot buy more than four stocks.

(c) Since stocks 3 and 5 belong to the same company, the broker recommends purchase of at most one of these.

(d) John's broker suggests the following two combinations: either choose two from stocks 1, 2, 3, and 4, or at least two from stocks 3, 4, 5, and 6.

(e) Stock 4 can only be purchased if stock 1 is bought.

3.7 Consider a system that uses binary digits to represent any possible value of all variables. For example, a decision of "yes" is represented by 1 and "no" is represented by 0. A die with six sides can be represented using three digits of binary values: 001 is 1, 010 is 2, 011 is 3, 100 is 4, 101 is 5, and 110 is 6. How many digits will be required to represent the alphabets, the states in the United States, and the days in a year?

3.8 Formulate the following scheduling problem as an IP problem. A set of n jobs are to be processed on m machines ($n > m$). Each job visits one and only one machine. Each job j has a release time (earliest time when the job is ready to be processed) r_j and a processing time p_{ij} on machine i, and no two jobs can be processed on the same machine simultaneously. But during the same time interval, different jobs can be processed on different machines. Schedule the processing order of the jobs so that the makespan z (the completion time of the last job) is minimized. (*Hint*: $z \geq x_{ij} + p_{ij}$, the completion time of job j on machine i.) Consider the following two cases:

(a) The m machines are identical, that is, $p_{ij} = p_j$ for each j

(b) The m machines are different

3.9 Consider the following single machine scheduling problem. A set of eight jobs, each with weight w_j and processing time p_j, are to be processed on a single machine. The precedence limitations of the jobs are depicted in Figure 3.2. No two jobs can be processed simultaneously. Schedule the processing order of the jobs so that the sum of the weighted completion times (the product of the completion time and weight) of all jobs is minimized.

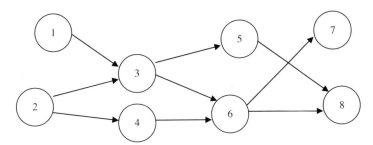

FIGURE 3.2 Job precedence.

Let x_j be the completion time of job j.

(a) Express the objective function using x_j.

(b) Express the constraints that job 1 must precede job 3, and job 4 must precede job 6.

(c) Express that no two jobs can be processed simultaneously. (*Hint*: For each pair of independent jobs, the schedule must be either i preceding j or j preceding i. Let $y_{ij} = 1$ if job i precedes job j, 0 otherwise.) Formulate the problem using y_{ij}'s.

3.10 A set of n jobs, each with processing time p_j and due date d_j, are to be processed on a single machine. If the completion time of a job exceeds its due date, then the job is said to be a "tardy" job. Otherwise, the job is not tardy. Formulate the scheduling problem to minimize the total number of tardy jobs as an IP.

3.11 The volume of a certain solid object changes with temperature. Let t be the temperature and $f(t)$ be its expansion index in percentage. The expansion index changes with temperature as follows:

$$f(t) = \begin{cases} 0.002t & 0 \le t \le 40°C \\ -0.001t + 0.12 & 40°C \le t \le 100°C \\ 0.003t - 0.28 & 100°C \le t \le 200°C \end{cases}$$

Transform this function into linear integer function(s).

3.12 Big Burger is a fast food restaurant with many chain stores all over the states. Customers to Big Burger can be divided into three types: kids, drivers drawn from highways, and workers nearby. Each type has different purchasing preference. The most popular food in Big Burger includes burgers, fries, and soft drinks. They can be purchased individually or as a combo meal, which includes all three items. The maximum price each customer group is willing to pay for each item or the combo, as well as monthly estimated number of customers from each group, is listed in Table 3.8. Determine the price to charge for individual and combo purchases so as to maximize revenue from sales.

TABLE 3.8 Big Burger Customers

Customer Type	Expected Monthly Customers	Maximum Price Customer is Willing to Pay			
		Burger	Fries	Soft Drink	Combo Meal
Kids	300	2.69	1.39	1.09	4.29
Drivers	240	2.99	0.99	1.29	4.89
Workers	600	2.59	0.99	1.19	4.19

3.13 Study the electrical circuit shown in Figure 3.3, where A, B, C, and D are four light bulbs.

 (a) Let the event "electrical current is through" be denoted by E, show the logic relationships between E and A, B, C, D.

 (b) Since each light bulb is either working or not working, we can define the possible value by 0–1 variable. If we want to maximize the probability of electrical current getting through, how would you formulate the objective function?

 (c) If we know the following constraints are enforced, formulate the problem as binary IP and transform it into linear IP:

 - A and C have the same performance.
 - The probability of both B and C are working is greater than the probability of two A's working.
 - The probability of D working is no larger than the probability of either A or C working.

 (d) Verify your work in part (3) using truth table.

3.14 The ABC University selects students out of a large population of applicants. To ease the selection process, the university makes a checklist of the students' qualifications. If a student satisfies at least four items in the list, then he/she is admitted. Formulate this as IP.

 (a) SAT score higher than 600
 (b) More than 10 A's in high school
 (c) GPA over 3.0

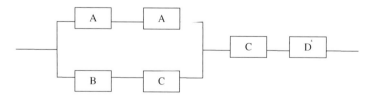

FIGURE 3.3 Electrical circuit schematic.

(d) Participated in at least one national subject competition

(e) Won at least one presidential prize

(f) Is considered talented in art

(g) Has been in gifted program for at least 5 years

(h) Is recommended by his/her high school counselor

3.15 Mandy is ordering a set of football tickets for the coming season. She plans to sell the tickets to make some money. There are two types of tickets: tickets for road games and tickets for home games. For each road game ticket, she could make a profit of $150, and for each home game ticket, the profit is $50 on average. The ticket office offers two price options:

(a) $5/home ticket, and no more than $50 purchase per person; $50/road ticket, and no more than $300 purchase per person.

(b) $7.5/home ticket, and no more than $100 purchase per person; $45/road ticket, and no more than $250 purchase per person.

How many tickets of each type should Mandy purchase so as to maximize the total profit she can make?

3.16 HomeMax offers the following promotion offer to customers: if a customer makes a purchase between $200 and $300, then he/she gets $25 off, and if the purchase if $300 or more, then he/she gets $40 off. Jimmy is buying some furniture at HomeMax. He can choose from the following items (not necessarily all): a gazebo, four chairs, one dining table, two long tables, one coffee table, and two TV stands. Price for each item is listed in Table 3.9. Help Jimmy decide which items to buy so as to achieve maximum saving, defined as the difference between the original price and the amount actually paid.

3.17 *Step function* (Taha, 1975. Used with permission). Show how the following step function can be represented as a 0–1 expression:

$$f(x) = b_i \quad a_{i-1} \leq x \leq a_i, \quad i = 1, 2, \ldots, n$$

where $b_i > b_{i-1}$ for all $i = 1, 2, \ldots, n$. In particular, show how the 0–1 variables relate to the variable x by specifying the appropriate constraints.

3.18 *Conditional constraints* (Taha, 1975. Used with permission). Consider the constraints $a_i \leq f_i(x)_i \leq b_i$, $i = 1, 2, 3$ and a_i and b_i are given constants for all

TABLE 3.9 HomeMax Prices

Item	Gazebo	Chair	Dining Table	Long Table	Coffee Table	TV Stand
Price ($)	119	69	199	29	49	69

i. Show how the following conditional constraints can be expressed as manageable forms by using 0–1 variables:

(i) $f_1(x) > 0 \Rightarrow f_2(x) \geq 0$

(ii) $f_1(x) > 0 \Rightarrow f_2(x) \geq 0$ and $f_1(x) < 0 \Rightarrow f_3(x) \geq 0$

3.19 *Absolute Value Constraint* (Taha, 1975. Used with permission). Convert the following constraint into a manageable form by simulating the effect of the absolute value using 0–1 variables:

$$\left| \sum_{j=1}^{n} a_{ij} x_j \right| \geq b > 0$$

4

BETTER FORMULATION
BY PREPROCESSING

For any given *integer programming* problem, there always exist many, possibly infinite, alternative formulations. Intuitively, a *better formulation* means that it is easier to solve. In this chapter, we make precise what is meant by a formulation, examine alternative formulations, and explain why some formulations might be better than others. Then we present some basic preprocessing techniques that can be used for transforming a given formulation to a better formulation for a *general* MIP program as well as for a *special* pure 0–1 program. Problem preprocessing has made a great contribution to the success of the modern branch-and-cut algorithms for solving combinatorial optimization problems and large-scale 0–1 integer programming problems.

4.1 BETTER FORMULATION

For a linear programming (LP) problem, the size of the *problem matrix* is commonly used for measuring the quality of a formulation, where problem size is a function of the number of constraints (or rows), number of variables (or columns), and number of nonzero elements in the problem matrix. In the absence of a sparse matrix, LP problem size can be approximated by the product of the number of variables and constraints. Thus, for an LP problem, a smaller problem matrix generally means a better formulation. But for an IP/MIP problem, the simple use of the problem matrix is no longer accurate for measuring its formulation quality for several reasons.

Applied Integer Programming: Modeling and Solution, By Der-San Chen, Robert G. Batson, and Yu Dang
Copyright © 2010 John Wiley & Sons, Inc.

First, integer and continuous variables should not be weighted equally in measuring IP formulation quality because a problem with integer variables is much more difficult to solve than that with continuous variables. Moreover, the degree of difficulty increases exponentially as the number of integer variables increases.

Second, an IP formulation with *more* constraints may be *easier* to solve (not *harder* to solve, contrary to an LP formulation) because extra constraints may often help "tighten" a continuous feasible region. The cutting plane methods (see Chapter 11) provide good evidence for this phenomenon, in which the latter iteration has more cutting constraints and a smaller feasible region. Another example arises in the alternative formulations for the uncapacitated facility location problem (Section 2.6.2), which will be detailed later in this section.

Therefore, measuring the quality of an IP formulation needs a different criterion than the one associated with an LP problem matrix. This criterion is based on the quality of the polyhedron (feasible region) of the linear programming relaxation of the given integer program. The key idea of preprocessing is to reformulate problems so as to make the difference in the objective function values between the solutions to the *linear programming relaxation* and the respective *integer program* as small as possible.

There is another important reason why the LP relaxation is commonly used for measuring the quality of IP formulations. This is due to the fact that the state-of-the-art methods used for solving *general* integer programs are *based* on linear programming, or the so-called LP-based methods. In fact, as can be seen in later chapters, all existing general IP methods (including branch-and-bound, cutting plane, and branch-and-cut) require solving a large number of LP relaxation problems, numbering in the thousands or even millions for a moderate or large-scale combinatorial optimization problem.

Before rigorously defining the terms such as polyhedron, formulation, and better formulation, let us consider three expository examples of *pure* IP constraints:

$$\text{IP1:} \quad 2y_1 + 2y_2 \leq 3$$
$$y_1, y_2 \text{ integer}$$

$$\text{IP2:} \quad 3y_1 + 2y_2 \leq 3$$
$$y_1, y_2 \text{ integer}$$

$$\text{IP3:} \quad y_1 + y_2 \leq 1$$
$$y_1, y_2 \text{ integer}$$

Plotting them in Figure 4.1, we see that each IP constraint set contains exactly the same set of feasible *integer* points, $S = \{(0, 0), (0, 1), (1, 0)\}$.

Relaxing the integer requirements of y_1 and y_2 from the given programs, we obtain the following LP relaxations:

$$\text{LP1:} \quad 2y_1 + 2y_2 \leq 3$$
$$y_1, y_2 \geq 0$$

$$\text{LP2:} \quad 3y_1 + 2y_2 \leq 3$$
$$y_1, y_2 \geq 0$$

$$\text{LP3:} \quad y_1 + y_2 \leq 1$$
$$y_1, y_2 \geq 0$$

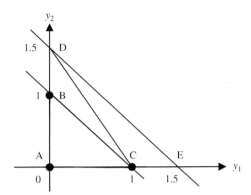

FIGURE 4.1 Integer feasible points.

Plotting LP1, LP2, and LP3, respectively, in Figure 4.1, we obtain three (continuous) feasible regions enclosed by triangles ADE, ADC, and ABC. Geometrically, each of these regions is called a *polyhedron*. Since each of these polyhedra contains the same set $S = \{(0, 0), (0, 1), (1, 0)\}$, they are *alternative formulations for set S.*

Now we rigorously define the terms "polyhedron," "formulation," and "better formulation." Let E^n and Z^p, respectively, denote n-dimensional real space and p-dimensional integer space. Thus, $\mathbf{x} \in E^n$, $\mathbf{y} \in Z^p$, and the set of mixed integer feasible solutions $S \subseteq E^n \times Z^p$.

Definition 4.1 The set of all points (or solutions) that satisfy a set of linear constraints, denoted by $P = \{\mathbf{x}: \mathbf{Ax} \le \mathbf{b}, \mathbf{x}$ continuous$\}$, is a *polyhedron.*

Recall that $\mathbf{Ax} \le \mathbf{b}$ represents the *standard* linear constraint set and those non-standard linear constraints in $\le$, $\ge$, and $=$ forms as well as nonnegativity restrictions $\mathbf{x} \ge \mathbf{0}$ can be converted to the standard form. For example, $\mathbf{x} \ge \mathbf{0}$ can be converted to $-\mathbf{x} \le \mathbf{0,}$ which in turn is a simple form of $\mathbf{Ax} \le \mathbf{b}$. This definition states that any feasible region formed by a linear program is a polyhedron.

For a pure integer program, a formulation for a pure integer feasible region S_y is defined below.

Definition 4.2 Given $S_y = \{\mathbf{y} \in Z^p: \mathbf{Gy} \le \mathbf{b}, \mathbf{y}$ integer$\}$. A polyhedron $P \subseteq E^p$ is a *formulation* for S_y if and only if $S_y = P \cap S_y$, which is the same as $S_y \subseteq P$.

Note that a formulation for a set of pure integer feasible solutions must satisfy two conditions: (1) It must be a polyhedron defined in the same p-dimensional real space. (2) It must contain *exactly* the same set of integer feasible points S_y (i.e., the polyhedron contains no more integer points than in S_y and no less integer points than in S_y). For example, polyhedron ABFC in Figure 4.2 is not a formulation for $S_y = \{(0,0), (0,1), (1,0)\}$ because it contains an extra integer point $(1, 1)$. Polyhedron AGC is not a formulation for S_y because point $(0, 1)$ is not in S_y.

For a mixed integer program, a formulation for a mixed feasible region S_{xy} is defined below.

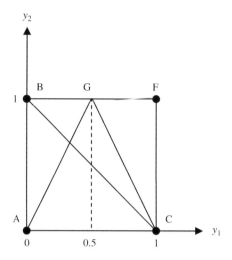

FIGURE 4.2   Formulations.

Definition 4.3   Given $S_{xy} = \{(\mathbf{x}, \mathbf{y}): \mathbf{A}\mathbf{x} + \mathbf{G}\mathbf{y} \leq \mathbf{b}, \mathbf{x} \in E^n, \mathbf{y} \in Z^p\}$. A polyhedron $P \subseteq E^{n+p}$ is a *formulation* for S_{xy} if and only if $S_{xy} = P \cap S_{xy}$.

For example, consider the constraint set of a mixed integer program

$$x + y \leq 1$$
$$x \geq 0$$
$$y \text{ integer}$$

Plotting in Figure 4.3, we see that the set of mixed integer feasible points S_{xy} contains point $(0, 1)$ and a line segment AC defined by $(x, 0)$, $0 \leq x \leq 1$. Relaxing the integer requirement on y, we obtain formulations ABC and ADC for S_{xy}.

We now define a *better formulation* for a pure integer program below. Definition of a better formulation for a mixed integer program is similar.

Definition 4.4   Given two formulations P_1 and P_2 for S_y. P_1 is a *better formulation* than P_2 if $P_1 \subset P_2$, that is, P_1 is a proper subset of P_2.

In Figure 4.1, since formulation LP3 $\subset$ LP2, LP3 is a *better formulation* than LP2. Similarly, LP2 is a better formulation than LP1. However, given any two formulations, we may not know whether one is better than the other. For example in Figure 4.4, we cannot tell whether P_3 is better than P_2 even though P_3 looks smaller in area than P_2. Note that just the size alone of a feasible region does not necessarily determine the quality of formulation.

Definition 4.5   Given $S = \{\mathbf{y}: \mathbf{G}\mathbf{y} \leq \mathbf{b}, \mathbf{y} \text{ integer}\}$. A formulation S is *ideal* if all extreme points of the polyhedron are integer.

In linear programming theory, if there exists a finite optimum solution (maximum or minimum) and if all the extreme points (or basic feasible solutions) of the

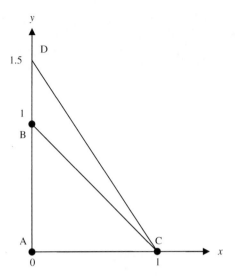

FIGURE 4.3   Better formulation.

polyhedron are integer, then one of the extreme points must be an integer optimum to a pure integer program. Therefore, in this special case, solving the relaxed LP problem will automatically solve the original integer program. It is instructive to draw several linear objective functions of various slopes on Figure 4.4. We can see that no matter what the coefficients (gradient vector) of the objective functions is, a maximum or minimum must always fall on one of the extreme points.

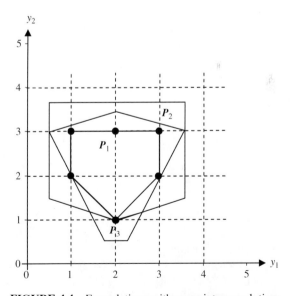

FIGURE 4.4   Formulations with same integer solutions.

Considering Figure 4.1, many other formulations are also possible for the set S, and readers are encouraged to draw some of them. You can see that there is a smallest formulation whose extreme points are integer. Geometrically, it is called the *convex hull* of the set S, denoted by Conv(S). This is an *ideal* formulation. Special classes of combinatorial optimization problems, such as the assignment, transportation, trans-shipment, maximum flow, and linear minimum cost flow, have the property that their LP relaxation is the convex hull of basic feasible integer solutions. We refer to this class as "easy integer programs" to be discussed in Chapter 10. However, the ideal formulation for a general integer program is very difficult to find.

In what follows, we give two real-world examples to show how one formulation is better than the other. One example is a knapsack problem for a pure 0–1 integer program and another is an uncapacitated facility location problem for a mixed integer program.

Example 4.1 (The Knapsack Problem)   The following two polyhedra, P_1 and P_2, are formulations for S because they satisfy $P \subseteq E^5$ and $S = P \cap S$, where $S = \{(0, 0, 0, 0, 0), (1, 0, 0, 0, 0), (0, 1, 0, 0, 0), (0, 0, 1, 0, 0), (0, 0, 0, 1, 0), (0, 0, 0, 0, 1), (1, 1, 0, 0, 0), (1, 0, 1, 0, 0), (1, 0, 0, 1, 0), (1, 0, 0, 0, 1), (0, 1, 1, 0, 0), (0, 1, 0, 1, 0), (0, 1, 0, 0, 1), (0, 0, 1, 1, 0), (0, 0, 1, 0, 1), (0, 0, 0, 1, 1), (1, 1, 1, 0, 0), (1, 1, 0, 0, 1), (1, 0, 1, 1, 0), (1, 0, 1, 0, 1), (1, 0, 0, 1, 1), (0, 1, 1, 1, 0), (0, 1, 1, 0, 1), (0, 1, 0, 1, 1,), (0, 0, 1, 1, 1), (1, 1, 1, 0, 1), (1, 0, 1, 1, 1), (0, 1, 1, 1, 1)\}$:

$$P_1 = \{\mathbf{y} \in E^5 : 13y_1 + 21y_2 + 4y_3 + 17y_4 + 4y_5 \le 47, \mathbf{0} \le \mathbf{y} \le \mathbf{1}\}$$
$$P_2 = \{\mathbf{y} \in E^5 : 3y_1 + 3y_2 + y_3 + 3y_4 + y_5 \le 8, \mathbf{0} \le \mathbf{y} \le \mathbf{1}\}$$

To show that formulation P_2 is better than P_1, we must show that $P_2 \subset P_1$, which is equivalent to show that $P_2 \subseteq P_1$ and $P_2 \ne P_1$. If we can show that all the points in P_2 are also in P_1, then $P_2 \subseteq P_1$. In addition, if we can also show there exists a point in P_1 but not in P_2, then $P_2 \subset P_1$.

First, we show that all the points in P_2 are also in P_1. Multiplying by 4 on both sides of the constraint in P_2, we obtain an equivalent constraint

$$12y_1 + 12y_2 + 4y_3 + 12y_4 + 4y_5 \le 32$$

The constraint in P_1 can be rewritten as

$$13y_1 + 21y_2 + 4y_3 + 17y_4 + 4y_5 \le 47$$
$$\text{or} \quad (12y_1 + 12y_2 + 4y_3 + 12y_4 + 4y_5) + (y_1 + 9y_2 + 5y_4) \le 47$$

If we can show that $y_1 + 9y_2 + 5y_4 \le 47 - 32$, then it implies that $P_2 \subseteq P_1$. The claim is true because $\mathbf{0} \le \mathbf{y} \le \mathbf{1}$.

Next, to show that there exists a point in P_1 but not in P_2, consider the point $\mathbf{y}^* = (0.02, 1, 1, 1, 1)$. We have $\mathbf{y}^* \in P_1$ since $13(0.02) + 21(1) + 4(1) + 17(1) + 4(1) = 46.26 < 47$, but $\mathbf{y}^* \notin P_2$ since $3(0.02) + 3(1) + 1(1) + 3(1) + 1(1) = 8.06 > 8$. Hence, we conclude that $P_2 \subset P_1$.

Example 4.2 (The Uncapacitated Facility Location Problem) There are m machines available to meet the production requirement from n workshops, each with a demand 1. Once a machine is set up, a fixed cost of f_i is incurred. Unit transportation cost of products from machine i to workshop j is c_{ij}. The objective is to find a production plan with the lowest cost while meeting the demands at all workshops.

To model this problem, we let x_{ij} be the fraction of demand from workshop j met by machine i. Also, let y_i be 1 if machine i is used, and 0 otherwise. Two alternative models are obtained.

$$(IP_1) \quad \text{Minimize} \quad \sum_{i=1}^{m}\sum_{j=1}^{n} c_{ij}x_{ij} + \sum_{i=1}^{m} f_i y_i$$

$$\text{subject to} \quad \sum_{i=1}^{m} x_{ij} = 1 \qquad j = 1, 2, \ldots, n \qquad (4.1)$$

$$\sum_{j=1}^{n} x_{ij} \le n y_i \qquad i = 1, 2, \ldots, m \qquad (4.2)$$

$$x_{ij} \ge 0 \qquad i = 1, 2, \ldots, m; j = 1, 2, \ldots, n$$

$$y_i = 0 \text{ or } 1 \qquad i = 1, 2, \ldots, m$$

$$(IP_2) \quad \text{Minimize} \quad \sum_{i=1}^{m}\sum_{j=1}^{n} c_{ij}x_{ij} + \sum_{i=1}^{m} f_i y_i$$

$$\text{subject to} \quad \sum_{i=1}^{m} x_{ij} = 1 \qquad j = 1, 2, \ldots, n$$

$$x_{ij} \le y_i \qquad i = 1, 2, \ldots, m; j = 1, 2, \ldots, n \qquad (4.3)$$

$$x_{ij} \ge 0 \qquad i = 1, 2, \ldots, m; j = 1, 2, \ldots, n$$

$$y_i = 0 \text{ or } 1 \qquad i = 1, 2, \ldots, m$$

Note that the two IP models are similar except for constraint sets (4.2) and (4.3). Considering the problem size, model IP_2 is *larger* than IP_1 because the number of constraints in (4.3) is n times that of (4.2). However, we claim that formulation P_2 is *better* than P_1 because $P_2 \subset P_1$, where these represent the LP relaxation of the respective integer programs, hence in both models y_i becomes *continuous* on [0, 1].

To show that $P_2 \subset P_1$, we simply need to show that any points in P_2 also lie in P_1, but not vice versa. Since the only difference in these two formulations is that (4.3) replacing (4.2) in P_2, showing $P_2 \subset P_1$ is equivalent to showing that any points satisfying (4.3) also satisfy (4.2), but not vice versa.

Clearly, if we sum the inequalities in (4.3) over the range of j, then we obtain (4.2). Hence, every point satisfying (4.3) must also satisfy (4.2). On the other hand, we can easily find an example that satisfies (4.2) but not (4.3).

Consider a special case where $m = n$. Let $y_i = (1/n)$ for all i, then $ny_i = 1$. The coefficient matrix of $\mathbf{x}$ is a diagonal matrix with all the elements on the diagonal equal to 1, and others 0:

$$\begin{bmatrix} 1 & 0 & \ldots & 0 & 0 \\ 0 & 1 & \ldots & 0 & 0 \\ \ldots & \ldots & \ldots & \ldots & \ldots \\ 0 & 0 & \ldots & 1 & 0 \\ 0 & 0 & \ldots & 0 & 1 \end{bmatrix}$$

We can see that in this case, $\sum_j x_{ij} = 1 = ny_i$, which satisfies (4.2), but $x_{ii} > y_i$ for each i, which violates (4.3). So we can conclude that $P_2 \subset P_1$, and P_2 is a better formulation. In fact, the example discussed earlier is not the only case where (4.2) is satisfied but (4.3) is violated. Readers are encouraged to make up other examples too.

4.2 AUTOMATIC PROBLEM PREPROCESSING

Building good formulations for a given IP problem is both art and science. It often depends on the creativity of a model builder as well as scientific techniques. Even for the same model builder, one can often expect that his/her original formulation can be improved, either artistically or scientifically. In the remaining sections, we will introduce some logical rules that can be used to automatically improve a given formulation. These rules routinely process a given problem formulation before it is actually solved by an MIP algorithm. These rules are bundled together to form the so-called *preprocessor* or *presolver*. The preprocessor has been proven very efficient in reducing the solution space and speeding up the solution time. In fact, nowadays most popular IP software has a built-in preprocessor. Most preprocessors cover the following basic functions:

1. Tightening bounds on variables
2. Fixing variables
3. Eliminating redundant constraints
4. Identifying infeasibility
5. Tightening constraints
6. Decomposing the problem into independent subproblems
7. Scaling the coefficient matrix

There are many preprocessing techniques available in the literature; the interested reader should refer to the note of this section. In Section 4.3, we introduce a basic preprocessing technique for tightening bounds, fixing variables, and identifying redundant constraints and infeasibility for *general* integer programs. Then in Section 4.4, we introduce basic techniques for the same functions specially designed for pure *0–1 integer* programs. Methods for decomposing the problem and scaling the coefficient matrix are given in Sections 4.5 and 4.6, respectively.

4.3   TIGHTENING BOUNDS ON VARIABLES

We introduce a basic technique based on tightening upper and lower bounds on variables in mixed integer programs. Three types of variables are considered in order: continuous, general integer, and 0–1 variables. First, we introduce a bounding technique on continuous variables as a foundation. Then we modify and simplify this bounding technique for the special treatment of general integer and 0–1 variables.

Basically, a preprocessor is *initiated* with the inputted IP model. It examines and computes possible tighter upper and lower bounds on all variables, one at a time, in the following order: constraint 1, constraint 2, ... , constraint i, ... , constraint m; and within each constraint, variable $x_1, x_2, \ldots, x_k, \ldots, x_n$. If a computed upper bound of a variable is lower than the best upper bound found so far, or a computed lower bound is higher than the best lower bound found so far, then the computed bound replaces the current bound.

After an entire constraint set is evaluated, the process is *terminated* if there are no bound improvements on any of the variables. If any bound is improved, then a smaller or better formulation is obtained and another round (pass) of preprocessing is repeated on the new formulation. The process is repeated until no improvements are possible on either lower or upper bounds for any variables of an entire formulation. Alternatively, a termination condition may be set to a maximum number of passes predetermined by the user.

4.3.1   Bounds on Continuous Variables

The *bounded* linear programming problem can be stated as

$$\text{Maximize} \quad z = \sum_j c_j x_j$$

$$\text{subject to} \quad \sum_j a_{ij} x_j \leq b_i \quad (i = 1, 2, \ldots, m) \tag{4.4}$$

$$l_j \leq x_j \leq u_j \quad (j = 1, 2, \ldots, n)$$

Separating positive and negative coefficients, the constraints can be rewritten as

$$\sum_{j:a_{ij}>0} a_{ij} x_j + \sum_{j:a_{ij}<0} a_{ij} x_j \leq b_i \quad (i = 1, 2, \ldots, m) \tag{4.5}$$

Isolating variable x_k, we have

$$a_{ik} x_k + \sum_{j \neq k. a_{ij}>0} a_{ij} x_j + \sum_{j \neq k. a_{ij}<0} a_{ij} x_j \leq b_i \quad (i = 1, 2, \ldots, m) \tag{4.6}$$

where k is the index of the variable to be computed for possible tighter bounds, j is the index of the remaining variables, u_j and l_j, respectively, denote the tightest upper and lower bounds found so far on variable j. If they are not specified, we may initially let $u_j = M$ (a big number) and $l_j = 0$. The upper and lower bounds on variable x_k can be computed based on (4.6).

If $a_{ik} > 0$, then an upper bound on x_k can be computed by

$$\hat{u}_k = \frac{1}{a_{ik}} \left(b_i - \sum_{j \neq k, a_{ij} > 0} a_{ij} l_j - \sum_{j \neq k, a_{ij} < 0} a_{ij} u_j \right) \tag{4.7}$$

If $a_{ik} < 0$, then a lower bound on x_k can be computed by

$$\hat{l}_k = \frac{1}{a_{ik}} \left(b_i - \sum_{j \neq k, a_{ij} > 0} a_{ij} l_j - \sum_{j \neq k, a_{ij} < 0} a_{ij} u_j \right) \tag{4.8}$$

The basic idea is that any potential tighter bound must not exclude any feasible solution even under the "worst" conditions. For a positive coefficient a_{ik} in (4.7), the worst possible conditions are l_j for positive coefficient a_{ij} and are u_j for negative a_{ij}. Similarly, for a negative coefficient a_{ik} in (4.8), the possible worst conditions are l_j for positive coefficient a_{ij} and are u_j for negative a_{ij}.

After calculations, the new best bounds are updated by setting $u_k = \hat{u}_k$ if $\hat{u}_k < u_k$ and setting $l_k = \hat{l}_k$ if $\hat{l}_k > l_k$.

4.3.2   Bounds on General Integer Variables

In the presence of integer variables, upper and lower bounds can be further tightened by rounding the fractional values. If an integer variable has a computed upper bound ($\hat{u}_k$) that is noninteger, then it can be further tightened by *rounding* it *down* to obtain the largest integer smaller than $\hat{u}_k$, or symbolically $x_k \leq \lfloor \hat{u}_k \rfloor$. For example, if $u_k = 2.47$, then 2 is a tighter upper bound for integer x_k.

If an integer variable has a lower bound ($\hat{l}_k$) that is noninteger, then it can be tightened by *rounding* it *up* to obtain the smallest integer greater than $\hat{l}_k$, or symbolically $x_k \geq \lceil \hat{l}_k \rceil$. For example, if $\hat{l}_k = 0.3$, then 1 is a lower bound for integer x_k. The current best bound will then be replaced (updated) if the computed bound is tighter.

Example 4.3 (MIP Problem)

$$4x_1 - 3x_2 - 2x_3 + y_4 + 2y_5 \leq 13$$
$$-3x_1 + 2x_2 - x_3 + 2y_4 + 3y_5 \leq -9$$
$$x_1 \geq 0$$
$$0 \leq x_2 \leq 3$$
$$1 \leq x_3 \leq 5$$
$$2 \leq y_4 \leq 4$$
$$y_5 \geq 0 \text{ and integer}$$

Initialization:

$$u_1 = M, \; l_1 = 0, \; u_2 = 3, \; l_2 = 0, \; u_3 = 5, \; l_3 = 1, \; u_4 = 4, \; l_4 = 2, \; u_5 = M, \; l_5 = 0$$

Iteration 1

Check constraint 1:

$x_1 : u_1 = (13-1(2)-2(0)+3(3)+2(5))/4 = 7.5 < M$, so u_1 is updated to 7.5

$x_2 : l_2 = (13-4(0)-1(2)-2(0)+2(5))/(-3) = -7 < 0$, so l_2 is not updated

$x_3 : l_3 = (13-4(0)-1(2)-2(0)+3(3))/(-2) = -10 < 1$, so l_3 is not updated

$y_4 : u_4 = \lfloor (13-4(0)-2(0)+3(3)+2(5))/1 \rfloor = 32 > 4$, so u_4 is not updated

$y_5 : u_5 = \lfloor (13-4(0)-2(0)+3(3)+1(4))/2 \rfloor = 13 < M$, so u_5 is updated to 13

Check constraint 2:

$x_1 : l_1 = (-9-2(0)-2(2)-3(0)+1(5))/(-3) = 2.67 > 0$, so l_1 is updated to 2.67

$x_2 : u_2 = (-9-2(2)-3(0)+3(7.5)+1(5))/2 = 7.25 > 3$, so u_2 is not updated

$x_3 : l_3 = (-9-2(0)-2(2)-3(0)+3(7.5))/(-3) = -3.17 < 1$, so l_3 is not updated

$y_4 : u_4 = \lfloor (-9-2(0)-3(0)+3(7.5)+1(5))/2 \rfloor = 9 > 4$, so u_4 is not updated

$y_5 : u_5 = \lfloor (-9-2(0)-2(2)+3(7.5)+1(5))/3 \rfloor = 4 < 13$, so u_5 is updated to 4

After the first iteration, we have

$$2.67 \le x_1 \le 7.5, \quad 0 \le x_2 \le 3, \quad 1 \le x_3 \le 5, \quad 2 \le y_4 \le 4, \quad 0 \le y_5 \le 4$$

Iteration 2

Check constraint 1:

$x_1 : u_1 = (13-1(2)-2(0)+3(3)+2(5))/4 = 7.5$, so u_1 is not updated

$x_2 : l_2 = (13-4(2.67)-1(2)-2(0)+2(5))/(-3) = -3.44 < 0$, so l_2 is not updated

$x_3 : l_3 = (13-4(2.67)-1(2)-2(0)+3(3))/(-2) = -3.11 < 1.67$, so l_3 is not updated

$y_4 : u_4 = \lfloor (13-4(2.67)-2(0)+3(3)+2(5))/1 \rfloor = 21 > 4$, so u_4 is not updated

$y_5 : u_5 = \lfloor (13-4(2.67)-2(0)+3(3)+1(4))/2 \rfloor = 7 > 4$, so u_5 is not updated

Check constraint 2:

$x_1 : l_1 = (-9-2(0)-2(2)-3(0)+1(5))/(-3) = 2.67$, so l_1 is not updated

$x_2 : u_2 = (-9-2(2)-3(0)+3(7.5)+1(5))/2 = 7.25 > 3$, so u_2 is not updated

$x_3 : l_3 = (-9-2(0)-2(2)-3(0)+3(7.5))/(-3) = -3.17 < 1$, so l_3 is not updated

$y_4 : u_4 = \lfloor (-9-2(0)-3(0)+3(7.5)+1(5))/2 \rfloor = 9 > 4$, so u_4 is not updated

$y_5 : u_5 = \lfloor (-9-2(0)-2(2)+3(7.5)+1(5))/3 \rfloor = 4$, so u_5 is not updated

Stop, because no bounds that can be further tightened were found in this iteration.

4.3.3   Bounds on 0–1 Variables

Recall that for a 0–1 variable, the worst lower bound is 0 and the worst upper bound is 1. Thus, initially we can set $l_j = 0$ and $u_j = 1$ for all j. Any fractional upper bound may be rounded down to 0 and any fractional lower bound may be rounded up to 1. By rounding, for example, if $u_1 = 0.37$, then 0 is a new upper bound for binary variable y_1; and if $l_2 = 0.42$, then 1 is a new lower bound for binary variable y_2.

Example 4.4 (BIP Problem)

$$8y_1 + 11y_2 - 9y_3 + 4y_4 \leq 0$$
$$y_1 - 4y_2 - 6y_3 + y_4 \quad \leq -5$$
$$\text{all } y_j = 0 \text{ or } 1$$

Iteration 1

Check constraint 1:

$$u_1 = \lfloor (0 - 11(0) - 4(0) + 9(1))/8 \rfloor = 1, \text{ so } u_1 \text{ is not updated}$$
$$u_2 = \lfloor (0 - 8(0) - 4(0) + 9(1))/11 \rfloor = 0, \text{ so } y_2 \text{ is fixed to } 0$$

Substituting $y_2 = 0$ to the given problem, we have a new constraint set

$$8y_1 - 9y_3 + 4y_4 \leq 0$$
$$y_1 - 6y_3 + y_4 \leq -5$$

Iteration 2

Check constraint 1:

$$u_1 = \lfloor (0 - 4(0) + 9(1))/8 \rfloor = 1, \text{ so } u_1 \text{ is not updated}$$
$$l_3 = \lfloor (0 - 8(0) - 4(0))/(-9) \rfloor = 0, \text{ so } l_3 \text{ is not updated}$$
$$u_4 = \lfloor (0 - 8(0) + 9(1))/4 \rfloor = 2 > 1, \text{ so } u_4 \text{ is not updated}$$

Check constraint 2:

$$u_1 = \lfloor (-5 - 1(0) + 6(1))/1 \rfloor = 1, \text{ so } u_1 \text{ is not updated}$$
$$l_3 = \lceil (-5 - 1(0) - 1(0))/(-6) \rceil = 1, \text{ so } y_3 \text{ is fixed at } 1$$

Substituting $y_3 = 1$, we have

$$8y_1 + 4y_4 \leq 9$$
$$y_1 + y_4 \leq 1$$

Iteration 3

Check constraint 1:

$$u_1 = \lfloor (9-4(0))/9 \rfloor = 1, \text{ so } u_1 \text{ is not updated}$$
$$u_4 = \lfloor (9-8(0))/4 \rfloor = 2 > 1, \text{ so } u_4 \text{ is not updated}$$

Check constraint 2:

$$u_1 = \lfloor (1-1(0))/1 \rfloor = 1, \text{ so } u_1 \text{ is not updated}$$
$$u_4 = \lfloor (1-1(0))/1 \rfloor = 1, \text{ so } u_4 \text{ is not updated}$$

Stop, because no tighter bounds can be obtained in both constraints. The final set of constraints remains the same as start of Iteration 3.

$$8y_1 + 4y_4 \leq 9$$
$$y_1 + y_4 \leq 1$$
$$y_2 = 0$$
$$y_3 = 1$$

4.3.4 Variables Fixing, Redundant Constraints, and Infeasibility

There are a number of variable fixing techniques available. We discuss three of them. First, for a maximization problem in the form given in (4.4), if $a_{ij} > 0$ for all $i = 1, 2, \ldots, m$ and $c_j < 0$, then fix x_j at l_j. If $a_{ij} < 0$ for all $i = 1, 2, \ldots, m$ and $c_j > 0$, then fix x_j at u_j. Second, if the best bounds on any variable obtained after applying the bound tightening routine having $l_k^* = u_k^*$, then variable x_k can be fixed at l_k^*. Third, based on bounds on the left-hand side of a constraint, we can fix variables under the condition described below. Once a variable is fixed, it can be removed by substituting its fixed value into the current formulation (model), resulting in a smaller feasible region.

For the *i*th constraint, define the following upper and lower *row bounds*:

$$U_i = \sum_{j:a_{ij}>0} a_{ij}u_j + \sum_{j:a_{ij}<0} a_{ij}l_j$$

$$L_i = \sum_{j:a_{ij}>0} a_{ij}l_j + \sum_{j:a_{ij}<0} a_{ij}u_j$$

Note that U_i is an upper bound for the left-hand-side of the ith constraint (or row) and L_i is a lower bound for the left-hand-side of the ith constraint (or row). Comparing with the right-hand side b_i, these row bounds can be used to (a) identify a redundant constraint, (b) identify an infeasible constraint, and (c) fix variables. Normally, we have

$$L_i \leq b_i \leq U_i$$

Consider the following three cases outside the above bounds:

(a) If $b_i \geq U_i$, then the ith constraint is redundant and can be removed from the problem.

(b) If $b_i < L_i$, then the ith constraint cannot be satisfied and no feasible solution exists.

(c) If $b_i = L_i$, then all x_j with $a_{ij} > 0$ can be fixed at $x_j = l_j$, and all x_j with $a_{ij} < 0$ can be fixed at $x_j = u_j$.

Example 4.5   Consider the following constraint set and bounds on variables

$$x_1 + x_2 + x_3 - 2x_4 \leq -6$$
$$-x_1 - 3x_2 + 2x_3 - x_4 \leq 4$$
$$-x_1 + x_2 + x_4 \leq 0$$
$$0 \leq x_1 \leq 2$$
$$0 \leq x_2 \leq 1$$
$$1 \leq x_3 \leq 2$$
$$2 \leq x_4 \leq 3$$

Compute $U_1 = 1(2) + 1(1) + 1(2) - 2(2) = 1$
$L_1 = 1(0) + 1(0) + 1(1) - 2(3) = -5 > -6$

Constraint 1 is infeasible since $b_1 < L_1$.

Example 4.6   Consider the following constraint set and bounds on variables:

$$x_1 + x_2 + x_3 - 2x_4 \leq -1$$
$$-x_1 - 3x_2 + 2x_3 - x_4 \leq 4$$
$$-x_1 + x_2 + x_4 \leq 0$$
$$0 \leq x_1 \leq 2$$
$$0 \leq x_2 \leq 1$$
$$1 \leq x_3 \leq 2$$
$$2 \leq x_4 \leq 3$$

Compute $U_1 = 2 + 1 + 2 - 2(2) = 1$
$L_1 = 0 + 0 + 1 - 2(3) = -5$

No action is taken since $-5 \leq -1 \leq 1$.

Compute $U_2 = -0 - 3(0) + 2(2) - 2 = 2$
$L_2 = -2 - 3(1) + 2(1) - 3 = -6$

Thus, constraint 2 is redundant since $b_2 > U_2$. Remove constraint 2 and continue.

Compute $U_3 = -0 + 1 + 3 = 4$
$L_3 = -2 + 0 + 2 = 0$

Since $b_3 = L_3$, we can fix variables: $x_1 = u_1 = 2$, $x_2 = l_2 = 0$, and $x_4 = l_4 = 2$. Substituting these fixed values, constraint 3 reduces to

$$2 + 0 + x_3 - 2(2) \leq -1$$
$$\text{or} \quad x_3 \leq 1$$

Combined with the given bound $x_3 \geq 1$, we have $x_3 = 1$. Since all variables are fixed, the problem is solved.

To illustrate the first-mentioned variable fixing technique, let us assume $x_3 \leq 2$ instead of $x_3 \leq 1$, which leads to $1 \leq x_3 \leq 2$. Then we can determine the value of x_3 using the given objective function. For the maximization problem, if the associated $c_3 > 0$, then $x_3 = u_3 = 2$. If $c_3 < 0$, then $x_3 = l_3 = 1$.

Although the *row* bounding technique can also be applied even before the variable bounding routine, the power of this technique depends on the tightness of bounds on variables.

4.4 PREPROCESSING PURE 0–1 INTEGER PROGRAMS

Problem preprocessing is most effective when a given model is a pure 0–1 integer program, which arises frequently in combinatorial optimization problems (see Chapters 5 and 6). Problem preprocessing includes the following functions for pure 0–1 integer programs:

- Fixing 0–1 variables
- Detecting redundant constraints and infeasibility
- Tightening constraints (coefficients reduction)
- Generating cutting planes (from minimum cover)
- Rounding by division with GCD

For distinction within an MIP problem, in this section we shall use y_j instead of x_j to denote a 0–1 variable in a pure 0–1 integer program.

4.4.1 Fixing 0–1 Variables

Isolating a variable y_k and separating positive and negative coefficients of the other variables, we can rewrite the standard form of the constraint set as

$$a_{ik}y_k + \sum_{j \neq k, a_{ij} > 0} a_{ij}y_j + \sum_{j \neq k, a_{ij} < 0} a_{ij}y_j \leq b_i \quad (i = 1, 2, \ldots, m) \qquad (4.9)$$

Any constraint of $\geq$ form can be converted to a corresponding constraint of $\leq$ form by multiplying by (-1). Note that the right-hand side constant may be negative. For fixing a 0–1 variable, the following two rules are applied to each constraint i:

Rule 1: Identify the variable (say y_k) with the *largest positive coefficient* (say $a_{ik} > 0$). If the sum of a_{ik} and all $a_{ij} < 0$ exceeds b_i, then constraint i is violated at $y_k = 1$ and hence y_k should be *fixed at* 0.

Rule 2: Identify the variable (say y_k) with the *most negative coefficient* (say $a_{ik} < 0$). If the sum of all $a_{ij} < 0$ ($j \neq k$) exceeds b_i, then constraint i is violated at $y_k = 0$ and hence y_k should be *fixed at* 1.

Note that in rule 1, the sum of a_{ik} and all $a_{ij} < 0$ is equivalent to setting $y_k = 1, y_j = 1$ if its coefficient $a_{ij} < 0$, and $y_j = 0$ if $a_{ij} > 0$ for all $j \neq k$ in (4.9). In rule 2, the sum of all $a_{ij} < 0$ is equivalent to setting $y_k = 0$, $y_j = 1$ if $a_{ij} < 0$, and $y_j = 0$ if $a_{ij} > 0$ for all $j \neq k$ in (4.9).

Consider the following example in $\leq$ form:

$$6y_1 + 2y_2 - 2y_3 - y_4 \leq 2$$

Identify y_1 as the variable having the largest positive coefficient and apply rule 1. Since $6 + (-2) + (-1) = 3 > 2$ violates the constraint, y_1 must be fixed at 0. Identify y_3 as the variable having the most negative coefficient. Rule 2 cannot be applied because $-1 \leq 2$.

Consider the following constraint in $\geq$ form:

$$3y_1 + y_2 - 3y_3 \geq 2$$

Multiplying (-1) through the constraint, we obtain

$$-3y_1 - y_2 + 3y_3 \leq -2$$

Identify y_3 as the variable with the largest positive coefficient and apply rule 1. Since $3 + (-3) + (-1) > -2$ violates the constraint, y_3 must be fixed at 0. Identify y_1 as the variable with the most negative coefficient and apply rule 2. Since $-1 > -2$, y_1 must be fixed at 0.

Once a variable is fixed at 0 or 1 using a certain constraint, the fixed value can be substituted into the other constraints, which results in problem reduction.

Example 4.4 (Continued)

$$8y1 + 11y_2 - 9y_3 + 4y_4 \leq 0$$
$$y_1 - 4y_2 - 6y_3 + y_4 \leq -5$$
$$\text{all } y_j = 0 \text{ or } 1$$

By applying rules 1 and 2 to constraint 1, we can fix $y_2 = 0$ and $y_3 = 1$ resulting in the formulation

$$8y_1 + 4y_4 \leq 9$$
$$y_1 + y_4 \leq 1$$
$$y_2 = 0$$
$$y_3 = 1$$

which, coincidently, is the same set of reduced constraints as obtained by the bound tightening technique described in the previous section.

Moreover, fixing a variable from one constraint can sometimes generate a *chain reaction* of fixing other variables from other constraints. Example 4.7 presents an extension of Example 4.4 in which two constraints and three variables are added.

Example 4.7   The set of constraints include two constraints in Example 4.4 plus the following constraints:

$$y_3 + y_4 + y_5 \leq 1$$
$$y_5 - y_6 \geq 0$$

Variable fixing in Example 4.4 yields the reduced constraints

$$8y_1 + 4y_4 \leq 9$$
$$y_1 + y_4 \leq 1$$

by fixing $y_2 = 0$ and $y_3 = 1$. Next, continue the fixing process for two additional constraints.

$$y_3 + y_4 + y_5 \leq 1 \text{ implies } y_4 = y_5 = 0, \text{ and } y_5 - y_6 \geq 0 \text{ implies } y_6 = 0$$

Fixing variables can achieve a drastic reduction on the size of a pure 0–1 integer program. Crowder et al. (1983) reported that a problem of 2756 variables has been reduced to a problem of 1415 variables.

4.4.2   Detecting Redundant Constraints and Infeasibility

There are many techniques that can be used to detect a redundant constraint. The technique presented in the previous section is based on row bounding. Here we present another one that is based on a similar idea for variable fixing as presented in rules 1 and 2 in Section 4.4.1.

Rule 3: For a $\leq$ constraint, assign a value of 1 to the variables with positive coefficients and 0 otherwise. If the constraint is still satisfied, then it is redundant and can be dropped from further consideration.

Again, if a constraint is in $\geq$ form, convert it to one in $\leq$ form by multiplying by (-1). For example,

$$2x_1 + x_2 + 3x_3 \leq 7$$

is redundant, since $2(2) + 1(1) + 3(1) = 6 < 7$. As another example,

$$3x_1 - 2x_2 - x_3 \leq 0$$

is redundant, since $3(1) - 2(0) - 1(0) = 3 \leq 3$.

Very often redundant constraints are not detected from the original model, but are detected from the reduced models after fixing some variables. Note that the two techniques presented in this and the last sections, as well as other techniques, do not ensure detecting *all* redundant constraints. There are many more techniques in the literature for detecting redundant constraints.

4.4.3 Tightening Constraints (or Coefficients Reduction)

We use a flowchart in Figure 4.5 to demonstrate a constraint tightening procedure. Suppose we are given a constraint of the form

$$a_1 y_1 + a_2 y_2 + \cdots + a_n y_n \leq b \quad \text{where } y_j = 0 \text{ or } 1 \text{ for all } j$$

Example 4.8   Tighten the following constraint

$$6y_1 + 3y_2 - 5y_3 + 2y_4 + 7y_5 - 4y_6 \leq 15$$

Iteration 1

Calculate $M = 6 + 3 + 2 + 7 = 18$, $M - b = 18 - 15 = 3$.

$S = \{a_1, a_3, a_5, a_6\}$.

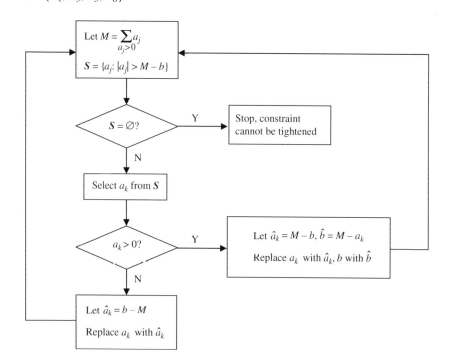

FIGURE 4.5   Process for coefficient reduction.

Pick a_1 to begin with. Since $a_1 > 0$, calculate $\hat{a}_1 = M - b = 3$, $\hat{b} = M - a_1 = 12$. Thus, the given constraint is tightened as

$$3y_1 + 3y_2 - 5y_3 + 2y_4 + 7y_5 - 4y_6 \leq 12$$

Iteration 2

$M = 3 + 3 + 2 + 7 = 15$. $M - b = 15 - 12 = 3$.

$S = \{a_3, a_5, a_6\}$.

Pick a_3 to start with. Since $a_3 < 0$, calculate $\hat{a}_3 = b - M = -3$.

The tightened constraint becomes

$$3y_1 + 3y_2 - 3y_3 + 2y_4 + 7y_5 - 4y_6 \leq 12$$

Iteration 3

$M = 3 + 3 + 2 + 7 = 15$. $M - b = 15 - 12 = 3$.

$S = \{a_5, a_6\}$.

Pick a_5 to start with. Since $a_5 > 0$, calculate $\hat{a}_5 = M - b = 3$, $\hat{b} = M - a_5 = 8$.

The tightened constraint becomes

$$3y_1 + 3y_2 - 3y_3 + 2y_4 + 3y_5 - 4y_6 \leq 8$$

Iteration 4

$M = 3 + 3 + 2 + 3 = 11$. $M - b = 11 - 8 = 3$.

$S = \{a_6\}$.

Since $a_6 < 0$, calculate $\hat{a}_6 = b - M = -3$.

The tightened constraint becomes

$$y_1 + 3y_2 - 3y_3 + 2y_4 + 3y_5 - 3y_6 \leq 8$$

Iteration 5

$S = \emptyset$. Stop, the constraint cannot be further tightened.

4.4.4 Generating Cutting Planes from Minimum Cover

A cutting plane (or cut) for an IP problem is a derived constraint that reduces the feasible region for the LP relaxation without eliminating any feasible solution for the IP problem. Here we will see a particular type of cutting planes for pure 0–1 integer programs. Such a cut is generated from a constraint in $\leq$ form with all coefficients and the right-hand side positive,

$$a_1 y_1 + a_2 y_2 + \cdots + a_n y_n \leq b$$

where $b > 0$, $a_j > 0$, and $y_j = 0$ or 1 for all j. Recall that this type of constraint appears in the knapsack problem. The procedure for generating cutting planes is as follows:

Step 1. Find a group of variables (called a *minimum cover* of the constraint, or *knapsack cover*) such that (a) the constraint is violated if every variable in the group is set to 1 and all other variables are set to 0, and (b) the constraint becomes satisfied if the value of every one of these variables is changed from 1 to 0. Let n_C denote the number of variables in the group.

Step 2. The resulting cutting plane has the form $\sum_{j=1}^{n_c} y_j \leq n_C - 1$.

Applying this procedure to the constraint $2y_1 + 3y_2 + 5y_3 + 6y_4 \leq 10$, we see that the group of variables $\{y_1, y_2, y_4\}$ is a minimum cover because (a) (1, 1, 0, 1) violates the constraint and (b) the constraint becomes satisfied if the value of every one of these three variables is changed from 1 to 0. Since $n_C = 3$, the resulting cutting plane is

$$y_1 + y_2 + y_4 \leq 2$$

This same constraint also has another minimum cut $\{y_3, y_4\}$ because (0, 0, 1, 1) violates the constraint, but both (0, 0, 1, 0) and (0, 0, 0, 1) satisfy the constraint. Thus, the resulting cutting plane is $y_3 + y_4 \leq 1$.

These cutting planes are very effective in tightening the LP relaxation. For example, for the Crowder et al. (1983) test problem with 2756 binary variables considered, 3326 cutting planes were generated. The result narrows the gap between the optimal objective value for the LP relaxation of the entire 0–1 integer program and the optimal objective value for this problem by 98%. The integration of this cutting plane and the branch-and-bound techniques provides a powerful, effective approach for solving binary integer programs.

4.4.5   Rounding by Division with GCD

Consider a constraint of the form

$$a_1 y_1 + a_2 y_2 + \cdots + a_n y_n \leq b$$

where y_j is a 0–1 variable and a_j is an integer constant. Denote GCD as the greatest common divisor of $a_1, a_2, \ldots, a_n$. The constraint can be tightened by dividing all terms by the GCD, then rounding b/GCD down to largest integer $\leq b/\text{GCD}$:

$$\frac{a_1}{\text{GCD}} y_1 + \frac{a_2}{\text{GCD}} y_2 + \cdots + \frac{a_n}{\text{GCD}} y_n \leq \left\lfloor \frac{b}{\text{GCD}} \right\rfloor$$

If coefficients are not all integer, they can be made integer by the following procedure:

Find k such that $k = \min\{p: a_j(10^p)$ are integer for all $j\}$. Then all coefficients can be made integer by computing $a'_j = a_j(10^k)$ and $b' = b(10^k)$ and substituting a'_j for a_j and b'_j for b_j, respectively.

To derive a tightening constraint, find $GCD = \max\{d : (a'_j)/d$ is integer for all $j\}$, then divide the transformed constraint by GCD, and round the right-hand side down to the next largest integer.

Example 4.9 Tighten the following constraint by dividing GCD and rounding

$$1.05y_1 + 0.35y_2 - 1.4y_3 + 0.63y_4 \leq 6$$

Make all coefficients integer by multiplying 100,

$$105y_1 + 35y_2 - 140y_3 + 63y_4 \leq 600$$

Since GCD $(105, 35, 140, 63) = 7$, the constraint can be tightened as

$$15y_1 + 5y_2 - 20y_3 + 9y_4 \leq \left\lfloor \frac{600}{7} \right\rfloor = 85$$

If the given constraint is in $\geq$ form, then reverse the inequality sign by multiplying by (-1).

4.5 DECOMPOSING A PROBLEM INTO INDEPENDENT SUBPROBLEMS

A large-scale IP problem contains many variables and/or constraints. However, sometimes, special structures in the set of constraints enable us to partition the problem into two or more subproblems that are independent of each other. Combining solutions to these subproblems will yield a solution to the original problem. In this way the problem is greatly simplified. Whether a problem can be decomposed can be determined by looking at the coefficient matrix of the constraints. To be specific, if the coefficient matrix $\mathbf{A}$ for the constraint set $\mathbf{Ay} \leq \mathbf{b}$, after rearrangement, takes the following form:

y_1	y_2	y_3	y_4	y_5	y_6	y_7	y_8
$\mathbf{M_1}$			0			0	
0			$\mathbf{M_2}$			0	
0			0			$\mathbf{M_3}$	

where $\mathbf{M_1}, \mathbf{M_2}, \mathbf{M_3} \neq \mathbf{0}$, then the IP problem can be decomposed into three independent subproblems. Subproblem 1 optimizes over variables y_1, y_2, and y_3. Subproblem 2 optimizes over variables y_4, y_5, and y_6. Subproblem 3 optimizes over variables y_7 and y_8.

Example 4.10   Consider the IP problem

$$\text{Minimize}\quad 2y_1 + 3y_2 + y_3 - 2y_4$$
$$\text{subject to}\quad y_1 + y_3 \le 7$$
$$2y_2 + y4 \le 11$$
$$-y_2 + 5y_4 \ge 3$$
$$y_1 \ge 1$$
$$\mathbf{y} \ge \mathbf{0} \text{ and integer}$$

Rearranging the coefficient matrix by exchanging columns 2 and 3, we obtain

$$
\mathbf{A} =
\begin{array}{cccc}
y_1 & y_2 & y_3 & y_4 \\
\end{array}
\begin{bmatrix}
1 & 1 & 0 & 0 \\
1 & 0 & 0 & 0 \\
0 & 0 & 2 & 1 \\
0 & 0 & -1 & 5
\end{bmatrix}
$$

Obviously, the problem can be decomposed (partitioned) into two independent subproblems:

Subproblem 1	Subproblem 2
Minimize $\quad 2y_1 + y_3$	Minimize $\quad 3y_2 - 2y4$
subject to $\quad y_1 + y_3 \le 7$	subject to $\quad 2y_2 + y_4 \le 11$
$\quad\quad\quad\quad y_1 \ge 1$	$\quad\quad\quad\quad -y_2 + 5y_4 \ge 3$

Subproblem 1 only involves variables y_1 and y_3, while subproblem 2 only involves variables y_2 and y_4.

Note that the decomposing technique discussed in this section is not what is known as "decomposition approach," which is the topic of Section 14.6.

4.6   SCALING THE COEFFICIENT MATRIX

When a practical problem is modeled, it is important to pay attention to the units in which quantities are measured. Great disparity in the sizes of the coefficients in an MIP model could make such a model difficult to solve and yield an inaccurate solution due to rounding and truncation errors.

If capacity constraints allow quantities in thousands of tons, it would be better to allow each variable to represent a quantity in thousands of tons rather than tons. In general, constraints concerning a given resource should share a common measure. But different sets of constraints may have big difference in measuring units. Ideally, one should choose units so that each nonzero coefficient in an MIP model is of a magnitude between 0.1 and 10. In practice, this may not always be possible. However, most commercial packages have procedures for automatically *scaling* the coefficients of a model before it is solved. Even with that, some software guides such as the one accompanying LINGO® suggest that the user define units of the objective function, right-hand sides, and decision variables so that no nonzero coefficients have absolute values of more than 100,000 or less than 0.0001. The solution is then automatically *unscaled* before being printed out.

4.7   NOTES

Section 4.1

The quality of an IP formulation defined in this section is primarily based on the constraint set (polyhedron) of the problem with little or no consideration of the objective function. Nevertheless, in Section 4.3, one of the three variable fixing techniques does take advantage of the objective function. Prior to solving the problem, the role of preprocessing is to rapidly reduce the problem dimension with little computational effort and leave the optimization step for the solution phase.

Section 4.3

Brearley et al. (1975) presented some preprocessing techniques *prior to* applying the simplex method and reported implementation of such techniques in the earlier software systems such as MPSX® of IBM and APEX II® of CDC under system procedures called REDUCE and ANALYZE, respectively. Earlier, Zionts (1968) derived upper and lower bounds on variables *during* the simplex iterations for linear and integer programs.

Section 4.4

Hoffman and Padberg (1991) presented various techniques for automatically improving the LP-representations of 0–1 linear programs for branch-and-cut.

   Savelsbergh (1994) presented various preprocessing and probing techniques for mixed integer programming problems.

Section 4.6

For scaling the coefficient matrix, see the notes in Williams (1993) and Winston (1994).

4.8   EXERCISES

4.1   Plot the feasible regions for the following two formulations from the same problem. Can you tell from the graphs, which is a better formulation? Why or why not?

Formulation 1	Formulation 2
$2y_1 + y_2 \leq 25$	$y_1 \leq 9$
$2y_1 - y_2 \leq 5$	$y_1 \geq 1$
$4y_1 + y_2 \geq 5$	$y_2 \leq 13$
$y_1 - 2y_2 \geq 2$	$y_2 \geq 0$
$\mathbf{y} \geq \mathbf{0}$ and integer	$\mathbf{y} \geq \mathbf{0}$ and integer

4.2 Does a better formulation imply that it has fewer variables and/or constraints? Why or why not? If not, give a counterexample.

4.3 Examine the following two formulations, P_1 and P_2, for a single machine scheduling problem to minimize the total weighted completion time. Which do you think is a better formulation? Why? Let w_j be the weight of job j, p_j be the integer processing time of job j, M be a large number, and x_j be the completion time of job j.

$$(P_1) \quad \text{Minimize} \quad \sum_{j=1}^{n} w_j x_j$$

$$\text{subject to} \quad x_j \geq p_j$$
$$-x_j + x_i + p_j \leq My$$
$$-x_i + x_j + p_i \leq M(1-y)$$
$$y = 0 \text{ or } 1$$

$$(P_2) \quad \text{Minimize} \quad \sum_{j=1}^{n} \sum_{t=0}^{t} w_j(t+p_j)y_{jt}$$

$$\text{subject to} \quad \sum_{t=1}^{l} y_{jt} = 1 \qquad \text{for all } j$$

$$\sum_{j=1}^{n} \sum_{k=\max(t-p_j,0)}^{t-1} y_{jk} = 1 \quad \text{for all } t$$

$$y_{jk} = 0 \text{ or } 1 \qquad \text{for all } j \text{ and } k$$

where $l = \sum_{j=1}^{n} p_j - 1$, $y_{jt} = 1$ if job j starts at time t (t is integer), 0 otherwise.

4.4 Find the best bounds for each variable in the following constraints:

$$2y_1 + 7y_2 - 3y_3 + 6y_4 - 9y_5 + y_6 \leq -12$$
$$y_1 - 2y_2 + y_3 + 4y_4 + 2y_5 - 3y_6 \leq 13$$
$$1 \leq y_1 \leq 4$$
$$0 \leq y_2 \leq 7$$
$$4 \leq y_3 \leq 10$$
$$2 \leq y_4$$
$$y_5 \leq 2$$
$$\mathbf{y} \geq \mathbf{0} \text{ and integer}$$

4.5 Show that tightening bounds on 0–1 variables is the same as fixing 0–1 variables.

4.6 Use the variable fixing technique to fix the values of the variables as many as possible in the following 0–1 constraints:

$$30x_1 - 20x_2 + 40x_3 + 17x_4 - 23x_5 + 11x_6 \leq 70$$
$$23x_1 + 15x_2 + 30x_3 - 27x_4 + 13x_5 - 21x_6 \geq 61$$

4.7 Tighten bounds of the following constraints by using one of the techniques mentioned in this chapter:

$$\frac{10}{13}x_1 + \frac{25}{9}x_2 - \frac{5}{4}x_3 + 3x_4 - x_5 \leq \frac{17}{3}$$

$$\frac{18}{5}x_1 - \frac{5}{6}x_2 + \frac{17}{10}x_3 - \frac{5}{7}x_4 + x_5 \geq 4$$

4.8 Fix the variables in Exercise 4.4 after the bounds are tightened.

4.9 Tighten the constraints in Exercise 4.3.

4.10 Tighten the constraints obtained in Exercise 4.5.

4.11 Consider the following set of constraints for an IP problem. Identify redundant or infeasible constraints, if any.

$$5x_1 + x_2 + 3x_3 - 2x_4 + x_5 - 3x_6 \leq 9$$
$$2x_1 - 2x_2 + x_3 + x_4 - 2x_5 + x_6 \leq 6$$
$$x_1 + x_2 - x_3 - x_4 + 2x_5 - x_6 \geq 2$$
$$2x_1 + x_2 - 2x_3 + 3x_4 - x_5 + x_6 \geq 8$$
$$\mathbf{x} \in (0, 1)$$

4.12 Show that the constraints in Example 4.3 can be further processed for redundancy.

4.13 Consider the set of possible solutions for a set covering problem defined as

$$S = \{y(0, 1) : \sum_j a_{ij}y_j \geq 1, \ a_{ij} = (0, 1) \text{ for all } (i, j)\}$$

(a) Under what condition(s) is the set S empty (constraint infeasible)?
(b) Under what condition(s) is each constraint redundant?
(c) Under what condition(s) can variable y_k be fixed to 0 and 1?

4.14 The following two inequalities are simultaneous constraints of some binary IP problem.

$$a_1y_1 + a_2y_2 + a_3y_3 + a_4y_4 \leq b$$
$$g_1y_1 + g_2y_2 + g_3y_3 + g_4y_4 \leq d$$

(a) Under what condition(s) is the second constraint redundant?

(b) Under what condition(s) is the problem infeasible (the two constraints contradict each other)?

4.15 Generate a cutting plane for the knapsack constraint below. (*Hint*: Transformation of the constraint is needed before generating the cutting plane.)

$$3y_1 - y_2 + 2y_3 + 4y_4 - 3y_5 \leq 5$$

4.16 Consider the following set of constraints for a binary IP problem. Preprocess the model using the techniques from this chapter. Identify the type of techniques you used.

$$37x_1 - 68x_2 + 78x_3 + x_4 - 21x_5 \leq 141$$
$$4x_1 + 7x_2 + 7x_3 - 2x_4 + 5x_5 \quad \leq 17$$
$$13x_1 + 11x_2 - 17x_3 + 6x_4 - x_5 \geq 10$$
$$\frac{3}{5}x_1 + \frac{8}{3}x_2 - \frac{27}{7}x_3 + \frac{27}{11}x_4 + \frac{1}{4}x_5 \geq \frac{7}{2}$$
$$\mathbf{x} \in (0, 1)$$

4.17 How would you rescale the following problem?

$$\begin{aligned}
\text{Minimize} \quad & 2x_1 + 0.003x_2 - x_3 \\
\text{subject to} \quad & 21x_1 - 0.005x_2 \leq 13 \\
& -11x_1 + x_3 \leq 9 \\
& 0.001x_2 + 4x_3 \geq 17 \\
& \mathbf{x} \geq \mathbf{0}
\end{aligned}$$

4.18 If you are using some software (designed specifically for mathematical programming) for solving IP problems, read the user's manual and identify the built-in preprocessing techniques.

5

MODELING COMBINATORIAL OPTIMIZATION PROBLEMS I

This chapter deals with important classes of combinatorial optimization problem (COP), introduced in Chapter 1. Other COPs have appeared in Chapter 2, and some are to appear in Chapters 6 and 10. This chapter will discuss the modeling and successful real-world applications of the following COPs: set covering, set partitioning, node covering, set packing, matching, and cutting stock.

5.1 INTRODUCTION

Recall that the combinatorial optimization problem is a class of optimization problem whose optimum solution(s) can be identified from a finite set of feasible solutions, which in principle can be obtained by complete enumeration of all possible combinations. In Chapter 2, we have seen the IP formulations of some COPs:

- The 0–1 knapsack problem (Section 2.3.1)
- The capital budgeting problem (Section 2.3.2)
- The uncapacitated lot sizing problem (Section 2.4.1)
- The workforce/staff scheduling problem (Section 2.5)
- The uncapacitated facility location problem (Section 2.6.2).

Applied Integer Programming: Modeling and Solution, By Der-San Chen, Robert G. Batson, and Yu Dang
Copyright © 2010 John Wiley & Sons, Inc.

Another group of COPs, the network optimization problems, can be found in most introductory OR or LP texts, and also in Chapter 10:

- The minimum cost network flow problem (Section 10.2.1).
- The assignment problem (Section 10.2.2, also Section 6.2)
- The transportation problem (Section 10.2.2)
- The transshipment problem (Section 10.2.3)
- The maximal flow problem (Section 10.2.4)
- The shortest path problem (Section 10.2.5)

The group of network optimization problems can be solved as if they were linear programs (by ignoring integer requirements) because their constraint matrices are totally unimodular. The minimum cost network flow problem is a general class of this group. By specifying appropriate values to the parameters of the minimum cost network flow problem, the other five problems are seen as special cases.

Furthermore, any pure IP problem with bounded variables, general or binary, can also be treated as a COP. The reader may verify this in the exercises of this chapter.

5.2 SET COVERING AND SET PARTITIONING

The set covering problem can be stated in a general way as follows. You are given a set of *requirements* or *characteristics* (say R) that must be satisfied entirely, a set of *activities* (say A_1, A_2, ..., A_n) whose union equals or "cover" the entire set of requirements, and a cost associated with each activity. Although an activity A_j may cover only a subset of R, a combination of some activities A_j's may cover R. The *set covering problem* is to determine a combination of activities A_j's that can collectively cover all the requirements while minimizing a certain objective function. For example, an airline company has a set of scheduled flights (set of requirements) to be covered entirely and has a set of crews (set of activities) available for flight assignments. Assuming each crew (activity) incurs a certain cost, the objective is to find a subset of crews that cover all flights and minimize the total cost of crew assignments. For example, using set notation, if

$$R = \{1, 2, 3, 4, 5\}$$

and

$$A_1 = \{1, 2, 5\}, \quad A_2 = \{3, 4\}, \quad A_3 = \{3, 4, 5\}, \quad A_4 = \{2, 4, 5\}$$

the selections $\{A_1, A_2\}$ and $\{A_1, A_3\}$ cover R because $R = A_1 \cup A_2 = A_1 \cup A_3$. But the selections $\{A_1, A_4\}$, $\{A_2, A_3\}$, and $\{A_3, A_4\}$ do not because $A_1 \cup A_4 \neq R$, $A_2 \cup A_3 \neq R$, and $A_3 \cup A_4 \neq R$.

5.2.1 Set Covering Problem

The set covering problem can be defined as follows:

1. Given set of requirements or characteristics that must be fully satisfied or covered.
2. Given set of activities (often very large), each of which can satisfy some requirements and incur a certain cost.
3. A feasible solution is defined as a select subset of activities that as a whole can satisfy all requirements.
4. An optimal solution is a feasible solution with a minimal total cost.

Many real-world problems have been modeled as the set covering and set partitioning problems, primarily in industries such as airlines, trucking, communication, hospitals, and manufacturing. Primary application areas include flight crew scheduling, facility location, truck/vehicle delivery and routing, and workforce scheduling. In what follows, we give two examples to show how to formulate a set covering problem. We begin with the identification of the sets of requirements and activities.

Example 5.1 (Location of Warehouses) A firm has five distribution centers and it is to be determined as to which subset of these distribution centers should be selected as a site for construction of warehouses. Suppose the goal is to build a minimal number of warehouses that can cover all distribution centers in a manner that every warehouse is located within 10 miles of each distribution center it services.

To solve this problem as a set covering problem, we first obtain a distance table as shown in Table 5.1. Each entry represents the distance (in miles) between two distribution centers. Based on this distance table and the distance limitation of 10 miles or less, we can construct a requirement–activity table as shown in Table 5.2. A requirement row corresponds to a distribution center. An activity column corresponds to the set of distribution centers that are located within 10 miles from each given center. The entries of this table are either 1 or 0. A "1" indicates that the corresponding requirement is covered, and a "0" otherwise. Activity column 1 represents that if a warehouse is built in center 1, then both centers 1 and 2 are covered. Activity column 2 indicates that centers 1, 2, and 5 are covered. Likewise, other activity columns have similar interpretation.

TABLE 5.1 Distance Between Distribution Centers

Center	1	2	3	4	5
1	0	10	15	20	18
2	10	0	20	15	10
3	15	20	0	8	17
4	20	15	8	0	5
5	18	10	17	5	0

TABLE 5.2   Requirements and Activities of Warehouse Location Problem

Requirement (Center)	Activity				
	1	2	3	4	5
1	1	1	0	0	0
2	1	1	0	0	1
3	0	0	1	1	0
4	0	0	1	1	1
5	0	1	0	1	1

At each distribution center, we must decide whether or not to build a warehouse there. Therefore, we have five activity columns, one for each center. The problem is to build a minimal number of warehouses that can cover all five distribution centers.

Now we are ready to formulate this problem as a set covering problem, a special 0–1 IP model.

Step 1

Input parameters: number of distribution centers (m), vectors of activity ($\mathbf{a}_j, j = 1, 2, \ldots, n$), requirement vector ($\mathbf{b} = \mathbf{1}$)

Decision variables: whether or not to build a warehouse at the ith distribution center ($y_i = 1$ or 0, $i = 1, 5$)

Constraint: each of the five distribution centers must be covered

Objective: minimize the number of warehouses built

Step 2. Let $y_i = 1$ if a warehouse is built at center i and 0 otherwise. To ensure at least one warehouse is within 10 miles of center 1, we have constraint

$$y_1 + y_2 \geq 1 \quad \text{(requirement 1 constraint)}$$

Likewise, we obtain constraints for all five distribution centers. Combining these constraints with the objective function, we obtain the following 0-1 IP model:

$$\text{Minimize } z = \sum_{i=1}^{5} y_i$$

$$\text{subject to } \begin{pmatrix} 1 & 1 & 0 & 0 & 0 \\ 1 & 1 & 0 & 0 & 1 \\ 0 & 0 & 1 & 1 & 0 \\ 0 & 0 & 1 & 1 & 1 \\ 0 & 1 & 0 & 1 & 1 \end{pmatrix} \begin{pmatrix} y_1 \\ y_2 \\ y_3 \\ y_4 \\ y_5 \end{pmatrix} \geq \begin{pmatrix} 1 \\ 1 \\ 1 \\ 1 \\ 1 \end{pmatrix}$$

$$y_i = 0 \text{ or } 1 \quad i = 1, 2, \ldots, m$$

Example 5.2 (Flight Crew Scheduling Problem) Flight crew scheduling (i.e., assigning crews to a given set of flights) is one of the most important problems faced by the airline industry. Almost all major airlines solve this problem by formulating it as an integer program in which the set covering problem is a core component. For example, American Airlines has over 8000 airplanes and 16,000 flight attendants to schedule. They estimate that their mathematical programming-based system (in which the covering problem is a major component) saves about $20 million per year (see Table 1.1). From this table, we can also see that many MIP applications in the airline industry involve crew scheduling.

The problem can be formulated as a set covering problem as follows: (1) List the set of flight legs that are required to be covered; (2) generate a set of flight sequences or tours that begin and end in the same city, subject to certain regulations and conditions; and (3) formulate a 0–1 integer program to find a subset of flight sequences that cover all flights at a minimal cost. A numerical example is given below.

Budget Airways is required to assign its crews based in New York to cover all the upcoming scheduled flights. There are many possible sequences of flights that are feasible for a crew to choose from, assuming one crew can only be assigned to one sequence. In Table 5.3, 10 flights and 8 feasible sequences of flights are viewed respectively as requirements and activities of the set covering problem. The associated cost of each sequence of flight is also listed in this table. The problem is to find crew assignment that covers all 10 flights at a minimal total cost.

In Table 5.3, the number in each column indicates the order of flight legs to be connected in a given sequence of flights. It is permissible to have more than one crew on a flight where the extra crews would fly as passengers and would get pay as if they were working.

To formulate this problem, we begin by forming a node–arc incidence matrix **A**, having an entry of "1" if a sequence of flight covers a certain flight and "0" otherwise. Then Table 5.3 can be converted to Table 5.4, which is the incidence matrix **A** used in this covering problem. We now formulate the flight crewing problem as a set covering problem.

TABLE 5.3 Feasible Sequences of Flights

Requirement (Flight)	Activity (Feasible Sequence of Flights)							
	1	2	3	4	5	6	7	8
1 New York to Buffalo	1			1			1	
2 New York to Cincinnati		1			1			
3 New York to Chicago			1			1		1
4 Buffalo to Chicago	2			2				
5 Chicago to Cincinnati			2	3		2		
6 Cincinnati to Pittsburgh		2		4		3		
7 Cincinnati to Buffalo			3		2			
8 Buffalo to New York			4		3		2	
9 Pittsburgh to New York		3		5		4		
10 Chicago to New York	3							2
Cost ($1000) for each sequence	5	4	4	9	7	8	3	3

TABLE 5.4   Requirements and Activities of Flight Crew Scheduling Problem

Requirement	Activity							
	1	2	3	4	5	6	7	8
1	1	0	0	1	0	0	1	0
2	0	1	0	0	1	0	0	0
3	0	0	1	0	0	1	0	1
4	1	0	0	1	0	0	0	0
5	0	0	1	1	0	1	0	0
6	0	1	0	1	0	1	0	0
7	0	0	1	0	1	0	0	0
8	0	0	1	0	1	0	1	0
9	0	1	0	1	0	1	0	0
10	1	0	0	0	0	0	0	1

Step 1

Input parameters: 0–1 incidence matrix (A) given in Table 5.4, cost for each feasible sequence of flight (c_j, $j = 1, 2, \ldots,$ 8), requirement vector ($b_i = 1$, $i = 1, 2, \ldots, 10$)

Decision variable: one 0–1 variable for each sequence of flight ($y_j = 1$ or 0, $j = 1, 2, \ldots, 8$)

Constraint: one constraint for each requirement or flight

Objective: minimize the total cost of assigning crews to the selected sequence of flights

Step 2. Let $y_j = 1$ if a crew is assigned to the jth sequence of flights and 0 otherwise. To ensure that at least one crew is assigned to the first flight, we have constraint

$$y_1 + y_4 + y_7 \geq 1 \quad \text{(requirement 1 constraint)}$$

Likewise, we obtain constraints for all 10 flights. Combining these constraints with the objective function, we obtain the following 0–1 IP model:

Minimize $z = 5y_1 + 4y_2 + 4y_3 + 9y_4 + 7y_5 + 8y_6 + 3y_7 + 3y_8$

subject to

$$
\begin{pmatrix}
1 & 0 & 0 & 1 & 0 & 0 & 1 & 0 \\
0 & 1 & 0 & 0 & 1 & 0 & 0 & 0 \\
0 & 0 & 1 & 0 & 0 & 1 & 0 & 1 \\
1 & 0 & 0 & 1 & 0 & 0 & 0 & 0 \\
0 & 0 & 1 & 1 & 0 & 1 & 0 & 0 \\
0 & 1 & 0 & 1 & 0 & 1 & 0 & 0 \\
0 & 0 & 1 & 0 & 1 & 0 & 0 & 0 \\
0 & 0 & 1 & 0 & 1 & 0 & 1 & 0 \\
0 & 1 & 0 & 1 & 0 & 1 & 0 & 0 \\
1 & 0 & 0 & 0 & 0 & 0 & 0 & 1
\end{pmatrix}
\begin{pmatrix}
y_1 \\ y_2 \\ y_3 \\ y_4 \\ y_5 \\ y_6 \\ y_7 \\ y_8
\end{pmatrix}
\geq
\begin{pmatrix}
1 \\ 1 \\ 1 \\ 1 \\ 1 \\ 1 \\ 1 \\ 1 \\ 1 \\ 1
\end{pmatrix}
$$

$$y_j = 0 \text{ or } 1 \quad j = 1, 2, \ldots, 8$$

In general, the set covering problem is defined as

$$\text{Minimize } z = \mathbf{c}^T\mathbf{y}$$
$$\text{subject to } \mathbf{Ay} \geq \mathbf{1}$$
$$\mathbf{y} \in (0, 1)$$

where $\mathbf{c}$ is a cost vector representing the costs associated with activities, $\mathbf{y}$ is a vector of 0–1 variables indicating whether the corresponding activity is chosen or not, $\mathbf{A}$ is a 0–1 matrix representing relationships between requirements and activities.

5.2.2 Set Partitioning and Set Packing

The *set partitioning problem* is the same as the set covering problem except that each requirement must be exactly satisfied. Mathematically, the $\geq$ constraints are replaced by $=$ constraints.

$$\text{Minimize } \quad z = \mathbf{c}^T\mathbf{y}$$
$$\text{subject to } \quad \mathbf{Ay} = \mathbf{1}$$
$$\mathbf{y} \in (0, 1)$$

As an example, consider the problem of delivering orders from a warehouse to n different stores by m trucks. Each store receives its order in exactly one delivery. A truck can deliver at most k $(k < n)$ orders (stores). Because a store may fall on more than one route, a truck may pass a store without delivery of that store's order. It is required that all orders (stores) must be delivered. Here, activity j represents a feasible delivery sequence of orders satisfying the truck capacity. The collection of feasible activities forms a matrix $\mathbf{A}$. The constraint set, $\mathbf{Ay} = \mathbf{1}$, ensures that every order is delivered exactly by one truck. In a busy day, it may be acceptable that some lower priority orders can be postponed to a later day. To represent this situation, the set of constraints becomes $\mathbf{Ay} \leq \mathbf{1}$. This problem is known as a *set packing problem*.

5.2.3 Set Covering in Networks

In the domain of an undirected network, the set covering problem can be posed as a *node covering problem*. The node covering problem is one of the simplest classes of combinatorial optimization problems. Consider an undirected network $G(V, E)$ of $n = |V|$ nodes and $m = |E|$ arcs, each arc joining a pair of nodes. A *cover* is a subset of arcs such that each of the n nodes is incident or connected to at least one arc of the subset. A *simple covering problem* is defined as finding a cover with a minimum number of arcs. Consider the undirected network in Figure 5.1, consisting of 7 nodes and 12 arcs. The subset of five arcs in Figure 5.2 is a cover because all the seven nodes are incident to these arcs. But this five-arc cover is not minimal because a cover using only four arcs can be obtained by dropping arc (4, 5) from the five-arc cover.

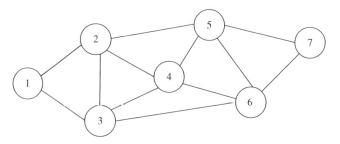

FIGURE 5.1   An example network.

We begin the IP formulation with constructing a node–arc incidence matrix of the network, in which row i corresponds to node i and column j corresponds to arc j. Let **A** denote node–arc incidence matrix, whose entries are a_{ij} ($i = 1, 2, \ldots, n; j = 1, 2, \ldots, m$). Let column vector $\mathbf{a}_j$ be the jth arc joining nodes p and q such that $a_{pj} = a_{qj} = 1$ and $a_{ij} = 0$ if $i \neq p, q$. Note that each arc column contains exactly two elements of 1's. The network in Figure 5.1 can be represented by the following node–arc incidence matrix:

$$\mathbf{A} = \begin{bmatrix} 1 & 1 & 0 & 0 & 0 & 0 & 0 & 0 & 0 & 0 & 0 & 0 \\ 1 & 0 & 1 & 1 & 1 & 0 & 0 & 0 & 0 & 0 & 0 & 0 \\ 0 & 1 & 0 & 0 & 1 & 1 & 1 & 0 & 0 & 1 & 0 & 0 \\ 0 & 0 & 0 & 1 & 0 & 1 & 0 & 1 & 0 & 0 & 0 & 0 \\ 0 & 0 & 1 & 0 & 0 & 0 & 0 & 1 & 1 & 0 & 1 & 0 \\ 0 & 0 & 0 & 0 & 0 & 0 & 1 & 0 & 1 & 1 & 0 & 1 \\ 0 & 0 & 0 & 0 & 0 & 0 & 0 & 0 & 0 & 0 & 1 & 1 \end{bmatrix}$$

Let variable y_j be 1 if the jth arc is selected and 0 otherwise. The node covering problem can be formulated as

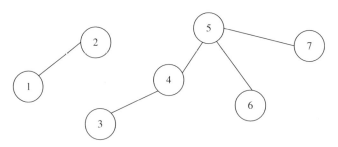

FIGURE 5.2   A node cover.

$$\text{Minimize} \quad z = \sum_{j=1}^{m} y_j$$

$$\text{subject to} \quad \sum_{j=1}^{m} a_{ij} y_j \geq 1 \quad i = 1, 2, \ldots, n$$

$$y_j = 0 \text{ or } 1 \quad j = 1, 2, \ldots, m$$

Or in matrix notation,

$$\text{Minimize} \quad z = \mathbf{c}^{\mathrm{T}} \mathbf{y}$$

$$\text{subject to} \quad \mathbf{Ay} \geq \mathbf{1}$$

$$\mathbf{y} \in (0, 1)$$

where $\mathbf{c}^{\mathrm{T}}$ is a row vector containing m elements of 1's, $\mathbf{y} = (y_1, y_2, \ldots, y_m)^{\mathrm{T}}$, $\mathbf{1} = (1, 1, \ldots, 1)^{\mathrm{T}}$, and $\mathbf{A}$ is an $n \times m$ matrix.

For example, the node covering problem of the graph in Figure 5.1 can be formulated as follows:

$$\text{Minimize} \quad z = (1, 1, \ldots, 1) \begin{pmatrix} y_1 \\ y_2 \\ \vdots \\ y_{12} \end{pmatrix}$$

$$\text{subject to} \quad \begin{bmatrix} 1 & 1 & \ldots & 0 \\ 1 & 0 & \ldots & 0 \\ 0 & 1 & \ldots & 0 \\ 0 & 0 & \ldots & 0 \\ 0 & 0 & \ldots & 0 \\ 0 & 0 & \ldots & 1 \\ 0 & 0 & \ldots & 1 \end{bmatrix} \begin{pmatrix} y_1 \\ y_2 \\ \vdots \\ y_{12} \end{pmatrix} \geq \begin{pmatrix} 1 \\ 1 \\ \vdots \\ 1 \end{pmatrix}$$

$$y_j = 0 \text{ or } 1 \quad j = 1, 2, \ldots, 12$$

Clearly, the node covering problem defined on an undirected network is a special case of the set covering problem. Note that each column of matrix $\mathbf{A}$ in the node covering problem contains exactly two 1's, while each column of $\mathbf{A}$ in the set covering problem may contain any number of 1's.

5.2.4 Applications of Set Covering Problem

Successful real-world applications of the set covering problem are ample, which can be classified in three major areas: (1) facility location, (2) scheduling or staffing of personnel, and (3) dispatching trucks/vehicles to routes/customers. For each area, we list below a few sample applications published in the literature.

1. *Facility location*
   - Determine the optimal location for a new fire station that can cover a given set of dispersed subdivisions, taking into account the average response time from a fire station to a fire in each subdivision.
   - Determine where emergency medical vehicles should be located in Austin, Texas, so that the number of people receiving adequate emergency service is maximized within a limited budget (Eaton et al., 1985).
   - Determine the least number of new supermarkets to be built to cover a number of geographically dispersed communities, taking into consideration the distance restriction and concentration of populations (Taha, 2007).
   - Determine which subset of a given number of potential transmission towers to be constructed that can cover a given number of contiguous geographical communities, taking into account their budgeted construction costs and maximization of potential population to be served (Guéret et al., 2002; Taha, 2007).
   - In an automated meter reading system for an electricity utility where meters from several customers are linked wirelessly to a single receiver, meters send monthly signals to designated receivers to report consumption of electricity and receivers send data to a central computer to general electricity bills. The problem is to determine the minimum number of receivers needed to cover a given number of customers (Taha, 2007).

2. *Scheduling or staffing of personnel*
   - Given a set of scheduled flights and a set of "preferred" flight crews, the staffing problem is to identify a subset of crews to cover all flights at a minimal cost. The problem has been modeled as a covering problem by most major airlines; recent references are Anbil et al. (1991), Hoffman and Padberg (1993), and Kontogiorgis and Acharya (1999). Other references dating back to 1957 are provided in the section notes.
   - For Pan American World Airways, determine optimal staffing levels for support staff for ticket counters, baggage loading and unloading, mechanical maintenance, and others, so that all work requirements are covered (Schindler and Semmel, 1993).
   - For a hospital, determine minimal number of nurses at various levels (RN, LPN, etc.) to cover the hourly requirements of various nursing functions, taking into account the upper limit on consecutive work hours.
   - Determine the minimal number of patrol police officers required to cover a given set of beats in San Francisco, taking into account response times (Taylor and Huxley, 1989).

3. *Dispatching trucks to routes/customers*
   - Determine emergency medical service vehicle deployment in Austin, Texas (Eaton et al., 1985).
   - Minimize the number of vehicles to meet a fixed periodic schedule (Orlin, 1982).

5.3 MATCHING PROBLEM

The *matching problem* belongs to another class of combinatorial optimization problems in which the constraint matrix **A** has exactly two 1's in each column. They deal with matching, pairing, or grouping objects such as selecting roommates, matching males to females, and assigning jobs to workers. The primary applications of this problem are for the development of matching-based algorithms. Examples include modeling network flow, routing, scheduling, spanning tree, and portfolio hedging and tracking.

5.3.1 Matching Problems in Network

The matching problem can be better described by the use of an undirected network. Consider an undirected network $G(V, E)$ of n $= |V|$ nodes and m $= |E|$ arcs, each arc joining a pair of nodes. The number of arcs incident to node i is called the *degree of node i*. The matching problem is to find a subset of arcs in the network such that at least a certain number of degrees, say b_i, are connected to node i, where b_i is a positive integer. This problem is called the **b**-*matching problem*, where b_i is an element of **b**.

The simplest matching problem is to find a matching in which each node can only be connected by at most one arc. This special case is called **1**-matching problem with b_i equal to 1 for all i. Consider the undirected network in Figure 5.1, the subset of arcs {(1, 2), (3, 4), and (5, 6)} is a **1**-matching, which is shown in Figure 5.3. A commonly used objective is to maximize the number of arcs selected.

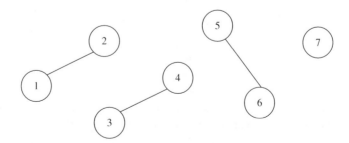

FIGURE 5.3 1-Matching for Figure 5.1.

The **1**-matching problem can be formulated as a 0–1 integer program:

$$\text{Maximize} \quad z = \sum_{j=1}^{m} y_j$$

$$\text{subject to} \quad \sum_{j=1}^{m} a_{ij} y_j \le 1 \qquad i = 1, 2, \ldots, n$$

$$y_j = 0 \text{ or } 1 \qquad j = 1, 2, \ldots, m$$

where a_{ij} is an element of a node–arc incidence matrix of the network and $y_j = 1$ if the jth arc is included in the matching, and 0 otherwise. Mathematically, the **1**-matching problem is different from the node covering problem in two ways: (1) The objective function is replaced by maximization, and (2) the constraint set is replaced by $\le$ inequalities. In other words, the node covering problem seeks to minimize the number of arcs used to cover all nodes in the network, while the **1**-matching problem seeks to maximize the number of arcs to connect nodes subject to at most one arc can be incident to each node. As a result, not all nodes are connected as shown in Figure 5.3.

5.3.2   Integer Programming Formulation

The **b**-matching problem is a generalization of the **1**-matching problem, which can be formulated as a 0–1 IP model:

$$\text{Max/min} \quad z = \mathbf{c}^{\mathrm{T}} \mathbf{y}$$

$$\text{subject to} \quad \mathbf{Ay} \le \mathbf{b}$$

$$\mathbf{y} \in (0, 1)$$

where $\mathbf{c} = (1, 1, \ldots, 1)^{\mathrm{T}}$, $\mathbf{y} = (y_1, y_2, \ldots, y_m)^{\mathrm{T}}$, $\mathbf{b} = (b_1, \ldots, b_n)$, and $\mathbf{A}$ is a node–arc incident matrix with each column containing exactly two elements of 1's. The matching problem can be a maximization or minimization problem. For example, find a maximum number of matching in the roommate selection problem, or find a minimum number of arcs forming a closed route in the postman problem.

When $\mathbf{c}$ is a vector of weights associated with variables, instead of 1's, the problem is called a *weighted b-matching*. When the constraint set is $\mathbf{Ay} = \mathbf{1}$, as in the set partitioning problem, the problem is called a *weighted perfect matching*.

In a weighted perfect matching problem, if the pairing objects are selected from two disjoint sets, then it becomes the classical assignment problem. Examples include assigning a set of workers (machines) to a set of jobs, assigning a set of plants to a set of potential locations, and assigning a set of tasks to a set of time slots. All of these assignments deal with pairing of objects from two disjoint sets. This problem can be represented by a *bipartite network*. In this bipartite network, a node in one set is connected to nodes in the other set, but the nodes of the same set are not connected.

5.4 CUTTING STOCK PROBLEM

Production activities in industries such as paper, textiles, plastic food wrap, aluminum foil, and steel sheet typically involve two stages. In the first stage, products are manufactured in large *standard sizes*, usually of a small variety due to economical and machinery considerations. In the second stage, these large standard sizes are cut into smaller *ordered* sizes, usually of a larger variety to satisfy diversified customer orders. The determination of how to cut the (larger) standard sizes into the (smaller) ordered sizes at minimum cost is called the *cutting stock problem*.

The cutting stock problem can be one or two dimensional. If all the ordered sizes are cut either horizontally or vertically, the problem is *one dimensional*. For example, a standard sheet of 72 in. width and 100 ft length can be slit horizontally into three pieces of 24 in. width and 100 ft length or vertically into four pieces of 72 in. width and 25 ft length. If an ordered size is made by both horizontal and vertical cuts, the problem is *two dimensional*.

For a given standard width (or standard length), usually there are many ways of cutting it into ordered widths (or ordered lengths). Each such way is called a *cutting pattern*. Figure 5.4 shows a possible cutting pattern from a roll of width W, which includes two rolls of width w_1, one roll of w_2, and one roll of trim loss T. Specifically, if $W = 12$ ft, $w_1 = 3$ ft, and $w_2 = 5$ ft, then there are seven possible cutting patterns as shown in Figure 5.5. However, patterns 4–7 that have trim loss greater than or equal to the smallest ordered width (i.e., ≥ 3 ft in this case) can be discarded. Therefore, only the first three patterns are effective.

5.4.1 One-Dimensional Case

Assume that there are available a sufficiently large number of rolls of a single standard width W, all having the same length L, that can be cut into at least b_i pieces of the ordered widths w_i ($i = 1, 2, \ldots, m$). A one-dimensional cutting stock problem is to determine how to cut rolls of the standard width into various ordered widths so that the required number of rolls (assuming trim pieces are useless) is minimal while satisfying the ordered quantity (b_i) of each width (w_i).

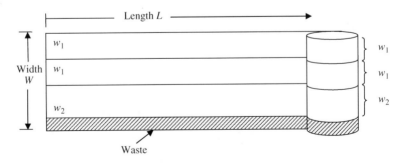

FIGURE 5.4 A cutting pattern from a roll of width.

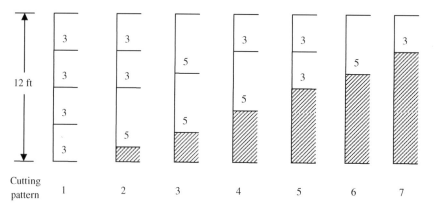

FIGURE 5.5   Another cutting pattern.

Let vector $\mathbf{a}_j$ denote the jth cutting pattern, whose component a_{ij} denotes the number of pieces in width w_i that can be generated from a standard width W. Mathematically,

$$\mathbf{a}_j = \begin{pmatrix} a_{1j} \\ \vdots \\ a_{ij} \\ \vdots \\ a_{mj} \end{pmatrix}$$

Note that a_{ij} is a nonnegative integer that must satisfy the following condition:

$$T_j = W - \sum_{i=1}^{m} w_i a_{ij} \geq 0$$

where $T_j \leq \min w_i$ is the trim loss associated with the jth cutting pattern. For example, the first three cutting patterns given in Figure 5.5 can be represented by

$$\mathbf{a}_1 = \begin{pmatrix} 4 \\ 0 \end{pmatrix}, \quad \mathbf{a}_2 = \begin{pmatrix} 2 \\ 1 \end{pmatrix}, \quad \mathbf{a}_3 = \begin{pmatrix} 0 \\ 2 \end{pmatrix}$$

with $T_1 = 0$, $T_2 = 1$, and $T_3 = 2$, respectively.

Following the modeling procedure described in Chapter 2, we formulate this problem.

Step 1

Input parameters:	standard width (W), ordered widths (w_i, $i = 1, 2,$ $\ldots, m$), required number of rolls of ordered widths (b_i, $i = 1, 2, \ldots, m$), all cutting patterns to be used ($\mathbf{a}_j$, $j = 1, 2, \ldots, n$)
Decision variables:	number of standard rolls (y_j, T_j) to be cut according to the jth pattern
State variables:	trim losses (T_j, $j = 1, 2, \ldots, n$)
Constraint:	total number of each ordered width w_i made must be at least b_i
Objective:	minimize the number of standard rolls needed

Step 2. The given cutting stock problem becomes

$$\text{Minimize} \quad \sum_{j=1}^{n} y_j$$

$$\text{subject to} \quad \sum_{j=1}^{n} a_{ij} y_j \geq b_i \qquad i = 1, 2, \ldots, m$$

$$y_j \geq 0 \text{ and integer} \quad j = 1, 2, \ldots, n$$

If a cost c_j is incurred with each cut using pattern $\mathbf{a}_j$, then the above objective function can be changed to minimize the total cost:

$$\text{Minimize} \quad \sum_{j=1}^{n} c_j y_j$$

The above formulation can be extended to one-dimensional problem with multiple standard widths (W^k, $k = 1, 2, \ldots, K$) with a fixed length L. For each standard width W^k, let n^k be the number of patterns, y_j^k be the number of the jth pattern to be cut, and c_j^k be the associated cost of cutting each jth pattern. Then the jth pattern can be represented by a vector $\mathbf{a}_j^k$, whose ith component is a_{ij}^k. We have the IP model for multiple standard widths:

$$\text{Minimize} \quad \sum_{k=1}^{K} \sum_{j=1}^{n^k} c_j^k y_j^k$$

$$\text{subject to} \quad \sum_{k=1}^{K} \sum_{j=1}^{n^k} a_{ij}^k y_j^k \geq b_i \qquad i = 1, 2, \ldots, m$$

$$y_j^k \geq 0 \text{ and integer} \quad j = 1, 2, \ldots, n^k; k = 1, 2, \ldots, K$$

Note that a_{ij}^k is a nonnegative integer that must satisfy the following condition:

$$T_j^k = W^k - \sum_{i=1}^{m} w_i a_{ij}^k \geq 0$$

The difficulty of cutting stock problem is that the number of possible cutting patterns n is usually too huge to enumerate them all. For example, with a roll of width 20 in. and demand for 40 different widths ranging from 20–80 in., the number of cutting patterns can exceed 100 million (Gilmore and Gomory, 1961). The number of cutting patterns is multiplied when there are multiple standard widths to be cut from.

Therefore, the IP model of a cutting stock problem is rarely solved exactly. In practice, its LP relaxation is solved by using Dantzig–Wolfe decomposition principle through a column generation technique. The details of such a solution approach will be discussed in Chapter 13.

5.4.2 Two-Dimensional Case

A two-dimensional cutting stock problem allows both horizontal and vertical cuts to get the ordered sizes. That is, rolls of standard width W and length L can be cut into b_i number of rectangular pieces of size $w_i \times l_i$ ($i = 1, 2, \ldots, m$). For example, given an unlimited number of standard rolls of size 4 ft × 10 ft and a demand of five rectangular pieces of size 2 ft × 4 ft and three pieces of size 3 ft × 7 ft, the problem is to determine the minimal number of rolls that satisfies the demand.

Similar to the one-dimensional problem, a set of suitable cutting patterns must be generated first to model the problem. Figure 5.6 shows two sample cutting patterns for the above example.

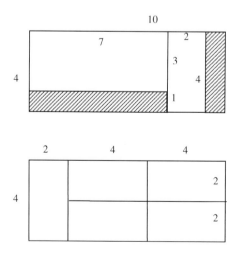

FIGURE 5.6 Examples of two-dimensional cutting patterns.

Let a_{ij} be the number of rectangular pieces of size $w_i \times l_i$ $(i = 1, 2, \ldots, m)$ generated by the jth cutting pattern. For example, the cutting patterns in Figure 5.6 are, respectively, represented by

$$\mathbf{a}_1 = \begin{pmatrix} 1 \\ 1 \end{pmatrix} \quad \text{and} \quad \mathbf{a}_2 = \begin{pmatrix} 5 \\ 0 \end{pmatrix}$$

Given the cutting patterns, the IP model of the two-dimensional case is the same as that of the one-dimensional case. However, due to the added dimension, the cutting pattern generation inequality and the column generating technique mentioned in the one-dimensional case cannot be used. Heuristics may be used to generate only the patterns that are likely to appear in the optimum solution.

In reality, the cost function of a cutting stock problem is more complex than the one described above. Schrage (2000) provides a list of additional cost considerations that are important in formulating a practical cutting stock model.

5.5 COMPARISONS FOR ABOVE PROBLEMS

The five problem types of the COP described in this chapter (namely, set covering, set partitioning, node covering, matching, and stock cutting) belong to a broader class of a pure IP model given below:

Find an integer vector $\mathbf{y}$
such that objective function $z = \mathbf{c}^T \mathbf{y}$ is minimized or maximized
subject to a set of equality or inequality constraints: $\mathbf{A}\mathbf{y} \, \{\geq \text{ or } = \text{ or } \leq\}\mathbf{b}$

The individual members of this class are distinguishable by their types of decision variables $\mathbf{y}$ (0–1 or general integer), types of optimization (min or max z), types of parameters ($\mathbf{c}^T$, $\mathbf{A}$, and $\mathbf{b}$), and types of relation ($\geq$, $=$, or $\leq$) between the left-hand and right-hand sides of the constraint set. For clarity, Table 5.5 provides comparisons for these problems.

To better understand the degree of difficulty for solving COB in particular, and IP in general, the basic concepts of the computational complexity of a problem and of an algorithm are explained next.

5.6 COMPUTATIONAL COMPLEXITY OF COP

A few people tend to overestimate the solution power of a computer (hardware) and underestimate the solution power of an algorithm (software). In fact, the solution power of an algorithm often exceeds the solution power of a computer. In what follows, we shall demonstrate how an algorithm affects the size of a problem that can be solved.

TABLE 5.5   Comparison of Five Problems

	Set Covering	Set Partitioning	Node Covering	Matching	Stock Cutting
y	0, 1	0, 1	0, 1	0, 1	Positive integer
z	Min	Min	Min	Min or max	Min
$\mathbf{c}^T$	Vector of real values	Vector of real values	$(1, 1, \ldots, 1)$	Vector of real values	Vector of real values
A	Any 0–1 matrix	Any 0–1 matrix	0–1 Matrix with two 1s in each column	0–1 Matrix with two 1s in each column	Positive integer matrix
b	$(1, 1, \ldots, 1)^T$	$(1, 1, \ldots, 1)^T$	$(1, 1, \ldots, 1)^T$	Positive integer vector	Positive integer vector
Relation	$\geq$	$=$	$\geq$	$\geq$ or $\leq$	$\geq$

In a practical sense, a combinatorial optimization problem can be viewed as a decision problem from which an optimum solution can be found among a finite set of feasible solutions that can be obtained by *explicit* enumeration of all possible combinations. Such a solution algorithm is known as *complete enumeration*. A grave shortcoming of complete enumeration is that the set of feasible solutions normally is too huge to handle, even for a small problem. In other words, solving a combinatorial optimization problem can be *algorithmically* simple but *computationally* intractable.

To show this, consider the knapsack problem discussed in Section 2.2. The problem deals with the optimal selection of a subset from a given set of items subject to a given knapsack capacity. The number of all possible feasible solutions is equal to the sum of combinations of selecting any one item from the given n items, any two items, and so on, which equals to

$$C_1^n + C_2^n + \cdots + C_{n-1}^n + C_n^n = 2^n - 1$$

Note that the number of all possible solutions increases exponentially in about 2^n as n increases. If $n = 20$, there are over 1 million (10^6) possibilities; if $n = 30$, over 1 billion (10^9); if $n = 40$, over 1 trillion (10^{12}); if $n = 50$, over 1000 trillion (10^{15}); and if $n = 60$, over 1 million trillion (10^{18}). Roughly, each additional 10 items will take nearly 1000-fold of additional computer time. On average, each such possible solution requires $2n$ arithmetic operations and comparisons.

Now suppose we have a computer that could calculate an arithmetic operation at a speed of light, that is, it can perform 8 trillion arithmetic operations per second. The complete enumeration algorithm would take over 4 years to solve a 70 item knapsack problem and would take over 4000 years to solve an 80 item problem.

In this section, we provide a brief, practical view of computational complexity for the combinatorial optimization problem.

5.6.1 Problem Versus Problem Instance

The word *problem* used in the domain of computational complexity is referred to as a class of problems having a common set of characteristics. For example, the knapsack problem is defined as a class of problems that is to determine an optimal subset of items within a prescribed knapsack capacity. The set covering problem is another class of problems that is to determine an optimal combination of activities covering all the prescribed requirements. In general, a problem is defined in terms of the types of its goal (min or max), decision variables, and constraints. To solve a particular COP problem, parameters such as $\mathbf{c}$, $\mathbf{A}$, $\mathbf{b}$, m, and n must be specified. A problem, after specification of its parameters, is called a *problem instance*. The warehouse location problem in Example 5.1 and the flight crew scheduling problem in Example 5.2 are problem instances.

5.6.2 Computational Complexity of an Algorithm

For a given algorithm, a mathematical function is often used to describe the growth of computational effort as problem size increases. Such a function is called *computational complexity of algorithm*. For example, the complete enumeration for the knapsack problem of size n has a computational complexity of 2^n, an exponential function.

Clearly, the size of an IP problem is a function of the number of integer variables n, the number of constraints, and the density of nonzero elements in a coefficient matrix. Traditionally and practically, however, only the number of integer variables is considered in the determination of the computational complexity (or complexity, for short) of an algorithm. Two reasons are behind this. First, an IP problem of more constraints does not necessarily require more computation time, based on Section 4.1. In fact, the reverse is often true. Second, the issue of the sparsity of matrix is problem specific or data specific, which is considered in the implementation of a given algorithm.

Ideally, an exact function in n is used to represent the computational complexity of an algorithm, sometimes even to a detailed level of counting the exact number of elementary operations $(+, -, \times, /,$ comparisons$)$ required to solve a problem instance. However, an exact function is often unattainable, in which case, an approximate function is used instead, and only the order of complexity for a large problem n is considered. For this purpose, the big O notation is commonly used in the theory of computational complexity to express an approximate upper bound computational effort for a given algorithm to solve a problem instance of size n. The expression

$$g(n) = O(f(n))$$

means that there exists a constant $k > 0$ and a small integer n_0 such that

$$g(n) \leq k \cdot O(f(n))$$

for all $n \geq n_0$. In other words, $f(n)$ gives a functional form of an upper bound on the value of $g(n)$ for a large n. The computational complexity of the given algorithm is then said to be $O(f(n))$, pronounced "a big-oh function of $f(n)$." For example, let

$$g(n) = an + b$$

Then the complexity is $O(n)$ for all $n \geq 1$ because

$$g(n) \leq an + bn = (a + b)n$$

where $k = (a + b) > 0$, $n_0 = 1$, and $f(n) = n$.

Note that the big O notation provides an upper bound on how poor an algorithm could be. It gives no information about how good an algorithm is.

Although the big O notation gives an approximate upper bound on the computational effort for a specified problem size, it should not be assumed that all problem instances of the same size require the same computational effort. Various problem instances of a given size may take different amounts of effort. Therefore, in practice, an average of computation times of problem instances is often taken to represent the performance. In addition, the worst-case and best-case analyses are also used.

5.6.3   Polynomial Versus Nonpolynomial Function

Unlike the complete enumeration algorithm, the computational complexity of most COP algorithms often cannot be *exactly* determined. In this case, an approximate function is estimated for the worst-case or average-case situations.

Based on the type of mathematical function, all algorithms are classified into two categories: polynomial time and nonpolynomial time. When the function is polynomial, the algorithm is said to be of *polynomial complexity* (or to be *polynomially bounded*) and such a polynomial algorithm is considered to be "easy" or efficient. Conversely, a nonpolynomial function is considered to be "hard" because the algorithm is capable of solving only a very small problem.

For example, if $g(n) = 6n^2 + 15n + 40$, then $O(f(n)) = O(n^2)$. Both $g(n)$ and $f(n)$ are polynomial functions except that the latter is a simplified form indicated by the order of the most significant term of the former. We say that the given algorithm has a polynomial function and the associated algorithm is considered to be easy.

Consider another example. The big O complexity of $g(n) = 2^{n-1} + n^2 + 10$ is $O(f(n)) = O(2^{n-1})$, or more loosely $O(2^n)$. We say that the algorithm has a nonpolynomial complexity and is considered to be hard.

Other forms of polynomial functions frequently used in the COP include $O(n^a)$ and $O(n^e)$, where n is the problem size (a variable), e is well-known constant approximately equal to 2.7183, and a is a positive constant. Other forms of nonpolynomial functions frequently used in the COP include $O(e^n)$ and $O(a^n)$, where e and a are positive constants and n is a variable. The forms of these big O functions can be further reduced or approximated to $O(n^2)$ for polynomial and $O(2^n)$ for nonpolynomial functions, respectively. The distinction of the two is that a

TABLE 5.6   Polynomial Versus Nonpolynomial Function

n	n^2	2^n
1	1	2
10	1×10^2	1.02×10^3
100	1×10^4	1.27×10^{30}
1000	1×10^6	1.07×10^{301}

nonpolynomial function has a constant base (2 or a) and a variable exponent (n), while a polynomial function has a variable base (n) and constant exponent (2 or a). To help the reader perceive how great is the difference on the computational complexity between the two functions, we give Table 5.6 and Figure 5.7. A problem is said to be a *polynomial problem* if there exists at least one polynomial–time algorithm. A problem is said to be a *nonpolynomial problem* if no polynomial–time algorithms have been found.

5.7   NOTES

Section 5.2

For the airline crew scheduling problems, see Anbil et al. (1991), Arabeyre et al. (1969), McCloskey and Hannsmann (1957), Baker and Fisher (1981), Hoffman and Padberg (1993), Kontogiorgis and Acharya (1999), Miller et al. (1976), Schindler and Semmel (1993), and Thirez (1968).

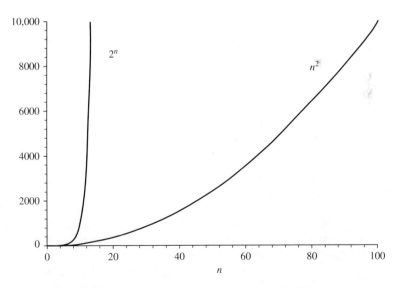

FIGURE 5.7   Polynomial versus nonpolynomial function.

For the staffing problem, see Agnihothri and Taylor (1991), Andrews and Parsons (1993), Aykin (1996), Rothstein (1973), Schindler and Semmel (1993), Taylor and Huxley (1989), and Warner (1976).

For the vehicle dispatching problems, see Agarwal et al. (1989), Balinski and Quandt (1964), Brown et al. (1987), Clarke and Wright (1964), Dantzig and Ramser (1960), Eaton et al. (1985), Lasky (1969), and Orlin (1982).

For assembly line balancing and information retrieval problems, see Salveson (1955) and Day (1965), respectively.

Section 5.3

For a survey on the matching problem, see Balinski and Quandt (1964) and Balinski (1965). For developing matching-based algorithms, see Edmonds and Johnson (1973) and Ball et al. (1983).

Section 5.4

For the cutting stock problem, see Dyckhoff (1981), Farley (1990), Gilmore and Gomory (1961, 1963), and Schrage (2003).

Section 5.6

See Ausiello et al. (1999) for a complete description on the computational complexity and approximation.

5.8 EXERCISES

5.1 Consider the pure integer program with bounded variables $y_j \leq K$ ($j = 1, 2, \ldots, n$), where K is a positive integer.

    (a) Show that the set of all possible solutions can be completely enumerated.

    (b) Show that the computational complexity of complete enumeration is nonpolynomial.

5.2 Given the graph below, (a) identify two different node covers; (b) set up its node–arc incidence matrix; and (c) formulate an IP model for finding a minimum cover of the graph (Figure 5.8).

5.3 Give two real-world applications of the set partitioning problem that have not been mentioned in this text.

5.4 Find a real-world application of cutting stock problem. (*Hint*: Search on Internet with appropriate keywords). Is it one dimensional or two dimensional? Is the problem formulated mathematically? If yes, give the formulation. If no, how will you formulate it? Can you think of an example of three or more dimensional problem?

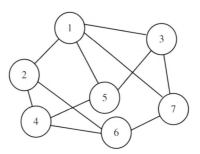

FIGURE 5.8 A simple graph.

5.5 (A Broker Model) A broker is placing m bids on n jobs. Each bid j can generate a possible profit of w_j, and each job i can only be bid at most once. Each bid either includes a job or not. The broker is trying to maximize the total profits generated by the bids.

(a) Formulate the problem as IP.

(b) What type of COP does this one belong to?

5.6 Based on Eaton et al. (1985) Gotham City has been divided into eight districts. The time (in minutes) that an ambulance takes to travel from one district to another is shown in Table 5.7. The population of each district (in thousands) is as follows—district 1: 40, district 2: 30, district 3: 35, district 4: 20, district 5: 15, district 6: 50, district 7: 45, and district 8: 60. Suppose Gotham City has n ambulance locations. Determine the locations of ambulances that maximize the number of people who live within 2 min of an ambulance. Do this separately for $n = 1$, $n = 2$, $n = 3$, and $n = 4$.

5.7 (Schrage, 2003) Suppose you manage your company's strategic planning department. There are eight analysts in the department. Your department is about to move into a new suite of offices. There are four offices in the new suite

TABLE 5.7 Travel Distances for Gotham City Problem

		To							
		1	2	3	4	5	6	7	8
	1	0	3	4	6	8	9	8	10
	2	3	0	5	4	8	6	12	9
	3	4	5	0	2	2	3	5	7
From	4	6	4	2	0	3	2	5	4
	5	8	8	2	3	0	2	2	4
	6	9	6	3	2	2	0	3	2
	7	8	12	5	5	2	3	0	2
	8	10	9	7	4	4	2	2	0

TABLE 5.8   Analysts' Incompatibility Ratings

Analysts	1	2	3	4	5	6	7	8
1	—	9	3	4	2	1	5	6
2	—	—	1	7	3	5	2	1
3	—	—	—	4	4	2	9	2
4	—	—	—	—	1	5	5	2
5	—	—	—	—	—	8	7	6
6	—	—	—	—	—	—	2	3
7	—	—	—	—	—	—	—	4

and you need to match up your analysts into four pairs, so that each pair can be assigned to one of the new offices. Based on past observations, you know some of the analysts work better together than they do with others. In the interest of departmental peace, you would like to come up with a pairing of analysts that results in minimal potential conflicts. To this goal, you have come up with a rating system for pairing your analysts. The scale runs from 1 to 10, with a 1 rating for a pair meaning the two get along fantastically, whereas all sharp objects should be removed from the pair's office in anticipation of mayhem for a rating of 10. The ratings appear in Table 5.8.

Since the pairing of analyst I with analyst J is indistinguishable from the pairing of J with I, we have only included the above diagonal elements in the table. Our problem is to find the pairings of analysts that minimizes the sum of the incompatibility ratings of the paired analysts.

5.8 Find a node cover and a 1-matching for the following network (Figure 5.9).

5.9 (Zionts, 1974. Used with permission) A lumber yard stocks 2 in. × 4 in. beams in three lengths: 8 ft, 14 ft, and 16 ft. The beams are sold by foot and no charge is made for cuts. The yard has an order for the following lengths:

80	12ft lengths
60	10ft lengths
200	8ft lengths
100	4ft lengths

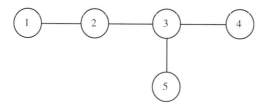

FIGURE 5.9   A simple network.

TABLE 5.9   Transmission Tower Data

Transmission Tower	Covered Population Centers	Cost ($M)
1	1, 2, 3	2.3
2	3, 5, 7	1.5
3	1, 6, 9	1.9
4	2, 4, 8, 9	3.1
5	4, 5, 7, 11, 12	2.7
6	10, 11, 12	2.0

The cost of the 2 × 4's to the lumber yard is $0.30 per 8 ft length, $0.60 per 14 ft length, and $0.70 per 16 ft length. Cutting costs can be assumed to be zero. Assuming that the lumber yard has enough of each of the three lengths in stock, what is the minimum cost method of filling the order?

5.10   Plywood is sold at Walls-are-Us in 48 in. × 96 in. rectangular sheets. A large job at a local construction site requires the following:

100	36in. × 48in. pieces
200	24in. × 35in. pieces
50	20in. × 48in. pieces
100	16in. × 30in. pieces

How many standard sheets of plywood should be purchased to minimize the cost to the contractor, assuming any cutting patterns are feasible and cutting is free of cost?

5.11   A cellular telephone service provider plans to offer service in a developing country, with 12 population centers (the rest is uninhabited mountainous terrain). The company has budgeted 10 million dollars to construct as many as 6 transmission towers to cover as much population as possible in the 12 population centers. The centers covered by each transmitter and the cost of construction are shown in Table 5.9.

The following table contains the population at each center:

Center	1	2	3	4	5	6	7	8	9	10	11	12
Population (in thousands)	5	4	17	7	8	10	8	3	6	15	9	10

Which of the proposed transmission towers should be constructed?

6

MODELING COMBINATORIAL OPTIMIZATION PROBLEMS II

This chapter deals with perhaps the most important class of combinatorial optimization problem: the traveling salesman problem (TSP) and its variants. Its purposes include (1) explain why the TSP is so important, (2) describe how to transform a variety of problems into a standard TSP, (3) provide a wide range of real-world applications of TSP, and (4) introduce several popular IP formulations for the TSP.

6.1 IMPORTANCE OF TRAVELING SALESMAN PROBLEM

The traveling salesman problem perhaps has been the most *well-studied* combinatorial optimization problem. Since the seminal paper published by Dantzig et al. (1954), the TSP has been actively and systematically studied by mathematicians, operations researchers, management scientists, and computer scientists for over five decades. During this period, thousands of refereed papers have been continuously published in the literature on TSP theories, formulations, applications, algorithms, and computations. In the two TSP books published in Gutin and Punnen (2002) and Applegate et al., (2006), the total number of distinct papers cited already exceeded 1000.

The TSP perhaps plays the most important role in the combinatorial optimization problem because over the decades it has been regarded as a *representative* or *typical model* of the combinatorial optimization problems whose computational complexity is of *nonpolynomial* (i.e., the problem is "hard" to solve). The TSP has been a primary driving force for the development of novel optimization concepts and solution

Applied Integer Programming: Modeling and Solution, By Der-San Chen, Robert G. Batson, and Yu Dang
Copyright © 2010 John Wiley & Sons, Inc.

TABLE 6.1   Milestones of TSP Instances Solved

Year	No. of Cities	Data Set	Research Team
1954	49	dantzig42	Dantzig, Fulkerson, Johnson
1971	64	random points	Held and Karp
1975	67	random points	Camerini, Fratta, Maffioli
1977	120	gr120	Grötschel
1980	318	lin318	Crowder and Padberg
1987	532	att532	Padberg and Rinaldi
1987	666	gr666	Grotschel and Holland
1987	2392	pr2392	Padberg and Rinaldi
1994	7397	pla7397	Applegate, Bixby, Chvátal, Cook
1998	13,509	usa13509	Applegate, Bixby, Chvátal, Cook
2001	15,112	d15112	Applegate, Bixby, Chvátal, Cook
2004	24,978	sw24978	Applegate, Bixby, Chvátal, Cook
2004	33,810	pla33810	Applegate, Bixby, Chvátal, Cook
2006	85,900	pla85900	Applegate, Bixby, Chvátal, Cook

Sources: www.tsp.gatech.edu (Applegate, 2007) and Applegate et al. (2006).

algorithms. For example, many AI algorithms, such as genetic algorithms, simulated annealing, and Tabu search, and many heuristic schemes, such as Lin–Kernighan's k-opt and the nearest-neighboring city, were developed at least in part to solve the TSP.

Commonly, the performance of an IP algorithm is measured by how large a TSP instance can be solved. Table 6.1 displays the milestones of the sizes of TSP instances that have been solved to optimality. The table marks the year, the problem size, and the contributor(s) when a particular sized TSP instance was solved to optimality. Many test data sets of the TSP are true distances on the road maps of the world's continents. For example, the data set d15112 is a map of 15,112 cities in Germany, usa13509 is a map of 13,509 cities in USA, and sw24978 is a map of 24,978 cities in Sweden. However, pla33810 and pla85900 are data sets derived from the application of TSP to integrated circuits (Applegate et al., 2006). These data sets can be found in TSPLIB, a library of sample instances for the TSP and related problems, maintained by Reinelt (1991, 2007). The largest six TSP instances in Table 6.1 were solved by a TSP solver called *Concorde*. The computer code is developed by the research team of Applegate et al. and written in the ANSI C programming language. "The full source code to the optimization package, as well as executables for various platforms, and a Windows graphical user interfaces to Concorde's traveling salesman solver are available for academic research use; for other uses, contact William Cook for licensing options" (www.tsp.gatech.edu). The user is suggested to download the Concorde package and try some TSP instances from the data sets of TSPLIB.

It is interesting to note that the ability for solving a large-scale TSP instance progressed very slowly in the first two and a half decades (1950s to mid-1970s) and progressed rapidly in the last three decades (mid-1970s to mid-2000s). The giant leap in computational capability is perhaps due to, among others, the introduction of novel solution approach called the *branch-and-cut*, which will be discussed in Chapter 12.

Another interesting observation is that prior to 1980s, the areas of TSP applications were limited, perhaps due to the small size of the TSP that could be solved. Most applications then were in the areas of machine sequencing in manufacturing and vehicle routing in transportation. After 1980s, areas of application are expanded to genome mapping of human and animals in life science, and circuit printing in the electronic industry. More details about these applications will be provided in the next section.

The statement of the traveling salesman problem is rather simple: A traveling salesman is to visit a number of cities and the distance connecting two cities are known; the problem is to find a shortest route that starts from a home city, visits other cities exactly once, and returns to the home city.

In graph theory, the TSP is commonly represented by a graph or network. A network is composed of a set of nodes (or vertices), a set of arcs (or edges) connecting nodes, and a known length (or distance) associated with each arc. An arc may be either undirected or directed. A directed arc allows travel only in the direction specified, while an undirected arc allows travel in either direction of the same length. Thus, any undirected or mixed network can be converted to a directed network by replacing any undirected arc by a pair of opposite directed arcs of the same length. Cities in a TSP are represented by nodes, and links between cities are represented by directed arcs. If a TSP is defined over a network (or digraph) of directed arcs, then it is called an *asymmetric TSP*. If a TSP is defined over a network consisting entirely of undirected arcs, it is called a *symmetric TSP*. Unless specified otherwise, we shall assume that the network is directed and replace any undirected arc by a pair of opposite directed arcs of the same length.

A *cycle* in a directed network is a sequence of nodes of the network such that it is possible to move from node to node, along directed arcs of the network, so that the selected nodes are encountered exactly once, except that the ending node is also the starting node. If a cycle contains *all* the nodes of the network, it is called a *Hamiltonian cycle*. For example, the directed network given in Figure 6.1 has a Hamiltonian cycle {(1, 2), (2, 5), (5, 6), (6, 4), (4, 3), (3, 1)} as depicted by the dotted lines. The TSP for a directed network with specified arc lengths is the problem of finding a Hamiltonian cycle of shortest length. A Hamiltonian cycle in the TSP is also called a *tour*. Any cycle that contains less than all the nodes in the network is a *subtour*.

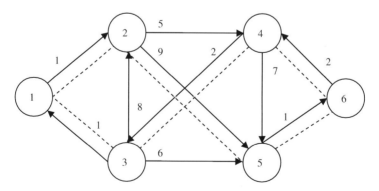

FIGURE 6.1   Hamiltonian cycle.

6.2 TRANSFORMATIONS TO TRAVELING SALESMAN PROBLEM

There are many problems that are minor variations of the TSP and can be easily transformed into the standard TSP. As a result, any efficient algorithm for solving the standard TSP can also be used to solve these variants, which extends the scope of the algorithmic application. In this section, we briefly describe and show how to transform the following problems into the standard TSP:

- Shortest Hamiltonian paths
- TSP with repeated city visits
- Multiple traveling salesmen problem
- Clustered TSP
- Generalized TSP
- Maximum TSP

Note that most variants can be formulated directly as integer programming models without using the transformations. The reader may attempt the models or see the cited references for formulations. Here, our purpose is to show how to transform each of these variants to a standard one. Note that the transformation may be reversed in some variants, for example, transforming a standard TSP to a shortest Hamiltonian path. Thus, knowing the reverse transformation can be beneficial if an efficient algorithm is available for finding a shortest Hamiltonian path.

6.2.1 Shortest Hamiltonian Paths

A path that starts from an arbitrary node, ends at another arbitrary node, and visits all other nodes exactly once in a given directed network is called a *Hamiltonian path* (or *H-path*). The shortest Hamiltonian path problem is a problem to find a *H*-path with the shortest distance.

A shortest *H*-path problem can be transformed into an equivalent TSP by constructing a new network G' from the original network G as follows. We add a new node (say $n + 1$) and new bidirectional dotted arcs (with distance 0) that, respectively, connect the new node with every node in the original network, as shown by the dotted arcs in Figure 6.2a. Now the shortest H-path problem on an n-node network G is equivalent to an $(n + 1)$-node TSP on network G'. Solving the TSP on network G', we obtain a shortest H-path, starting from node 5 and ending at node 2, as shown by the dotted arcs in Figure 6.2b.

Suppose the starting node is specified, say node 1. Then the shortest *H*-path problem can be transformed to the TSP by constructing a new network as follows. We add a directed arc with distance 0 to node 1 from every other node in the network, as shown by the dotted arcs in Figure 6.2c. Now the shortest *H*-path problem is equivalent to an n-node TSP on the new network.

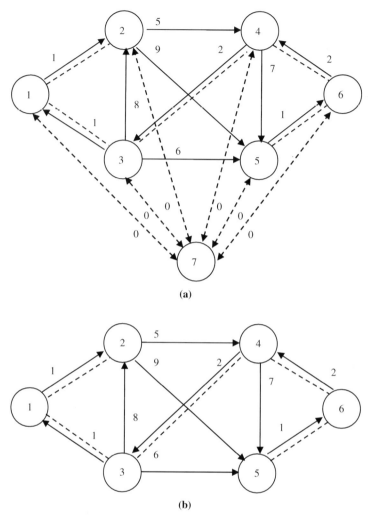

FIGURE 6.2   (a) Transformed TSP from H-path problem, (b) shortest Hamiltonian path, and (c) shortest H-path from node 1 to any node.

6.2.2   TSP with Repeated City Visits

Suppose that we have a standard TSP, except that it is required to visit each city *at least once* instead of exactly once. The challenge is: how to transform this problem to a standard TSP? We construct a new network with arcs representing the shortest paths between each pair of nodes. To show this construction, we consider a simpler network given in Figure 6.3a to obtain a new network in Figure 6.3b. Comparing these two figures, observe that in Figure 6.3a we have a shortest path from node 1 to node 4, {(1, 2), (2, 4)}, which has a shortest distance 7, shorter than the directed arc (1, 4) of distance 9. Now the TSP with multiple city visits is equivalent to a standard TSP

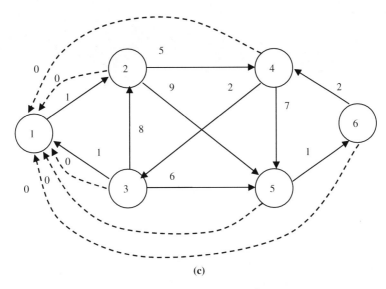

FIGURE 6.2 *(Continued)*

defined on the new network. In other words, we simply replace the given distance matrix by a new one whose elements are the distances of shortest paths between all pairs of nodes. Determination of shortest distances between all pairs of nodes can be easily computed in polynomial time $O(n^3)$ by a variety of shortest path algorithms.

6.2.3 Multiple Traveling Salesmen Problem

The multiple traveling salesmen problem (MTSP) can be stated as follows: Given a home city (node 0) and a set of $n-1$ customer cities (nodes $1, 2, \ldots, n-1$) to visit, the

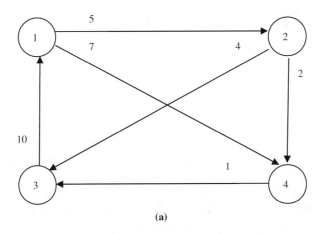

FIGURE 6.3 (a) TSP with repeated city visits; (b) transformed TSP from TSP with repeated city visits.

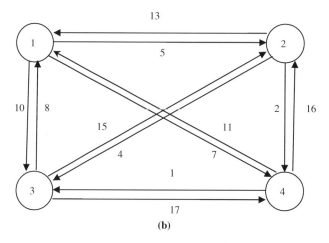

(b)

FIGURE 6.3 (*Continued*)

problem is to determine how many of the m salesmen should be utilized and to determine their respective routes that starts from and returns to the home city 0, so that the total distance traveled is minimized, subject to the constraint that each city (except home city) is visited once by one and only one salesman, where a fixed cost f_p ($p = 0$, $1, \ldots, m - 1$) is incurred if salesman p is activated and a distance cost $c(i, j)$ is incurred if arc (i, j) is traversed. Assuming the network is directed, the MTSP can be transformed to a standard asymmetric TSP by modifying the original network as follows:

1. Arrange the fixed costs of the salesman in ascending order:

$$f_0 \leq f_1 \leq f_2 \leq \cdots \leq f_{m-1}$$

2. Add dummy nodes, labeled by $-1, -2, \ldots, -(m-1)$, as a home city for salesman $2, 3, \ldots, m$, respectively.
3. Add a directed arc $(-i, j)$ for each $i = 1, 2, \ldots, (m-1)$ and each arc $(0, j)$ with distance $c'(-i, j) = c(0, j) + 1/2f_i$.
4. Add an directed arc $(j, -i)$ for each arc $(j, 0)$ with distance $c'(j, -i) = c(j, 0) + 1/2f_i$.
5. Add a directed arc $(-i, -(i-1))$ for every pair of $i = 1, 2, \ldots, (m-1)$ with distance $c'(-i, -(i-1)) = 1/2 f_{i-1} - 1/2 f_i$.

Applying the above procedure, we obtain a new network as shown in Figure 6.4. Suppose we apply a standard TSP algorithm to solve the new network and obtain the following optimal tour: {(0, 1), (1, 4), (4, −2), (−2, −1) (−1, 2), (2, 3), (3, 0)}. Then the corresponding optimal solution to the MTSP is interpreted as follows: Salesman 0 visits customer cities 1 and 4; salesman 2 visits no customer city; and salesman 1 visits customer cities 2 and 3. To obtain the total cost for the MTSP, we simply sum up the fixed and distance costs for salesmen 0 and 1.

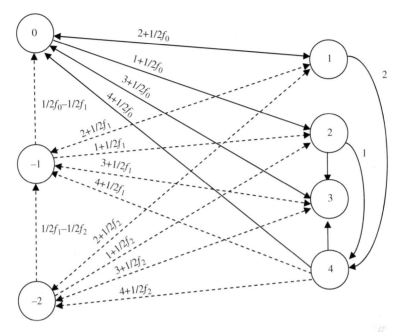

FIGURE 6.4 A multiple traveling salesmen problem.

Note that the transformed network is a directed network and the transformed TSP is an asymmetric TSP. Other transformation schemes are available (see Section 6.6 for references).

6.2.4 Clustered TSP

The clustered TSP can be stated as follows. Start with a directed network $G = (N, A)$ in which the set of nodes N is partitioned into k disjoint clusters of nodes $N_1, N_2, \ldots, N_k$. The problem is to find a least cost tour in G subject to the constraints that nodes within the same cluster must be visited consecutively. This problem can be transformed to the standard TSP by adding a large cost M to the cost of each intercluster arc.

6.2.5 Generalized TSP

The generalized TSP can be stated as follows. Start with a directed network $G = (N, A)$ in which the set of nodes N has been partitioned into k disjoint clusters of nodes $N_1, N_2, \ldots, N_k$, where $|N_i| \geq 1$ for $1 \leq i \leq k$. Then, the definition of the generalized TSP varies somewhat in the literature. Here, we define the generalized TSP as a problem, to find a least cost tour in G that passes through *exactly one node* from each cluster N_i, $1 \leq i \leq k$. In particular, if $|N_i| = 1$ for $1 \leq i \leq k$, then the generalized TSP is reduced to the standard TSP. Now we show how to transform a generalized TSP to a standard TSP according to Noon and Bean (1993). Basically, a new network is constructed from the given network by adding some new arcs and adjusting the connection of some old arcs. The procedure is given below:

1. For each cluster of $|N_i|>1$, we arbitrarily label the nodes as $n_{i1}, n_{i2}, \ldots, n_{ip_i}$ and form a cycle within cluster N_i, say $\{(n_{i1}, n_{i2}), (n_{i2}, n_{i3}), \ldots, (n_{i,p_i-1}, n_{i,p_i}), (n_{i,p_i}, n_{i1})\}$. For example, the dotted lines in Figure 6.5b shows a cycle within cluster N_4 of the give network in Figure 6.5a.

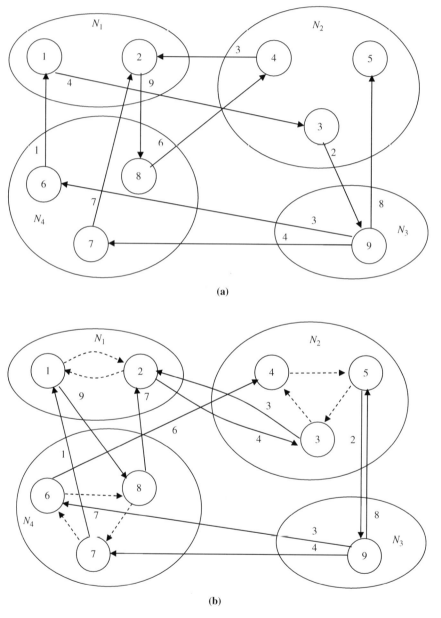

FIGURE 6.5 (a) Original network for generalized TSP; (b) transformed network for generalized TSP.

2. For each added arc $(n_{ij}, n_{i,j+1})$, we associate a cost of zero: $c'(n_{i1}, n_{i2}) = c'(n_{i2}, n_{i3}) = \cdots = c'(n_{i.p_i-1}, n_{i.p_i}) = c'(n_{i.p_i}, n_{i1}) = 0$.

3. For each outgoing arc (n_{ij}, n_{kq}) from a cluster, we replace it with arc $(n_{i,j-1}, n_{kq})$ such that $n_{i,j-1}, n_{ij} \in N_i, n_{kq} \in N_k$, and $k \neq i$. Its associated cost is adjusted as follows: $c'(n_{i,j-1}, n_{kq}) = c(n_{ij}, n_{kq})$ and $c'(n_{i1}, n_{kq}) = c(n_{i.p_i}, n_{kq})$.

4. For each remaining arc, its connection and cost remain unchanged.

Figure 6.5b depicts the transformed network G' created from the given network G in Figure 6.5a. Step 2 implies that once a TSP tour enters cluster N_i through n_{ij}, it will visit all other nodes in N_i following a predetermined cycle without additional cost. But the TSP tour must leave cluster N_i from node $n_{i,j-1}$ rather than from node n_{ij}.

Step 3 ensures that a generalized tour visits exactly one node in each cluster N_i. For example in Figure 6.5b, there are three arcs leaving from cluster N_4. If the entering node n_{ij} is node 6, then its preceding node $n_{i,j-1}$ is node 7. Thus, arc (6, 1) is replaced by arc (7, 1) with cost $c'(7, 1) = c(6, 1)$. Similarly, we have $c'(8, 2) = c(7, 2)$ and $c'(1, 4) = c(8, 4)$. This step ensure that if a TSP tour in G enters a cluster N_i through node n_{ij}, it visits nodes of cluster N_i in the order of $n_{ij}, n_{i,j+1}, \ldots, n_{i,p_i}, n_{i1}, \ldots, n_{i,j-1}$ and leaves the cluster N_i from the node $n_{i,j-1}$. Note that all outgoing arcs from node $n_{i,j-1}$ in G' correspond to the original outgoing arcs from node n_{ij} in G. As a result, all other nodes in the cluster are treated as dummy, and only the entering and leaving arcs of node n_{ij} are counted. For example, suppose there is a TSP tour that enters cluster N_4 from cluster N_3 by arc (9, 6), visits nodes 6, 8, and 7, and finally leaves cluster N_4 by using arc (7, 1) to node 1. This tour in Figure 6.5b corresponds to a tour in Figure 6.5a, which enters N_4 by arc (9, 3), visits node 6, and leaves cluster N_4 by arc (6, 1) to cluster N_1. Thus, the generalized TSP on G is equivalent to standard TSP on G'.

6.2.6 Maximum TSP

The *maximum TSP* can be stated as follows: Given a network G where the profit associated with each arc may be either positive or negative, the problem is to find a tour in G where the total profit of arcs of the tour is maximal. The problem can be transformed to a standard (minimum) TSP by replacing each arc profit by its negative value. If some resultant arc values are negative, then add a large constant M to each of the arc values to ensure that all arc values in G are nonnegative.

6.3 APPLICATIONS OF TSP

Real-world applications of the TSP and its variants are ample. Here, we give examples of four application areas reported in the literature: (1) machine sequencing problems in various manufacturing systems, (2) machine sequencing problems in electronic industry, (3) vehicle routing problems for delivery/dispatching, and (4) genome sequencing problems for genetic study. There is a wide variety of TSP applications. The interested reader may refer the survey papers mentioned in Section 6.6.

6.3.1   Machine Sequencing Problems in Various Manufacturing Systems

Perhaps the largest application area of the TSP is machine sequencing/scheduling problems arising in various manufacturing systems across multiple industries. In general, there are two types of sequencing problems that can be solved as a TSP. The first type of sequencing problem is scheduling n jobs on a single machine or on an assembly line. The second type is scheduling n jobs on m machines in the same order with no wait in process. In what follows, we give three systems for the first type and one system for the second type.

- *Job scheduling*: There are n jobs with known processing times to be processed sequentially on a single machine. The jobs can be processed in any order but their machine setup times are job dependent. That is, the setup time requiring for processing job j immediately follow job i may vary. The objective is to find a sequence of jobs so that all jobs are completed in the shortest possible time.
- *Assembly line*: In an *assembly line system*, jobs are often grouped together as a cluster so that the setup time, if any, between jobs within the same cluster is relatively small compared to the setup time between jobs in different clusters. This type of manufacturing system can be viewed as a *clustered TSP*.
- *Cellular manufacturing*: In a cellular manufacturing system, families of parts (products) that required similar processing are grouped and processed together in a specialized machine cell to achieve efficiency and cost reductions. This production philosophy is known as *group technology*. Aneja and Kamoun (1999) showed that the problem of sequencing jobs processed by a robot in a machine cell can be formulated as a TSP. Its objective is to find an optimal job sequence such that the robot's total time of movement is minimized.
- *Flow shop sequencing*: The problem of scheduling n jobs on m machines, with processing in the same order for each job, with no wait in process is also known as the *flow shop sequencing problem with no wait in process*. The sequencing problem can be described as follows. There are n jobs with known processing time that require processing by m machines in the same order. Each machine can work on at most one job at a time and once it begins work on a job it must continue working on it until completion without interruption. It is assumed that once a job is completed on a machine j, it must be immediately processed on machine $j + 1$ with no wait in process. The objective is to finish the last job as soon as possible. It can be shown that this sequencing problem can be formulated as an n-city shortest Hamiltonian path problem, which in turn can be transformed to an $(n + 1)$-city TSP by adding a dummy city.

6.3.2   Sequencing Problems in Electronic Industry

The electronic industry has utilized the TSP to solve the sequencing problems arising in design, production, and testing of integrated circuits (IC), also known as computer chips. In fact, the history of the TSP applications in the electronic industry paces the history of TSP applications. Such applications began as early as 1973 when Lin and

Kernighan (1973) introduced their famous heuristic algorithm for solving a 318-city problem arising in sequencing a numerically controlled drilling machine efficiently through a set of hole positions of an IC. The problem is called lin318 in the TSPLIB. As the IC technologies evolved over the decades, the use of the TSP in solving the sequencing problem still prevailed. The following are examples of the TSP uses in designing, manufacturing, and testing of ICs.

- *Drilling holes on printed IC boards*: A large number of holes are needed on printed IC boards for mounting chips and other hardware, or connecting layers to attain some specific functionality. Such holes are typically produced by auto-mated drilling machines that move to drill holes between specified locations. The TSP is to minimize the total traveling time of the drill, where the cities correspond to the hole locations. The hole drilling problems arise in production of both general and customized ICs.
- *Testing ICs via scan chain technology*: A scan chain is automatically generated test pattern for an IC in which components (scan points) in the IC are connected in a chain having input and output connections on the boundary of the chip. As stated by Applegate et al. (2006), "a scan chain permits test data to be loaded into the scan points through the input end, and after the chips performs a series of test operations [in pre-determined sequence] the data can be read and evaluated from the output end." In creating a scan chain, chip designers have naturally turned to the TSP in order to determine the minimum distance sequence of the scan points to save time in the testing phase. Pathways on ICs are only in the vertical and horizontal directions, so scan chains form paths that run from input to outputs in the manner of city street layouts, using the so-called "taxicab" or "Manhattan" metric: $d[(x_1, y_1), (x_2, y_2)] = |x_1 - x_2| + |y_1 - y_2|$.

6.3.3 Vehicle Routing for Delivery and Dispatching

Another common application of the TSP is the vehicle routing for delivery and dispatching services:

- *School bus routing*: Scheduling a fleet of school buses to pick up and transport waiting children to schools can be viewed as a multiple TSP if the constraints of time windows and bus capacities are of no concerns. Otherwise, the problem is a vehicle routing problem rather than a TSP.
- *Parcel/postal delivery/dispatching*: This type of problem is normally considered as a postman problem where a vehicle visits a given set of streets (or arcs) rather than a given set of locations (or nodes). However, recently the TSP software has been modified and adopted for use in these applications with successful reports.
- *Meals on wheels*: In many urban areas, dispatch a fleet of vehicles with meals to deliver to elderly and sickly people on a regular basis.
- *Clinic on wheels*: Dispatch a medical vehicle to service medical needs for a set of rural communities.

- *Maintenance on wheels*: Periodically, a maintenance vehicle with crew and equipments is dispatched to inspect and maintain the equipments in a number of bases or stations.

6.3.4 Genome Sequencing for Genetic Study

Mapping the genome of a species of animals is an ordering problem of huge proportion. For example, human have 23 chromosomes and each has to be mapped. As stated by Applegate et al. (2006), before finding the genome sequence of a species, the research team must determine "accurate placement of markers that serve as landmarks for the genome maps... A genome map has for each chromosome a sequence of markers with some estimate of the distance between adjacent markers. The markers in these maps are segments of DNA that appear exactly once in the genome under study... It is particularly useful to have accurate information on the order in which the markers appear on the genome, and this is where the TSP comes into play."

Many genome sequencing research projects that employ the TSP approach have been reported in the literature in the 2000s. Species under study include human, macaque, horse, dog, cat, mouse, rat, and cow (see Section 6.6 for references).

6.4 FORMULATING ASYMMETRIC TSP

The *asymmetric* TSP is defined on a directed network in which travels are allowed only in the directions specified. The problem is to find a *directed tour* with a minimal distance. This section presents the formulation of the asymmetric TSP as a 0–1 integer program. All of IP formulations in this chapter are based on the assignment problem. We follow the modeling procedure described in Chapter 2. Although any city can be a starting city, for simplicity we assume it is city 1.

Step 1

Input parameters:	all directed arcs (i, j) and associated distances (c_{ij})
Decision variables:	whether or not each directed arc (i, j) is in the tour $(y_{ij} = 1$ or $0)$
Constraints:	each city j must be entered exactly once, each city i must be exited exactly once, a tour that starts from a starting city must travel exactly $n - 1$ cities and return to the starting city, and no subtours are allowed
Objective:	total distance traveled on a tour must be minimum

Step 2. The asymmetric TSP can be formulated as

$$\text{Minimize} \quad \sum_{(i,j) \in A} c_{ij} y_{ij} \qquad (6.1)$$

$$\text{subject to} \quad \sum_{\{i:(i,j)\in A\}} y_{ij} = 1 \quad \text{for all cities } j = 1 \text{ to } n \tag{6.2}$$

$$\sum_{\{j:(i,j)\in A\}} y_{ij} = 1 \quad \text{for all cities } i = 1 \text{ to } n \tag{6.3}$$

$$y_{ij} = 0 \text{ or } 1 \quad \text{for all arcs } (i,j) \in A \tag{6.4}$$

A set of subtour elimination constraints

Constraints (6.2) ensure that each city j must be entered exactly once. Constraints (6.3) ensure that each city i must be exited exactly once. Note that constraints (6.1)–(6.4) correspond to the classical *assignment problem*. However, any solution satisfying (6.2)–(6.4) is not sufficient to define a tour because it may also define some disjoint subtours, for example, of $n = 6$ cities as shown in Figure 6.6. The solution $y_{12} = y_{25} = y_{51} = y_{43} = y_{36} = y_{64} = 1$ (and $y_{ij} = 0$ for all others) satisfies constraints (6.2)–(6.4) but forms two disjoint subtours: $1 \to 2 \to 5 \to 1$ and $3 \to 6 \to 4 \to 3$ as shown in Figure 6.7. Thus, the assignment problem is a relaxation of the TSP, and the TSP is a restriction of the assignment problem.

Many existing IP formulations for the TSP are relaxations of the assignment problem. Their difference is in the formulation of subtour elimination constraints. In this text, we present two popular ones for the asymmetric TSP and one for the symmetric TSP.

6.4.1 Subtour Elimination by Dantzig–Fulkerson–Johnson Constraints

This subtour elimination scheme is based on the fact that for every tour or subtour, the number of nodes must be equal to the number of arcs. Therefore, to prevent from forming subtours but allow forming a tour, the number of arcs must be less than the number of cities for every subset that consists of 2 to $n-1$ cities. Mathematically,

$$\sum_{i\in S}\sum_{j\in S} y_{ij} \leq |S|-1 \quad \text{for all subsets } |S| = 2, 3, \ldots, n-1 \tag{6.5}$$

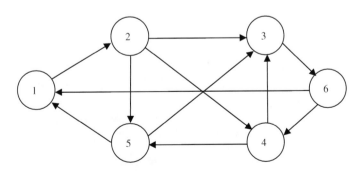

FIGURE 6.6   Directed graph.

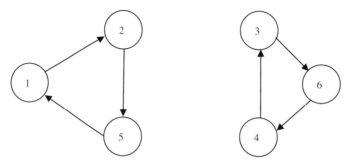

FIGURE 6.7   Directed subtours.

where S is a nonempty proper subset of all nodes V, and $|S|$ is the number of cities in S. The total number of constraints in (6.5) is nearly 2^n because there are 2^n possible subsets for n cities. For example, the subtour $1 \rightarrow 2 \rightarrow 5 \rightarrow 1$ in Figure 6.7 could not be satisfied because $\sum_{i \in S} \sum_{j \in S} y_{ij} = y_{12} + y_{25} + y_{51} = 3$, $|S| - 1 = 2$, and thus a corresponding constraint in (6.5) would be violated. Likewise, the subtour $3 \rightarrow 6 \rightarrow 4 \rightarrow 3$ in Figure 6.7 would also violate a constraint in (6.5).

6.4.2   Subtour Elimination by Miller–Tucker–Zemlin (MTZ) Constraints

This set of subtour elimination constraints is introduced by Miller et al. (1960). This set is derived based on the following observation. Consider that a tour is just a sequence of all cities. If we define u_j as the sequence number of city j in a tour, we obtain the following set of subtour elimination constraints:

$$u_i - u_j + ny_{ij} \le n-1 \quad \text{for } (i,j) \in A, \ i \ne, \ j \ne 1, \ j \ne i \qquad (6.6)$$

Consider the first subtour in Figure 6.7, for example. We have $y_{36} = y_{64} = y_{43} = 1$. Because city 3 is the first city in the sequence, we have $u_3 = 1$, $u_6 = 2$, and $u_4 = 3$. From (6.6), we obtain three corresponding constraints:

$$u_3 - u_6 + 6y_{36} = 1-2+6 = 5 \le 5 \quad \text{(satisfied)}$$

$$u_6 - u_4 + 6y_{64} = 2-3+6 = 5 \le 5 \quad \text{(satisfied)}$$

$$u_4 - u_3 + 6y_{43} = 3-1+6 = 8 > 5 \quad \text{(violated)}$$

Clearly, this subtour violates the third constraint. Similarly, we can show that every subtour that does not contain city 1 will violate a constraint in (6.6).

Now if there is a set of y_{ij}'s that does not contain a subtour, then we can define a set of u_i's, starting from city 1, that does not violate any constraint in (6.6). Let $u_i = k$ indicate that city i is the kth city visited in the tour, where $k = 2, 3, \ldots, n$. If $y_{ij} = 1$, we then have $u_j = k + 1$ and the left-hand side of (6.6) is

$$u_i - u_j + ny_{ij} = k - (k+1) + n = n-1$$

which satisfies the right-hand side of (6.6) for every k. If $y_{ij}=0$ and $u_j=k'$ ($k'=2$, $3, \ldots, n$, $k' \neq k + 1$), then the left-hand side of (6.6) is

$$u_i - u_j + n y_{ij} = u_i - u_j = k - k'$$

Clearly, the largest difference for $k - k'$ occurs when $k = n$ and $k' = 2$, which is $n - 2 < n - 1$. We have shown that (6.6) can be satisfied for all cases of $y_{ij} = 1$ and $y_{ij} = 0$ if no subtours are involved. For example, consider a tour of $1 \rightarrow 2 \rightarrow 4 \rightarrow 5 \rightarrow 3 \rightarrow 6 \rightarrow 1$ for the above six-city problem. We have $y_{12} = y_{24} = y_{45} = y_{53} = y_{36} = y_{61} = 1$, $y_{23} = y_{25} = y_{43} = y_{51} = y_{64} = 0$, and set $u_1 = 1$, $u_2 = 2$, $u_4 = 3$, $u_5 = 4$, $u_3 = 5$, and $u_6 = 6$. Check the corresponding constraints in (6.6) for $i \neq 1$ and $j \neq 1$:

$$y_{24} = 1, \quad u_2 - u_4 + 6y_{24} = 2 - 3 + 6 = 5$$
$$y_{45} = 1, \quad u_4 - u_5 + 6y_{45} = 3 - 4 + 6 = 5$$
$$y_{53} = 1, \quad u_5 - u_3 + 6y_{53} = 4 - 5 + 6 = 5$$
$$y_{36} = 1, \quad u_3 - u_6 + 6y_{36} = 4 - 5 + 6 = 5$$

which satisfy all the constraints in (6.6). Next, we consider $y_{23} = y_{25} = y_{43} = y_{64} = 0$

$$y_{23} = 0, \quad u_2 - u_3 + 6y_{23} = 2 - 5 = -3 < 5$$
$$y_{25} = 0, \quad u_2 - u_5 + 6y_{25} = 2 - 4 = -2 < 5$$
$$y_{43} = 0, \quad u_4 - u_3 + 6y_{43} = 3 - 5 = -2 < 5$$
$$y_{64} = 0, \quad u_6 - u_4 + 6y_{64} = 6 - 3 = 3 < 5$$

which also satisfy all the constraints in (6.6). Now we have shown that every subtour will violate at least one of the constraints in (6.6) and that no complete tour can be excluded by (6.6).

The MTZ formulation, when compared with clique packing, adds only n variables to the model (the u's), but dramatically decreases the number of constraints to approximately n^2 from nearly 2^n. At first glance, this huge reduction in the number of constraints could mean great reduction in time for finding an optimum tour. On the contrary, the set of type (6.5) constraints is much tighter than the set of type (6.6) constraints. In fact, the clique packing formulation is better than the MTZ formulation in the sense that the polyhedron (say P_1) of the LP relaxation of the first formulation is a subset of the polyhedron (say P_2) of the second formulation. Nemhauser and Wolsey (1988) provided the following example to show that $P_2 \not\subset P_1$. If $n \geq 4$, The point $u_2 = u_3 = u_4 = 0$ and $y_{23} = y_{34} = y_{42} = (n - 1)/n > 2/3$ satisfies constraint set (6.6) but not (6.5).

The above two formulations for TSP show that the computability of IP reverses the rule of the computability of LP—the computational time increases as the number of constraints increases. Therefore, any attempts to finding compact IP formulations with a small number of constraints are often counterproductive for efficiently solving large-scale TSPs.

6.5 FORMULATING SYMMETRIC TSP

The *symmetric* traveling salesman problem is defined on an undirected network in which travel is allowed in either direction of each undirected arc or edge. The problem is to find an *undirected tour* with a minimal traveling distance. To formulate the symmetric TSP, one must note that $c_{ij} = c_{ji}$ and $y_{ij} = y_{ji}$. Thus, each edge can be identified by a single index k, each decision variable by y_k, and each distance by c_k. As a result, the symmetric problem can be formulated with only one-half the number of 0–1 variables as required in the asymmetric problem. To find a tour in an undirected network $G(E, A)$, one must select a subset of undirected arcs such that every node j is connected to exactly two of the undirected arcs selected. As in the asymmetric problem, the symmetric problem requires additional constraints to eliminate all subtours, but not any tour.

Again, following the modeling procedure described in Chapter 2, we now formulate the symmetric TSP as a 0–1 integer program.

Step 1

    Input parameters: a list of undirected arcs indexed by k and their associated distances, c_k

    Decision variables: whether or not each undirected arc k is in the tour ($y_k = 1$ or 0)

    Constraints: each city in the tour must have exactly two undirected arcs incident to it, and all subtours must be eliminated

    Objective: total distance traveled in a tour must be minimal

Step 2. Let E be the set of all undirected arcs, E_j be the set of all undirected arcs connected to city j, and E_S be the set of all undirected arcs connecting the cities in any proper subset S. Also let $y_k = 1$ if undirected arc $k \in E$ is in the tour, and $y_k = 0$ otherwise. Then the symmetric TSP can be formulated as a 0–1 integer program:

$$\text{Minimize} \quad \frac{1}{2} \sum_{j=1}^{n} \sum_{k \in Ej} c_k y_k \tag{6.7}$$

$$\text{subject to} \quad \sum_{k \in Ej} y_k = 2 \quad \text{for all cities } j = 1, 2, \ldots, n \tag{6.8}$$

$$\sum_{j \in E_S} y_j = |S| - 1 \quad \text{for all } |S| = 2, 3, \ldots, n-2 \tag{6.9}$$

$$y_j = 1 \text{ or } 0 \quad \text{for all } j \in E \tag{6.10}$$

Constraints (6.8) ensure that every node in the tour must have exactly two undirected arcs connected to it. As in the asymmetric case, (6.9) is a set of subtour elimination constraints equivalent to (6.5).

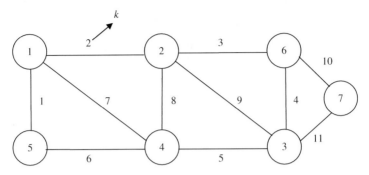

FIGURE 6.8 Undirected network.

For example, consider the undirected network given in Figure 6.8 that has 7 cities and 11 undirected arcs labeled by $k = 1, 2, \ldots, 11$. The solutions $y_1 = y_2 = y_8 = y_6$ $y_4 = y_{10} = y_{11} = 1$ and $y_3 = y_5 = y_7 = y_9 = 0$ satisfy constraints in (6.8) but not in (6.9). As a result, we obtain two disjoint subtours, 1–2–4–5–1 and 3–6–7–3, as shown in Figure 6.9.

First, we examine the subtour 1–2–4–5–1 using the constraints in (6.8) and (6.9). In this subtour, we have

$$S = \{1, 2, 4, 5\}, \quad |S| = 4$$
$$E_1 = \{1, 2, 7\}, \quad E_2 = \{2, 8, 3, 9\}, \quad E_4 = \{6, 7, 8, 5\}, \quad E_5 = \{1, 6\}$$

The corresponding constraints in (6.8) are

$$\text{For } j = 1, \quad \sum_{k \in E_1} y_k = y_1 + y_2 + y_7 = 2$$

$$\text{For } j = 2, \quad \sum_{k \in E_2} y_k = y_2 + y_8 + y_3 + y_9 = 2$$

$$\text{For } j = 4, \quad \sum_{k \in E_4} y_k = y_6 + y_7 + y_8 + y_5 = 2$$

$$\text{For } j = 5, \quad \sum_{k \in E_5} y_k = y_1 + y_6 = 2$$

which satisfy all the constraints in (6.8).

Next, we examine the constraints in (6.9). We have

$$E_S = \{1, 2, 6, 7, 8\}$$
$$y_1 = y_2 = y_6 = y_8 = 1$$
$$y_7 = 0$$

Thus, the left-hand side of (6.9) is

$$\sum_{j \in E_S} y_j = y_1 + y_2 + y_6 + y_7 + y_8 = 4$$

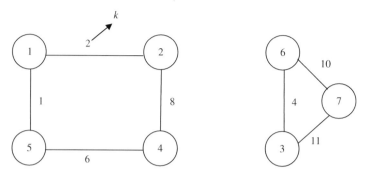

FIGURE 6.9   Two subtours in undirected network.

But the right-hand side of (6.10) is

$$|S| - 1 = 4 - 1 = 3$$

Hence, a constraint (6.9) is violated and the subtour 1–2–4–5–1 cannot happen. Likewise, we can show that the subtour 3–6–7–3 and all other subtours cannot occur because each subtour would violate a constraint in (6.9).

6.6   NOTES

Much of the material in this chapter is based on the TSP books by Applegate et al. (2006), Gutin and Punnen (2002), and Lawler et al. (1985). Related material is available on www.tsp.gatech.edu.

Section 6.1

Much of the data sets of TSP instances given in Table 6.1 can be found in TSPLIB (Reinelt, 1991), a traveling salesman problem library. Gutin and Punnen (2002) and Applegate et al. (2006), respectively, have 838 and 581 counts of TSP references. After removing the duplicated references, the total count exceeds 1000.

Section 6.2

See Lawler et al. (1985) for transforming a TSP to a shortest Hamiltonian path problem. See Hong and Padberg (1977) and Rao (1980) for transforming a symmetric multiple traveling salesmen problem to a standard TSP. See Jongens and Volgenant (1985) for transforming the symmetric cluster TSP to the standard TSP. See Noon (1988) and Noon and Bean (1993) for transforming generalized TSP to standard TSP. See Barvinok et al. (2002) for more detail about the maximum TSP.

Section 6.3

For scheduling jobs on a machine with sequence-dependent setup times, see Gilmore and Gomory (1964) and Bianco et al. (1988). For knowledge of scheduling theory and manufacturing systems, see Pinedo (2002). For more TSP applications, see survey papers by Garfinkel (1985), Lenstra and Rinnooy Kan (1975), and the book by Reinelt (1994).

6.7 EXERCISES

6.1 Visit "http://www.tsp.gatech.edu/index.html" or any other Web site containing information on TSP. Give at least three real-world applications of TSP that are not mentioned in this book.

6.2 For the asymmetric TSP, show that the following set of constraints is equivalent to the set of subtour elimination constraints in (6.5):

$$\sum_{i \in S} \sum_{j \in \bar{S}} y_{ij} \geq 1 \qquad \text{for all subsets } |S| = 2, 3, \ldots, n-1$$

where $\bar{S} = \mathbf{V} \backslash \mathbf{S}$.

6.3 For the symmetric TSP, show that each of the following sets of constraints is equivalent to the set of subtour elimination constraints in (6.10):

(a) $\sum_{j \in E\bar{S}} y_j \leq |\mathbf{V}| - |\mathbf{S}| - 1$ for all $|S| = 2, 3, \ldots, n-2$
    where $\bar{S} = \mathbf{V} \backslash \mathbf{S}$ and $E_{\bar{S}}$ is the set of all edges in $\bar{S}$.

(b) $\sum_{j \in E\bar{S}} y_j \geq 2$ for all subsets $|S| = 2, 3, \ldots, n-2$
    where $\bar{S} = V \backslash \mathbf{S}$ and $E_{S\bar{S}}$ is the set of all edges between S and $\bar{S}$.

6.4 Consider the directed network given in Figure 6.10:

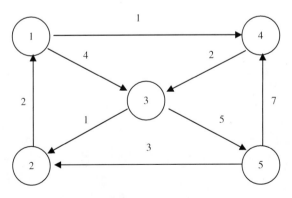

FIGURE 6.10 A directed network.

    (a) Find a Hamiltonian path starting from node 1. Find a Hamiltonian path starting from node 5.

    (b) Find a shortest Hamiltonian path by inspection.

    (c) Transform the given network G into a network G' such that a shortest Hamiltonian path can be found by a TSP algorithm.

    (d) Construct the cost matrix corresponding to the network G'.

6.5 Consider the network in Problem 6.5. Suppose nodes are allowed to visit more than once.

    (a) By inspection find the shortest distance between each pair of nodes.

    (b) Construct a transformed network that can be solved by a standard TSP algorithm/software.

    (c) Construct the distance matrix for the transformed network.

6.6 There are two trucks available at the warehouse (node 1) to be dispatched for delivering goods to all the customers (the remaining nodes). Assume each customer can be delivered by a truck only.

    (a) Suppose the costs of trucks and drivers vary: $100 and $120, respectively. Construct a transformed network that can be solved by a standard TSP algorithm/software.

    (b) In the transformed network, let nodes 1 and -1 be the starting nodes for trucks 1 and 2, respectively. Suppose an optimal TSP tour for the transformed network is found: $\{(1, 4), (4, 3), (3, 5), (5, 2), (2, -1), (-1, 1)\}$. Determine the number of trucks needed, tours, and the total cost.

6.7 In cellular manufacturing, similar parts (products) are grouped in a same cell to reduce costs. Consider the network of seven parts grouped in two cells N_1 and N_2

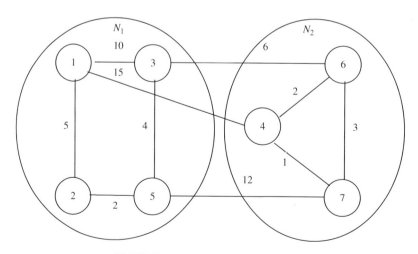

FIGURE 6.11   Parts assigned to cells.

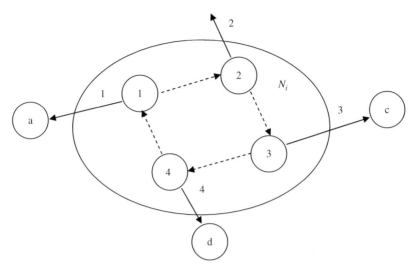

FIGURE 6.12   Subnetwork in a TSP.

shown in Figure 6.11. Assume that once a cell is entered, all the parts in the cell must be processed before moving out of the cell. Show how this network can be transformed into one that can be solved by a standard TSP.

6.8   Consider the generalized TSP discussed in this text. Suppose you are given a subnetwork for cluster N_i (Figure 6.12). There are four outgoing arcs from this cluster, (1, a), (2, b), (3, c), and (4, d), with respective costs 1, 2, 3, and 4. Draw a transformed subnetwork with appropriate links and costs so that the transformed network can be solved by a standard TSP. Show that if a TSP tour enters cluster N_i through node 4, it will leave N_i via a new arc with cost equal to 4.

6.9   Given a complete directed network of 10 nodes.

   (**a**) Determine the number of subtour elimination constraints required by equation (6.5).

   (**b**) Determine the number of subtour elimination constraints required by equation (6.6).

6.10   Given a complete undirected network of 10 nodes, determine the number of subtour elimination constraints required by equation (6.9).

PART II

REVIEW OF LINEAR PROGRAMMING AND NETWORK FLOWS

7

LINEAR PROGRAMMING— FUNDAMENTALS

This chapter reviews the basic linear algebra and linear programming theory essential to the understanding of the LP and IP solutions and methods to be discussed in the remainders of this book.

7.1 REVIEW OF BASIC LINEAR ALGEBRA

An n vector is a column or row array of n numbers. Throughout this text, we shall use a lowercase boldface letter to denote a column vector such as $\mathbf{b}$ or $\mathbf{d}$, and use the transpose of a column vector to denote a row vector such as $\mathbf{b}^T$ or $\mathbf{d}^T$, where superscript T stands for *transpose*.

7.1.1 Euclidean Space

Definition 7.1 An n-dimensional *Euclidean space*, denoted by E^n, is the collection of all vectors of dimension n having the following properties:

- *Addition of vectors*—for any two vectors $\mathbf{a}$ and $\mathbf{b}$ in E^n, vector $\mathbf{a} + \mathbf{b}$ is in E^n.
- *Scalar multiplication*—for any vector $\mathbf{a}$ and scalar k in E^n, vector $k\mathbf{a}$ is in E^n.

Applied Integer Programming: Modeling and Solution, By Der-San Chen, Robert G. Batson, and Yu Dang
Copyright © 2010 John Wiley & Sons, Inc.

- *Vector multiplication*—any two vectors **a** and **b** in E^n can be multiplied. The result of this vector multiplication is a real number called *inner (dot) product of the two vectors*, defined by $\mathbf{a}^T\mathbf{b} = a_1b_1 + a_2b_2 + \cdots + a_nb_n$.
- The *length (norm) of a vector* in the space, denoted by $\|\mathbf{a}\|$, is defined

$$\|\mathbf{a}\| = \sqrt{\sum_{i=1}^{n} a_i^2}$$

7.1.2 Linear and Convex Combinations

Definition 7.2   Given p vectors, $\mathbf{a}^1, \mathbf{a}^2, \ldots, \mathbf{a}^p$ in E^n, and p scalars $\lambda_1, \lambda_2, \ldots, \lambda_p$, the expression $\lambda_1\mathbf{a}^1 + \lambda_2\mathbf{a}^2 + \cdots + \lambda_p\mathbf{a}^p$ is called a *linear combination*. The scalars are real numbers that can be positive, negative, or zero. The linear combination becomes a *convex combination* when $\lambda_1 + \lambda_2 + \cdots + \lambda_p = 1$ and $0 \leq \lambda_1, \lambda_2, \ldots, \lambda_p \leq 1$.

7.1.3 Linear Independence

Definition 7.3   A collection of vectors $\mathbf{a}^1, \mathbf{a}^2, \ldots, \mathbf{a}^p$ of dimension n is called *linearly dependent* if there exist constants $\alpha_1, \alpha_2, \ldots, \alpha_p$, with at least one $\alpha_j \neq 0$, such that their linear combination equals to an n-dimensional null vector, or

$$\sum_{j=1}^{p} \alpha_j\mathbf{a}^j = \mathbf{0}$$

and the set of vectors is said to be *linearly independent* if the only solution is $\alpha_1 = \alpha_2 = \cdots = \alpha_p = 0$.

Note that in E^n, any set of $p > n$ vectors is always linearly dependent, but a set of $p \leq n$ vectors may or may not be linearly independent.

7.1.4 Rank of a Matrix

A matrix is a rectangular array of numbers. Thus, the size of a matrix is represented by the number of rows crossed with the number of columns, denoted by $m \times n$. In this text, we shall use an uppercase boldface letter to denote a matrix, for example, **A** or **B**.

Definition 7.4   Let **A** be an $m \times n$ matrix. The *row rank* of the matrix is equal to the maximum number of linearly independent rows of **A**. The *column rank* of **A** is the maximum number of linearly independent columns of **A**. It can be shown that the row rank of a matrix is always equal to its column rank, and hence the rank of $\mathbf{A}$, denoted by rank($\mathbf{A}$), is equal to the number of linearly independent rows (or columns) of $\mathbf{A}$.

7.1.5 Basis

A collection of vectors $\mathbf{a}^1, \mathbf{a}^2, \ldots, \mathbf{a}^p$ in $\boldsymbol{E}^n$ is said to *span* $\boldsymbol{E}^n$ if any vector in this space can be represented as a linear combination of this set of vectors. In other words, given any vector $\mathbf{b}$ in this space, we can always find scalars $\alpha_1, \alpha_2, \ldots, \alpha_p$ such that

$$\mathbf{b} = \sum_{j=1}^{p} \alpha_j \mathbf{a}^j$$

Definition 7.5   A collection of n linearly *independent row vectors* forms a *basis* of $\boldsymbol{E}^n$.

Thus, given a basis of $\boldsymbol{E}^n$, say $\mathbf{a}^1, \mathbf{a}^2, \ldots, \mathbf{a}^n$, any vector $\mathbf{b}$ in $\boldsymbol{E}^n$ is *uniquely* represented in terms of this basis. Note that a basis must be an $n \times n$ square matrix.

7.1.6   Matrix Inversion

Definition 7.6   Given an $m \times m$ matrix $\mathbf{A}$ if there exists an $m \times m$ matrix $\mathbf{B}$ such that their product is an identity matrix (i.e., $\mathbf{AB} = \mathbf{I}$ and $\mathbf{BA} = \mathbf{I}$), then $\mathbf{B}$ is called the *inverse* of $\mathbf{A}$. The *inverse matrix* is *unique* and usually denoted by $\mathbf{A}^{-1}$. Also, the inverse of $\mathbf{B}$ is $\mathbf{A}$, denoted by $\mathbf{B}^{-1}$.

Note that a square matrix $\mathbf{A}$ can have an inverse if and only if the *row* vectors of $\mathbf{A}$ are linearly independent, or if and only if the *column* vectors of $\mathbf{A}$ are linearly independent. There is a simple condition to check for existence of $\mathbf{A}^{-1}$, called nonsingularity.

7.1.7   Determinant of a Matrix

Definition 7.7   Every $n \times n$ square matrix $\mathbf{A}$ has a real number associated with it called the *determinant* of the matrix, denoted by $\det \mathbf{A}$, which is defined as follows:

$$\det \mathbf{A} = \sum_{i=1}^{n} a_{i1} A_{i1}$$

where A_{i1} is the cofactor of a_{i1} defined as $(-1)^{i+1}$ times the determinant of the submatrix of $\mathbf{A}$ obtained by deleting the ith row and the first column. The matrix $\mathbf{A}$ is *singular* if $\det \mathbf{A} = 0$ and is *nonsingular* if $\det \mathbf{A} \neq 0$.

To illustrate how to calculate the determinant of a matrix, consider the following example:

$$\det \begin{pmatrix} 2 & 1 & -1 \\ 1 & 3 & 1 \\ -2 & 1 & 1 \end{pmatrix}$$

$$= 2A_{11} + 1A_{21} - 2A_{31}$$

$$= 2(-1)^{1+1}\det\begin{pmatrix} 3 & 1 \\ 1 & 1 \end{pmatrix} + 1(-1)^{2+1}\det\begin{pmatrix} 1 & -1 \\ 1 & 1 \end{pmatrix} - 2(-1)^{3+1}\det\begin{pmatrix} 1 & -1 \\ 3 & 1 \end{pmatrix}$$

$$= 2(1)2 + 1(1)2 - 2(1)(4)$$

$$= -6$$

Thus, matrix $\mathbf{A}$ is nonsingular.

If a matrix $\mathbf{A}$ is nonsingular, then the inverse $\mathbf{A}^{-1}$ exists and can be calculated by

$$\mathbf{A}^{-1} = \frac{\text{adj } \mathbf{A}}{\det \mathbf{A}}$$

where adj $\mathbf{A}$ is called the *adjoint matrix* of $\mathbf{A}$, which is defined as the transpose of the matrix whose ij element is A_{ij}, the cofactor of a_{ij}. However, this method for finding the inverse of a matrix is not as effective as the method using the elementary row operations to be given in the following section.

Consider the system $\mathbf{A}\mathbf{x} = \mathbf{b}$ where $\mathbf{A}$ is a $n \times n$ nonsingular matrix, $\mathbf{b}$ is an n vector, and $\mathbf{x}$ is an n vector of unknowns. Then according to the *Cramer's rule*, the unique solution to this system is given by

$$x_j = \frac{\det \mathbf{A}_j}{\det \mathbf{A}} \quad \text{for } j = 1, 2, \ldots, n$$

where $\mathbf{A}_j$ is obtained from $\mathbf{A}$ by replacing the jth column of $\mathbf{A}$ by $\mathbf{b}$. However, this method is not as effective as that of the elementary row operations.

7.1.8 Upper and Lower Triangular Matrices

Definition 7.8   A square matrix $\mathbf{A} = (a_{ij})$ is called an *upper triangular matrix* if $a_{ij} = 0$ for all $i > j$. Matrix $\mathbf{A}$ is called a *lower triangular matrix* if $a_{ij} = 0$ for all $i < j$. A square matrix $\mathbf{D}$ is called a *diagonal matrix* with diagonal elements $d_{11}, d_{22}, \ldots, d_{nn}$ if all other elements (d_{ij}, $i \neq j$) are equal to 0. The diagonal matrix is denoted by $\mathbf{D} = \text{diag}\{d_{11}, d_{22}, \ldots, d_{nn}\}$.

The triangular matrix (either upper or lower) with nonzero diagonal elements has an inverse. Note that in solving a system of equations, finding an *equivalent* triangular coefficient matrix requires about one-half the computational operations as in finding the inverse of a basis matrix. Two systems of linear equations are said to be *equivalent* if their solution sets (including empty set) are the same.

A diagonal matrix $\mathbf{D}$ with diagonal elements $d_{11}, d_{22}, \ldots, d_{nn}$ is a special form of a triangular matrix (either upper or lower) in which $d_{ij} = 0$ for all $i \neq j$. Let $\mathbf{D}$ be the diagonal matrix obtained from matrix $\mathbf{A}$ by the elementary row operations, it can be shown that $\det \mathbf{A} = \det \mathbf{D}$ and therefore

$$\det \mathbf{A} = d_{11}, d_{22}, \ldots, d_{nn}$$

If any diagonal element $d_{ii} = 0$, then $\det \mathbf{A} = 0$ and hence matrix $\mathbf{A}$ is *singular*.

7.2 USES OF ELEMENTARY ROW OPERATIONS

Given a matrix $\mathbf{A}$ (or a collection of *row vectors*), we can perform some *elementary row operations* (or *row operations*, for short). In the context of this textbook, these operations have the following uses:

- Finding the rank of $\mathbf{A}$ (or the number of linearly independent row vectors)
- Calculating the inverse of $\mathbf{A}$ if exists
- Converting to an upper or a lower triangular matrix
- Converting to a diagonal matrix
- Calculating the determinant of a matrix
- Solving a system of linear equations

A row operation on a matrix is one of the following operations:

1. Interchange of any two rows ($\mathbf{R}_i \leftrightarrow \mathbf{R}_j$).
2. Multiply any row $\mathbf{R}_i$ by a scalar $k \neq 0$ and use the resultant row $k\mathbf{R}_i$ to replace $\mathbf{R}_i$.
3. Addition to any row $\mathbf{R}_j$ of a nonzero scalar multiple of another row $\mathbf{R}_i$, or use $\mathbf{R}_j + k\mathbf{R}_i$ to replace $\mathbf{R}_j$.

We can also view a matrix $\mathbf{A}$ as a collection of *column vectors* and can perform some *elementary column operations* (or *column operations*, for short) on a matrix in a way similar to row operations except that rows are replaced by columns.

7.2.1 Finding the Rank of a Matrix

To find the rank of a given matrix, we either apply row operations to find the row rank or apply column operations to find the column rank. The following example is to show an application of row operations.

Example 7.1   Find the rank of **A** defined by

$$\begin{pmatrix} 2 & 1 & 2 & 3 \\ 1 & 3 & 1 & 9 \\ 1 & 1 & 1 & 3 \end{pmatrix}$$

We apply row operations 2 and 3 to **A**, and *attempt* to make as many columns as possible to become *distinct unit vectors*. For this example, we attempt to obtain three distinct unit vectors in the following order:

$$\begin{pmatrix} 1 \\ 0 \\ 0 \end{pmatrix}, \quad \begin{pmatrix} 0 \\ 1 \\ 0 \end{pmatrix}, \quad \begin{pmatrix} 0 \\ 0 \\ 1 \end{pmatrix}$$

To obtain the first unit column vector, we multiply row R_1 by 0.5 and use $0.5R_1$ to replace R_1. Then use $R_2 - 1R_1$ to replace R_2, and use $R_3 - 1R_1$ to replace R_3. To obtain the second unit column vector, we multiply R_2 by 0.4 and use $0.4R_2$ to replace R_2. We then use $R_1 - 0.5R_2$ to replace R_1 and use $R_3 - 0.5R_2$ to replace R_3:

$$\mathbf{A} \rightarrow \begin{pmatrix} 1 & 0.5 & 1 & 1.5 \\ 0 & 2.5 & 0 & 7.5 \\ 0 & 0.5 & 0 & 1.5 \end{pmatrix} \rightarrow \begin{pmatrix} 1 & 0 & 1 & 0 \\ 0 & 1 & 0 & 3 \\ 0 & 0 & 0 & 0 \end{pmatrix}$$

After making two distinct unit columns, we find that the third row becomes a zero vector. This indicates that row 3 is a linear combination of rows 1 and 2. In other words, the maximum number of linearly independent vectors is 2, or the rank of **A** is 2.

7.2.2   Calculating the Inverse of a Matrix

An $m \times n$ matrix **A** is *invertible* or *nonsingular* if it contains m linearly independent rows. To calculate *the inverse* of a square matrix, we construct an augmented matrix $(\mathbf{A}|\mathbf{I})$ and perform row operations on it until the augmented matrix becomes $(\mathbf{I}|\mathbf{B})$. Then $\mathbf{B} = \mathbf{A}^{-1}$.

Example 7.2   Find the inverse of matrix A defined by

$$\begin{pmatrix} 2 & 1 & 2 \\ 1 & 3 & 1 \\ 2 & 1 & 1 \end{pmatrix}$$

Construct the matrix by augmenting an identity matrix of the same size and perform row operations 2 and 3 repeatedly until an identity matrix is obtained on the left-hand part of the augmented matrix.

$$\left(\begin{array}{ccc|ccc} 2 & 1 & 2 & 1 & 0 & 0 \\ 1 & 3 & 1 & 0 & 1 & 0 \\ 2 & 1 & 1 & 0 & 0 & 1 \end{array}\right) \rightarrow \left(\begin{array}{ccc|ccc} 1 & 1/2 & 1 & 1/2 & 0 & 0 \\ 0 & 5/2 & 0 & -1/2 & 1 & 0 \\ 0 & 0 & -1 & -1 & 0 & 1 \end{array}\right) \rightarrow$$

$$\left(\begin{array}{ccc|ccc} 1 & 0 & 1 & 3/5 & -1/5 & 0 \\ 0 & 1 & 0 & -1/5 & 2/5 & 0 \\ 0 & 0 & -1 & -1 & 0 & 1 \end{array}\right) \rightarrow \left(\begin{array}{ccc|ccc} 1 & 0 & 0 & -2/5 & -1/5 & 1 \\ 0 & 1 & 0 & -1/5 & 2/5 & 0 \\ 0 & 0 & 1 & 1 & 0 & -1 \end{array}\right)$$

Therefore, we obtain

$$\mathbf{A}^{-1} = \left(\begin{array}{ccc} -2/5 & -1/5 & 1 \\ -1/5 & 2/5 & 0 \\ 1 & 0 & -1 \end{array}\right)$$

and check that $\mathbf{AA}^{-1} = \mathbf{I}$.

7.2.3 Converting to a Triangular Matrix

Example 7.3   Find the upper triangular, lower triangular, and diagonal matrices for

$$\mathbf{A} = \left(\begin{array}{ccc} 1 & 1 & 1 \\ 1 & 3 & 3 \\ 1 & 1 & 2 \end{array}\right)$$

Apply the elementary row operations on A.

For upper triangular matrix

$$\mathbf{A} \rightarrow \left(\begin{array}{ccc} 1 & 1 & 1 \\ 0 & 2 & 2 \\ 0 & 0 & 1 \end{array}\right)$$

For lower triangular matrix

$$\mathbf{A} \to \begin{pmatrix} 2/3 & 0 & 0 \\ -1 & 3 & 0 \\ 2/3 & 0 & 1 \end{pmatrix} \to \begin{pmatrix} 2/3 & 0 & 0 \\ -1 & 3 & 0 \\ 2/3 & 0 & 1 \end{pmatrix}$$

From upper triangular to diagonal

$$\begin{pmatrix} 1 & 1 & 1 \\ 0 & 2 & 2 \\ 0 & 0 & 1 \end{pmatrix} \to \begin{pmatrix} 1 & 0 & 0 \\ 0 & 2 & 2 \\ 0 & 0 & 1 \end{pmatrix} \to \begin{pmatrix} 1 & 0 & 0 \\ 0 & 2 & 0 \\ 0 & 0 & 1 \end{pmatrix}$$

From lower triangular to diagonal matrix

$$\begin{pmatrix} 2/3 & 0 & 0 \\ -1 & 3 & 0 \\ 2/3 & 0 & 1 \end{pmatrix} \to \begin{pmatrix} 2/3 & 0 & 0 \\ 0 & 3 & 0 \\ 0 & 0 & 1 \end{pmatrix}$$

7.2.4   Calculating the Determinant of a Matrix

The determinant of a matrix $\mathbf{A}$ can be obtained by calculating the product of the diagonal elements of any one of its equivalent upper triangular, lower triangular, or diagonal matrix. From Example 6.3, we may compute

$$\det \mathbf{A} = 1 \times 2 \times 1 = \frac{2}{3} \times 3 \times 1 = 2$$

7.2.5   Solving a System of Linear Equations

There are two commonly used methods for solving a system of linear equations: (1) matrix inversion and (2) matrix triangularization. The former method, called *Gauss–Jordan reduction*, is well known in introductory operations research texts. The latter method, called *Gaussian reduction*, is more efficient and converts a given augmented matrix to a triangular matrix, either upper or lower triangular.

We shall first use the first method to solve a system of equations. Consider the system of linear equations $\mathbf{Ax} = \mathbf{b}$ where $\mathbf{A}$ is $m \times n$, $\mathbf{x}$ is $n \times 1$, and $\mathbf{b}$ is $m \times 1$. Such a system has one of three cases when solved: (1) a *unique solution*, (2) an *infinite number of solutions*, and (3) *no solution*. To know which case holds for a given system, we must use the concept of rank and the *augmented matrix* of the system denoted $(\mathbf{A}|\mathbf{b})$.

1. The system has a unique solution if $\text{rank}(\mathbf{A}) = \text{rank}(\mathbf{A}|\mathbf{b}) = n$.
2. The system has an infinite number of solutions if $\text{rank}(\mathbf{A}) = \text{rank}(\mathbf{A}|\mathbf{b}) < n$.
3. The system has no solution if $\text{rank}(\mathbf{A}) < \text{rank}(\mathbf{A}|\mathbf{b})$.

Example 7.4 (Unique Solution) Solve the following system:

$$x_1 + x_2 + x_3 = 4$$
$$x_1 + 3x_2 + 3x_3 = 2$$
$$x_1 + x_2 + 2x_3 = 6$$

$$\mathbf{A} = \begin{pmatrix} 1 & 1 & 1 \\ 1 & 3 & 3 \\ 1 & 1 & 2 \end{pmatrix} \rightarrow \begin{pmatrix} 1 & 1 & 1 \\ 0 & 2 & 2 \\ 0 & 0 & 1 \end{pmatrix} \rightarrow \begin{pmatrix} 1 & 0 & 0 \\ 0 & 1 & 1 \\ 0 & 0 & 1 \end{pmatrix} \rightarrow \begin{pmatrix} 1 & 0 & 0 \\ 0 & 1 & 0 \\ 0 & 0 & 1 \end{pmatrix} \Rightarrow \text{rank}(\mathbf{A}) = 3$$

$$(\mathbf{A}|\mathbf{b}) = \left(\begin{array}{ccc|c} 1 & 1 & 1 & 4 \\ 1 & 3 & 3 & 2 \\ 1 & 1 & 2 & 6 \end{array} \right) \rightarrow \left(\begin{array}{ccc|c} 1 & 1 & 1 & 4 \\ 0 & 2 & 2 & -2 \\ 0 & 0 & 1 & 2 \end{array} \right) \rightarrow \left(\begin{array}{ccc|c} 1 & 0 & 0 & 5 \\ 0 & 1 & 1 & -1 \\ 0 & 0 & 1 & 2 \end{array} \right) \rightarrow \left(\begin{array}{ccc|c} 1 & 0 & 0 & 5 \\ 0 & 1 & 0 & -3 \\ 0 & 0 & 1 & 2 \end{array} \right)$$

Because $\text{rank}(\mathbf{A}) = \text{rank}(\mathbf{A}|\mathbf{b}) = 3 = n$, the system has a unique solution

$$x_1 = 5, \quad x_2 = -3, \quad x_3 = 2$$

Example 7.5 (Infinite Number of Solutions) Solve the following system:

$$x_1 + x_2 + x_3 + x_4 = 4$$
$$x_1 + 3x_2 + 3x_3 = 2$$
$$x_1 + x_2 + 2x_3 - x_4 = 6$$

$$\mathbf{A} = \begin{pmatrix} 1 & 1 & 1 & 1 \\ 1 & 3 & 3 & 0 \\ 1 & 1 & 2 & -1 \end{pmatrix} \rightarrow \begin{pmatrix} 1 & 1 & 1 & 1 \\ 0 & 2 & 2 & -1 \\ 0 & 0 & 1 & -2 \end{pmatrix} \rightarrow \begin{pmatrix} 1 & 0 & 0 & 3/2 \\ 0 & 1 & 1 & -1/2 \\ 0 & 0 & 1 & -2 \end{pmatrix} \rightarrow \begin{pmatrix} 1 & 0 & 0 & 3/2 \\ 0 & 1 & 0 & 3/2 \\ 0 & 0 & 1 & -2 \end{pmatrix}$$

$$\Rightarrow \text{rank}(\mathbf{A}) = 3$$

$$(\mathbf{A}|\mathbf{b}) = \left(\begin{array}{cccc|c} 1 & 1 & 1 & 1 & 4 \\ 1 & 3 & 3 & 0 & 2 \\ 1 & 1 & 2 & -1 & 6 \end{array} \right) \rightarrow \left(\begin{array}{cccc|c} 1 & 1 & 1 & 1 & 4 \\ 0 & 2 & 2 & -1 & -2 \\ 0 & 0 & 1 & -2 & 2 \end{array} \right) \rightarrow \left(\begin{array}{cccc|c} 1 & 0 & 0 & 3/2 & 5 \\ 0 & 1 & 0 & -1/2 & -1 \\ 0 & 0 & 1 & -2 & 2 \end{array} \right)$$

$$\rightarrow \left(\begin{array}{cccc|c} 1 & 0 & 0 & 3/2 & 5 \\ 0 & 1 & 0 & 3/2 & -3 \\ 0 & 0 & 1 & -2 & 2 \end{array} \right)$$

Because $\text{rank}(\mathbf{A}) = \text{rank}(\mathbf{A}|\mathbf{b}) = 3 < n = 4$, the system has an infinite number of solutions. Note that the final equivalent system of equations reads

$$x_1 = 5 - \frac{3}{2} x_4$$

$$x_2 = -3 - \frac{3}{2} x_4$$

$$x_3 = 2 + 2x_4$$

Clearly, the system has an infinite number of solutions because an infinite number of possible values can be assigned to x_4.

Example 7.6 (No Solution)   Consider the following system:

$$2x_1 + x_2 + 2x_3 = 6$$
$$x_1 + 3x_2 + x_3 = 9$$
$$x_1 + x_2 + x_3 = 3$$

Applying the row operations to matrix $\mathbf{A}$, we have

$$\mathbf{A} = \begin{pmatrix} 2 & 1 & 2 \\ 1 & 3 & 1 \\ 1 & 1 & 1 \end{pmatrix} \rightarrow \begin{pmatrix} 1 & 1/2 & 1 \\ 0 & 5/2 & 0 \\ 0 & 1/2 & 0 \end{pmatrix} \rightarrow \begin{pmatrix} 1 & 0 & 1 \\ 0 & 1 & 0 \\ 0 & 0 & 0 \end{pmatrix} \Rightarrow \text{rank}(\mathbf{A}) = 2$$

$$(\mathbf{A}|\mathbf{b}) = \begin{pmatrix} 2 & 1 & 2 & | & 6 \\ 1 & 3 & 1 & | & 9 \\ 1 & 1 & 1 & | & 3 \end{pmatrix} \rightarrow \begin{pmatrix} 1 & 1/2 & 1 & | & 3 \\ 0 & 5/2 & 0 & | & 6 \\ 0 & 1/2 & 0 & | & 0 \end{pmatrix} \rightarrow \begin{pmatrix} 1 & 0 & 1 & | & 24/5 \\ 0 & 1 & 0 & | & 12/5 \\ 0 & 0 & 0 & | & -6/5 \end{pmatrix} \rightarrow \begin{pmatrix} 1 & 0 & 1 & | & 0 \\ 0 & 1 & 0 & | & 0 \\ 0 & 0 & 0 & | & 1 \end{pmatrix}$$

$$\Rightarrow \text{rank}(\mathbf{A}|\mathbf{b}) = 3$$

Because $\text{rank}(\mathbf{A}) < \text{rank}(\mathbf{A}|\mathbf{b})$, the system has no solution. Note that the third row of the last equivalent system gives the equation

$$0x_1 + 0x_2 + 0x_3 = 1$$

Clearly, the left-hand side of this equation never equals to the right-hand side. Hence, the system has no solution.

Now let us use the matrix triangularization method to solve a system of linear equations. Basically, the procedure performs a sequence of row operations on the augmented matrix $(\mathbf{A}|\mathbf{b})$ until a triangular matrix appears on the left-hand side. Then for the lower triangular matrix, use forward substitutions to find the solution; or for the upper triangular matrix, use backward substitutions to find the solution.

Example 7.7   Solve the following system by matrix triangularization

$$x_1 + x_2 + x_3 = 4$$
$$x_1 + 3x_2 + 3x_3 = 2$$
$$x_1 + x_2 + 2x_3 = 6$$

Perform row operations on the augmented matrix until an upper triangular matrix is obtained:

$$(\mathbf{A}|\mathbf{b}) = \begin{pmatrix} 1 & 1 & 1 & 4 \\ 1 & 3 & 3 & 2 \\ 1 & 1 & 2 & 6 \end{pmatrix} \rightarrow \begin{pmatrix} 1 & 1 & 1 & 4 \\ 0 & 2 & 2 & -2 \\ 0 & 0 & 1 & 2 \end{pmatrix}$$

The equivalent system of equations becomes

$$x_1 + x_2 + x_3 = 4$$
$$2x_2 + 2x_3 = -2$$
$$x_3 = 2$$

Applying backward substitutions, we have $x_3 = 2 \rightarrow x_2 = -3 \rightarrow x_1 = 5$.

Likewise, we perform row operations on the augmented matrix until a lower triangular matrix is obtained:

$$(\mathbf{A}|\mathbf{b}) = \begin{pmatrix} 1 & 1 & 1 & 4 \\ 1 & 3 & 3 & 2 \\ 1 & 1 & 2 & 6 \end{pmatrix} \rightarrow \begin{pmatrix} 2/3 & 0 & 0 & 10/3 \\ 1 & 3 & 3 & 2 \\ 1 & 1 & 2 & 6 \end{pmatrix} \rightarrow \begin{pmatrix} 2/3 & 0 & 0 & 10/3 \\ -1/2 & 3/2 & 0 & -7 \\ 1 & 1 & 2 & 6 \end{pmatrix}$$

Applying forward substitutions, we have $x_1 = 5 \rightarrow x_2 = -3 \rightarrow x_3 = 2$.

7.3   THE DUAL LINEAR PROGRAM

Every linear program, whether expressed in standard form or not, has another linear program associated with it called the "*dual*". In this context, the *given* or *original* LP problem will be referred to as the "*primal*". The dual problem complements its *primal* in many ways. In problem formulation, for example, if the primal (P) problem is a maximization, then the dual (D) is a minimization, and vice versa. Moreover, both problems share all the data (parameters) found in **A**, **c**, and **b**.

Knowing the relations between an LP and its dual is vital to understanding advanced topics in linear and nonlinear programming such as economic interpretation, sensitivity analysis, and development of dual and primal–dual simplex methods.

After we explain the formulation of the dual problem from the primal, first in standard form and then in *arbitrary* form, we will provide an economic interpretation of the dual problem using an insurance portfolio problem as an example. Then we will review the duality theory as to the relations between their respective feasible solutions and their respective optimal solutions in this chapter. Other contributions from duality theory will be introduced in later chapters when needed.

7.3.1 The Linear Program in Standard Form

First, let us define the linear programming problem in standard form:

$$(P) \quad \text{maximize} \quad z = \sum_j c_j x_j$$

$$\text{subject to} \quad \sum_j a_{ij} x_j \leq b_i \quad (i = 1, 2, \ldots, m)$$

$$x_j \geq 0 \qquad (j = 1, 2, \ldots, n)$$

or in matrix form,

$$(P') \quad \text{maximize} \quad z = \mathbf{c}^T \mathbf{x}$$

$$\text{subject to} \quad \mathbf{Ax} \leq \mathbf{b}$$

$$\mathbf{x} \geq \mathbf{0}$$

Throughout this text, a linear program is said to be in *standard form* if (1) the objective function is maximized, (2) all the constraints are of $\leq$ form, and (3) all continuous variables are ≥ 0 with no finite upper bound. However, all parameters b_i, c_j, and a_{ij} may be positive, negative, or zero.

Any LP problem that does not conform to conditions (1)–(3) is in *nonstandard form*, which can be converted to standard by simple substitutions. Various nonstandard forms are as follows:

- Minimization problem
- Inequality of $\geq$ form
- Equation (equality constraint)
- Unrestricted variable (continuous or integer)
- Variable with a lower bound other than 0
- Variable with a finite upper bound

The conversion procedures for an LP problem are identical to those for an MIP problem described in Section 1.2. For ease of presentation, we shall use the standard LP form for the remainder of the text unless specified otherwise.

The above mathematical definition of an LP problem implies the following assumptions:

- *Divisibility assumption* for each continuous variable $(x_j \geq 0)$
- *Certainty (constant) assumption* for each input parameter (c_j, a_{ij}, b_i)
- *Proportionality assumption* for each term in the constraint and objective function $(a_{ij}x_j, c_jx_j)$
- *Additivity and separability assumption* for each combined function in the objective and constraints $(\sum_j c_j x_j, \sum_j a_{ij} x_j)$
- *Single-objective assumption*(max or min z $= \sum_j c_j x_j$)
- *Simultaneousness (conjunction) assumption* for the system of all constraint equations and inequalities $(\sum_j a_{ij} x_j \leq b_i, i = 1, 2, \ldots, m)$

Note that these assumptions are the same as those imposed on an MIP problem except for the absence of the integrality assumption. The implications of *all* other assumptions are the same as those described in Section 2.1. We shall not reiterate here.

7.3.2   Formulating the Dual Problem

To formulate the dual of P, we first detach variables from the coefficients and form the following augmented matrix by combining $\mathbf{A}$, $\mathbf{b}$, and $\mathbf{c}^T$.

Dual variable

$$
\begin{array}{c}
\begin{array}{cccc} x_1 & x_2 & \cdots & x_n \quad \leq \end{array} \\
\left(\frac{\mathbf{A} \mid \mathbf{b}}{\mathbf{c}^T \mid 1}\right) =
\left(
\begin{array}{cccc|c}
a_{11} & a_{12} & \cdots & a_{1n} & b_1 \\
a_{21} & a_{22} & \cdots & a_{2n} & b_2 \\
\vdots & \vdots & \vdots & \vdots & \vdots \\
a_{m1} & a_{m2} & \cdots & a_{mn} & b_m \\
\hline
c_1 & c_2 & \cdots & c_n & 1
\end{array}
\right)
\begin{array}{c}
u_1 \geq 0 \\
u_2 \geq 0 \\
\vdots \\
u_m \geq 0 \\
\end{array}
\end{array}
$$

The dual problem is formulated by the following procedure:

1. Assign a *nonnegative dual variable* to each corresponding constraint, denoted by $u_1, u_2, \ldots, u_i, \ldots, u_m \geq 0$, or $\mathbf{u} \geq \mathbf{0}$.

2. Construct *dual constraints* with respect to variables x_j:

$$a_{1j}u_1 + a_{2j}u_2 + \cdots + a_{mj}u_m = \sum_{i=1}^{m} a_{ij}u_i \geq c_j, \quad j = 1, 2, \ldots, n$$

or, $\mathbf{A}^T\mathbf{u} \geq \mathbf{c}$

3. Construct *dual* objective function

$$\text{minimize } w = b_1u_1 + b_2u_2 + \cdots + b_iu_i + \cdots + b_mu_m = \sum_{i=1}^{m} b_iu_i$$

or, minimize $w = \mathbf{b}^T\mathbf{u}$

The dual problem is recapped as follows:

$$(D) \quad \text{Minimize} \quad w = \sum_{i=1}^{m} b_iu_i$$

$$\text{subject to} \quad \sum_{i=1}^{m} a_{ij}u_i \geq cj \quad j = 1, 2, \ldots, n$$

$$u_i \geq 0 \qquad i = 1, 2, \ldots, m$$

$$(D') \quad \text{Minimize} \quad w = \mathbf{b}^T\mathbf{u}$$

$$\text{subject to} \quad \mathbf{A}^T\mathbf{u} \geq \mathbf{c}$$

$$\mathbf{u} \geq \mathbf{0}$$

Basically, the formulation rule state that if a given (primal) problem is a maximization problem with all constraints of $\leq$ form and all variables ≥ 0, then the dual problem must be a minimization problem with all dual constraints of $\geq$ form and all dual variables ≥ 0. Both primal and dual problems share the same set of coefficient matrix and vectors of objective coefficients and right-hand side constants, while the dual problem takes on the *transpose* of the original matrix and vectors given in the primal problem. That is, in matrix notation, the dual problem uses $\mathbf{A}^T$, $\mathbf{c}$, and $\mathbf{b}^T$ as the coefficient matrix, right-hand side column, and objective coefficients, respectively. In other words, the transpose of a matrix implies that the objective coefficients of the primal become the right-hand sides of the dual, and vice versa. The number of variables in the primal equals the number of constraints in the dual, and vice versa. Relative to the dual problem, *primal* variables and *primal constraints* are used for referring the variables and constraints defined in the given problem.

Clearly, the dual of the dual is the primal because the transpose of the transpose of a matrix is itself. This implies that we may also refer the minimization problem as the primal and the maximization problem as the dual. Therefore, the format in problem D or D' will also be treated as the *standard minimization problem*.

Example 7.8   If the primal problem is

$$\text{Maximize}\quad z = x_1 + 2x_2 - 8x_3$$
$$\text{subject to}\quad x_1 + 3x_2 + 5x_3 \leq 8$$
$$2x_1 - 5x_3 \leq 7$$
$$x_1, x_2, x_3 \geq 0$$

then the dual problem is

$$\text{Minimize}\quad w = 8u_1 + 7u_2$$
$$\text{subject to}\quad u_1 + 2u_2 \geq 1$$
$$3u_1 \geq 2$$
$$5u1 - 5u_2 \geq -8$$
$$u_1, u_2 \geq 0$$

What if the given problem is not in standard form? Straightforwardly, we may first convert the given problem to a standard problem as usual, then formulate the dual from the converted standard problem, and finally convert this dual problem back to the original format. It can be shown that the result of this three-step procedure corresponds to the primal–dual formulation rules listed in Table 7.1. The reader is encouraged to verify these rules in exercises by the three-step procedure.

Example 7.9   Formulate the dual problem of the following LP:

$$\text{Maximize}\quad z = 2x_1 - x_2 + 5x_3 + 3x_4$$
$$\text{subject to}\quad x_1 + 2x_2 + 3x_3 - x_4 \geq 5$$
$$2x_1 - 3x_2 + x_3 + 2x_4 \leq 12$$
$$x_2 - x_3 + x_4 = -3$$
$$x_1, x_2, x_4 \geq 0, x_3 \leq 0$$

TABLE 7.1   Correspondence of Primal–Dual Formulation

Maximization Problem	Minimization Problem
Constraint i	Variable I
$\leq$	≥ 0
$\geq$	≤ 0
$=$	Unrestricted in sign
Variable j	Constraint j
≥ 0	$\geq$
≤ 0	$\leq$
Unrestricted in sign	$=$
Objective row	Right-hand side column
Right-hand side column	Objective row

The dual problem is

$$\text{Minimize} \quad w = 5u_1 + 12u_2 - 3u_3$$

$$\text{subject to} \quad u_1 + 2u_2 \geq 2$$

$$2u_1 - 3u_2 + u_3 \leq -1$$

$$3u_1 + u_2 - u_3 \leq 5$$

$$-u_1 + 2u_2 + u_3 \geq 3$$

$$u1 \leq 0, u_2 \geq 0, u_3 \text{ unrestricted in sign}$$

7.3.3 Economic Interpretation of the Dual

Suppose the primal problem represents a resource allocation problem at a manu-facturing plant; that is, x_j is the quantity of product j to produce in a given period (say one month) and the limited availability of raw material i is represented by $a_{i1}x_1 + a_{i2}x_2 + \cdots + a_{in}x_n \leq b_i$, for each $i = 1, \ldots, m$. The plant manager wants to maximize profit in the particular production mix $(x_1, x_2, \ldots, x_n)$ found to represent optimal utilization of available raw materials $b_1, b_2, \ldots, b_m$.

Now, consider the company's risk manager, who reports to the VP of finance. He is interested in insuring the raw materials on hand against loss, but only wants to insure them (place a per unit valuation on each) up to their value in producing products that result in sales and profit (represented by $c_1, c_2, \ldots, c_n$ for each potential product manufactured).

If we let u_i be the per unit insured value of raw material i, the total valuation of resource i on-hand would be $b_i u_i$. We shall assume that $u_i \geq 0$, and otherwise no cost to dispose of unused resource i. The objective function for minimizing the cost of insurance is to minimize $w = b_1 u_1 + b_2 u_2 + \cdots + b_m u_m$.

Furthermore, the combined values of the various raw materials used to make say one unit of product j must be at least c_j, the profit from producing one unit of product j; in mathematical notation: $a_{1j}u_1 + a_{2j}u_2 + \cdots + a_{mj}u_m \geq c_j$.

Because the above relationship must hold for each product $j = 1, \ldots, n$, the risk manager's linear program to choose the least cost insurance portfolio is

$$\text{Minimize} \quad \mathbf{b}^T \mathbf{u}$$

$$\text{subject to} \quad \mathbf{A}^T \mathbf{u} \geq \mathbf{c}$$

$$\text{and} \quad \mathbf{u} \geq \mathbf{0}$$

which is recognized as the dual problem to the primal production mix problem.

7.3.4 Importance of the Dual

We have just seen that the duality relationship for the standard form LP explained by the total insurance valuation of on-hand raw materials (**b**) should be precisely equal to the maximum dollar value that can be extracted from them, using the company's current technology (**A**) and profit per unit for each of n products (**c**). There are many more such insights to be gained from the dual LP, regardless of whether the primal represents the classic resource allocation model or some other model. Duality plays a central role in linear programming. The primal–dual relationships presented here and in the following section help establish the simplex method, and then develop an alternative version known as the dual simplex algorithm. It turns out that whether one is solving the primal or dual LP via the simplex (or related) algorithm, one is automatically solving the dual LP as well. More theory is needed to justify this statement, presented next.

7.4 RELATIONSHIPS BETWEEN PRIMAL AND DUAL SOLUTIONS

There are a series of primal–dual relationships the reader needs to know. Let $\mathbf{x}^0 = (x_1^0, x_2^0, \ldots, x_n^0)$ represent any feasible solution to the maximizing (primal) problem, with objective value $z^0 = \sum_{j=1}^{n} c_j x_j^0 = \mathbf{c}^T\mathbf{x}^0$. Let $\mathbf{u}^0 = (u_1^0, u_2^0, \ldots, u_m^0)$ be feasible for the minimizing (dual) problem, with objective function value $w^0 = \sum_{i=1}^{m} b_i u_i = \mathbf{b}^T\mathbf{u}^0$.

7.4.1 Relationships Between All Primal and All Dual Feasible Solutions

Since the primal problem can be either maximization or minimization and the dual problem can be either maximization or minimization too, we shall simply use the maximization problem or minimization problem in this section to ease the presentation of primal–dual relations.

The Weak Duality Theorem   If $\mathbf{x}^0$ is feasible a maximization problem, and $\mathbf{u}^0$ is feasible for the associated minimization problem, then the objective value of the maximization problem is a lower bound of the objective value of the associated minimization problem, or mathematically $z^0 = \sum_{j=1}^{n} c_j x_j^0 \leq w^0 = \sum_{i=1}^{m} b_i u_i^0$.

Proof   Since $\mathbf{x}^0$ is feasible to the maximizing problem, we have

$$z^0 = \sum_{j=1}^{n} c_j x_j^0 \tag{7.1}$$

$$\text{subject to} \quad \sum_{j=1}^{n} a_{ij} x_j^0 \leq b_i \quad (i = 1, 2, \ldots, m) \tag{7.2}$$

$$x_j^0 \geq 0 \quad (j = 1, 2, \ldots, n) \tag{7.3}$$

Since $\mathbf{u}^0$ is feasible to the minimizing problem, we have

$$w^0 = \sum_{i=1}^{m} b_i u_i^0 \tag{7.1'}$$

$$\text{subject to} \quad \sum_{i=1}^{m} a_{ij} u_i^0 > c_j \quad (j = 1, 2, \ldots, n) \tag{7.2'}$$

$$u_i^0 \geq 0 \quad (i = 1, 2, \ldots, m) \tag{7.3'}$$

Premultiplying (7.2) by (7.3') and summing over i, we obtain

$$\sum_{i=1}^{m} u_i^0 \sum_{j=1}^{n} a_{ij} x_j^0 \leq \sum_{i=1}^{m} b_i u_i^0 \tag{7.4}$$

Premultiplying (7.2') by (7.3) and summing over j, we obtain

$$\sum_{j=1}^{n} x_j^0 \sum_{i=1}^{m} a_{ij} u_i^0 \geq \sum_{j=1}^{n} c_j x_j^0 \tag{7.4'}$$

Rearranging and combining (7.4) and (7.4'), we get

$$\sum_{j=1}^{n} c_j x_j^0 \leq \sum_{i=1}^{m} \sum_{j=1}^{n} a_{ij} x_j^0 u_i^0 \leq \sum_{i=1}^{m} b_i u_i^0 \tag{7.5}$$

or, by definition, $z^0 \leq w^0$. Expressing (7.5) in matrix notation, we have

$$\mathbf{c}^T \mathbf{x}^0 \leq (\mathbf{u}^0)^T \mathbf{A} \mathbf{x}^0 \leq \mathbf{b}^T \mathbf{u}^0 \tag{7.6}$$

■

7.4.2 Relationship Between Primal and Dual Optimum Solutions

The Duality Theorem   There are three possible relationships between the primal and dual problems:

1. If one problem has a feasible solution $\mathbf{x}^*$ with a bounded objective value z^*, then the other problem has a feasible solution $\mathbf{u}^*$ with a bounded objective value w^*. Furthermore, both problems have a finite optimum solution with the relation $\mathbf{c}^T \mathbf{x}^* = (\mathbf{u}^*)^T \mathbf{A} \mathbf{x}^* = \mathbf{b}^T \mathbf{u}^*$.

2. If one problem has a feasible solution with an unbounded objective value, then the other problem has no feasible solution.

3. If one problem has no feasible solution, then the other problem has either no feasible solution or an unbounded solution.

Example 7.10 The reader should verify graphically that both primal and dual problems below are infeasible.

$$(P) \quad \text{maximize} \quad z = x_1 + 2x_2 \qquad (D) \quad \text{minimize} \quad w = -u_2$$

$$\text{subject to} \quad x_1 - x_2 \leq 0 \qquad\qquad \text{subject to} \quad u_1 - u_2 \geq 1$$

$$-x_1 + x_2 \leq -1 \qquad\qquad\qquad -u_1 + u_2 \geq 2$$

$$x_1, x_2 \geq 0 \qquad\qquad\qquad\qquad u_1, u_2 \geq 0$$

7.4.3 Relationships Between Each Complementary Pair of Variables at Optimum

We now convert problem P to *equality* constraints by *adding* slack variables x_{s_i} ($i = 1, 2, \ldots, m$).

$$\text{Maximize} \quad z = \sum_j c_j x_j$$

$$\text{subject to} \quad \sum_j a_{ij} x_j + x_{s_i} = b_i \quad (i = 1, 2, \ldots, m)$$

$$x_j \geq 0 \qquad\qquad (j = 1, 2, \ldots, n)$$

Likewise, convert problem D to dual equality constraints by *subtracting* surplus variables u_{s_j} ($j = 1, 2, \ldots, n$).

$$\text{Minimize} \quad w = \sum_{i=1}^{m} b_i u_i$$

$$\text{subject to} \quad \sum_{i=1}^{m} a_{ij} u_i - u_{s_j} = c_j \quad j = 1, 2, \ldots, n$$

$$u_i \geq 0 \quad j = 1, 2, \ldots, m$$

Complementary Slackness Theorem Let $x_1^*, x_2^*, \ldots, x_n^*$ be an optimum solution to the primal problem P and $x_{s_1}, x_{s_2}, \ldots, x_{s_m}$ be the associated slack variables. Also, let $u_1^*, u_2^*, \ldots, u_m^*$ be an optimum solution to the dual problem D and $u_{s_1}, u_{s_2}, \ldots, u_{s_n}$

be the associated surplus variables. Then, the following relation holds for each of the complementary pairs of variables:

$$x_{s_i}^* u_i^* = 0 \quad \text{for } i = 1, 2, \ldots, m$$
$$u_{s_j}^* x_j^* = 0 \quad \text{for } j = 1, 2, \ldots, n$$

Proof From the duality theorem, we have

$$\sum_{j=1}^{n} c_j x_j^* = \sum_{i=1}^{m} \sum_{j=1}^{n} a_{ij} x_j^* u_i^* = \sum_{i=1}^{m} b_i u_i^*$$

Subtracting the middle term from the rightmost term, we have

$$\sum_{i=1}^{m} \left(b_i - \sum_{j=1}^{n} a_{ij} x_j^* \right) u_i^* = 0$$

or

$$\sum_{i=1}^{m} x_{s_i}^* u_i^* = 0$$

But because $u_i^* \geq 0$ for all i, this equality implies that $x_{s_i}^* u_i^* = 0$ for $i = 1, 2, \ldots, m$. Likewise, subtracting the leftmost term from the middle term, we have

$$\sum_{j=1}^{n} \left(\sum_{i=1}^{m} a_{ij} u_i^* - c_j \right) x_j^* = 0$$

or

$$\sum_{j=1}^{n} u_{s_j}^* x_j^* = 0$$

But because $x_j^* \geq 0$ for all j, this equality implies $u_{s_j}^* x_j^* = 0$ for $j = 1, 2, \ldots, n$, completing the proof. ■

Example 7.11   Consider the following LP problem (P) and its dual (D) (with slack variables and surplus variables).

(P) maximize $z = x_1 - 3x_2 + x_3$

    subject to $-x_1 + 2x_2 + x_3 + x_{s_1} = 10$

                $3x_1 - 2x_2 + 3x_3 + x_{s_2} = 20$

                $\mathbf{x} \geq \mathbf{0}$

(D) minimize $w = 10u_1 + 20u_2$

    subject to $-u_1 + 3u_2 - u_{s_1} = 1$

                $2u_1 - 2u_2 - u_{s_2} = -3$

                $u_1 + 3u_2 - u_{s_3} = 1$

                $\mathbf{u} \geq \mathbf{0}$

Suppose we know that the optimal solution to D is $u_1 = 0$, $u_2 = 1/3$, $w^* = 20/3$, which implies that $u_{s_1} = 0, u_{s_2} = -7/3, u_{s_3} = 0$. By the duality and the complementary theorems, we can obtain that $z^* = w^* = 20/3, x_{s_2} = 0$, and $x_2 = 0$.

Let us look at one more example.

Example 7.12 Consider the following LP problem (P) and its dual (D).

(P) maximize $z = 7x_1 + 11x_2$

    subject to $x_1 + x_2 + x_{s_1} = 11$

               $2x_1 - x_2 + x_{s_2} = 5$

               $-3x_1 - 2x_2 + x_{s_3} = -20$

               $\mathbf{x} \geq \mathbf{0}$

(D) minimize $w = 11u_1 + 5u_2 - 20u_3$

    subject to $u_1 + 2u_2 - 3u_3 - u_{s_1} = 7$

               $u_1 - u_2 - 2u_3 - u_{s_2} = 11$

               $\mathbf{u} \geq \mathbf{0}$

Given that the optimal solution to P is $x_1 = 0, x_2 = 11, z^* = 121$, which implies that $x_{s_1} = 0, x_{s_2} = 16$, and $x_{s_3} = 2$. By strong duality and the complementary rule, we can obtain that $w^* = z^* = 121$, $u_2 = 0, u_3 = 0$, and $u_{s_2} = 0$, which implies $u_1 = 11$ and $u_{s_1} = 4$. Hence, all information about the dual optimal solution is also obtained.

7.5 NOTES

Section 7.2

In some discussions of linear programming, the converted LP with each constraint expressed as an equality is called "standard form," and the version with all $\leq$ constraints (maximizing objective) is called "canonical form."

Section 7.4

Duality theory for linear programming was a major focus of applied mathematicians in the early 1950s. The main duality theorem was originally stated by John von Neumann, and proof first appeared in an article by Goldman and Tucker (1956). The interested reader should review the papers by Farkas (1902), the paper by Gale et al. (1951), and then the paper by Goldman and Tucker (1956).

There is a much broader duality theory for convex programming. The classic reference is Rockafellar (1970). For a brief introduction to duality for integer programming, see Section 2.5 of Wolsey (1998). For a comprehensive survey of duality theory and its relation to the concept of relaxation in integer programming, see Chapter II.3 in Nemhauser and Wolsey (1988).

7.6 EXERCISES

7.1 Under what conditions the following expressions are (a) affine combinations and (b) convex combinations?

(1) $\lambda^2 \mathbf{x}^1 + 2\lambda \mathbf{x}^2 - \lambda \mathbf{x}^3 + 3\lambda^2 \mathbf{x}^4$

(2) $2\lambda^3 \mathbf{x}^1 + 3\lambda \mathbf{x}^2 + \lambda^2 \mathbf{x}^3 + 2\lambda \mathbf{x}^4$

7.2 Calculate the determinant of each of the following matrices using the two methods introduced in this chapter.

(1) $\mathbf{A} = \begin{bmatrix} 0 & 3 & 2 & 1 \\ 2 & -1 & 6 & 4 \\ 1 & 4 & -1 & 3 \\ 5 & 2 & 3 & 0 \end{bmatrix}$

(2) $\mathbf{B} = \begin{bmatrix} -3 & 0 & 2 \\ 4 & 7 & 3 \\ 2 & 1 & -5 \end{bmatrix}$

7.3 Generate upper and lower triangular matrices for each of the matrix below. Show your steps.

(1) $\mathbf{A} = \begin{bmatrix} 6 & 9 & 5 \\ -3 & 7 & 4 \\ 4 & 0 & -3 \\ 8 & -1 & 7 \end{bmatrix}$

(2) $\mathbf{B} = \begin{bmatrix} 2 & 5 & -3 & 0 \\ 0 & 4 & 6 & 5 \\ -1 & 2 & 5 & 3 \\ 3 & 3 & 7 & 4 \end{bmatrix}$

7.4 Apply elementary row operations to the matrices in Exercise 7.3 to generate new matrices. Show your steps.

7.5 Prove the following two statements:

(1) If $\mathbf{A}$ and $\mathbf{B}$ are nonsingular $n \times n$ matrices, then $(\mathbf{AB})^{-1} = \mathbf{B}^{-1}\mathbf{A}^{-1}(n \times n)$.

(2) If $\mathbf{A}$ is nonsingular, then $\mathbf{AB} = \mathbf{AC}$ implies $\mathbf{B} = \mathbf{C}$.

7.6 Determine the rank of each of the matrices in Exercises 7.2 and 7.3.

7.7 Determine whether the following linear system is feasible or not by applying the upper or lower triangular matrix method.

$$2x_1 + x_2 - 3x_3 + x_4 = 9$$
$$-x_1 + 3x_2 + x_3 + 2x_4 = 11$$
$$x_1 - x_2 + 4x_3 = 7$$
$$2x_2 + x_3 - 2x_4 = 5$$
$$x_1, x_2, x_3, x_4 \geq 0$$

7.8 Determine if the following linear system has (a) no solution, (b) unique solution, or (c) multiple solutions.

$$7x_1 - 3x_2 + x_3 + 4x_4 = 23$$
$$3x_1 + 2x_2 - 4x_3 - x_4 = 19$$
$$x_1 + x_2 + x_3 + 3x_4 = 15$$
$$-x_1 + 3x_2 + 8x_3 + 2x_4 = 29$$
$$x_1, x_2, x_3, x_4 \geq 0$$

7.9 Determine the feasibility of the following LP system without solving it. Use at least two methods in this chapter.

$$-x_1 + 3x_2 - 2x_3 = 7$$
$$2x_1 + x_2 - x_3 = 6$$
$$x_1, x_2, x_3 \geq 0$$

7.10 Test if the solution $(125/92, 4/23, 91/92)$ is optimal to the following LP. Why or why not?

$$\text{Maximize} \quad z = 2x_1 - 3x_2 + 10x_3$$
$$\text{subject to} \quad -3x_1 + x_2 + 9x_3 \leq 5$$
$$x_1 - 2x_2 + x_3 \leq 2$$
$$6x_1 + 5x_2 + 2x_3 \leq 11$$
$$\mathbf{x} \geq \mathbf{0}$$

7.11 Formulate the dual of the following LP problem.

$$\text{Maximize} \quad z = 11x_1 - 13x_2 + 7x_3 + 9x_4$$
$$\text{subject to} \quad -2x_1 + x_2 + 4x_3 - 5x_4 \le -5$$
$$5x_1 + 4x_2 - x_3 \le 17$$
$$2x_1 + x_3 - x_4 \le 5$$
$$\mathbf{x} \ge 0$$

7.12 Formulate the dual of the following LP problem.

$$\text{Minimize} \quad z = 11x_1 - 13x_2 + 7x_3 - 9x_4$$
$$\text{subject to} \quad 2x_1 - x_2 + 4x_3 - 5x_4 \le 5$$
$$5x_1 + 4x_2 - x_3 \le 17$$
$$-2x_1 + x_3 - x_4 = 5$$
$$x_1, x_2 \ge 0, x_3 \le 0, x_4 \text{ unrestricted in sign}$$

7.13 Consider the following LP problem and its unique optimal solution. Formulate its dual and figure out as much information about the dual optimal solution from the information given about the primal.

$$\text{Maximize} \quad z = 4x_1 + 3x_2 + x_3 + 7x_4 + 6x_5$$
$$\text{subject to} \quad x_1 + 2x_2 + 3x_3 + x_4 - 3x_5 \le 9$$
$$2x_1 - x_2 + 2x_3 + 2x_4 + x_5 \le 10$$
$$-3x_1 + 2x_2 + x_3 - x_4 + 2x_5 \le 11$$
$$\mathbf{x} \ge 0$$

optimal solution is $(7, 10, 0, 0, 6)$ with $z^* = 94$

7.14 Consider the following LP problem and its dual. Given the optimal solution to the dual, figure out as much information about the optimal solution to the primal.

Primal	Dual
maximize $z = 25x_1 - 2x_2 + 16x_3$	minimize $w = 5u_1 + 2u_2 + u_3 + 4u_4 + 3u_5$
subject to $3x_1 + x_2 + 9x_3 \le 5$	subject to $3u_1 + 5u_2 - 6u_3 + 2u_4 + 2u_5 \ge 25$
$5x_1 + 2x_2 - 4x_3 \le 2$	$u_1 + 2u_2 + 3u_3 - 7u_4 + 3u_5 \ge -2$
$-6x_1 + 3x_2 + 2x_3 \le 1$	$9u_1 - 4u_2 + 2u_3 + 5u_4 - u_5 \ge 16$
$2x_1 - 7x_2 + -5x_3 \le 4$	$\mathbf{u} \ge 0$
$2x_1 + 3x_2 - x_3 \le 3$	optimal solution :
$\mathbf{x} \ge 0$	$(60/19, 59/19, 0, 0, 0)$ with $w^* = 22$

7.15   Consider the following LP problem. Formulate its dual. Given that the optimal
solution to the primal problem is $(3, 4, 0, 3)$ with $z^* = 22$, is the optimal solution
to the dual degenerate? Why or why not?

$$\text{Maximize} \quad z = 3x_1 + x_2 + 2x_3 + 3x_4$$

$$\text{subject to} \quad -x_1 + 3x_2 + x_3 - 2x_4 \leq 17$$

$$7x_1 + 3x_3 + x_4 \leq 23$$

$$x_1 + 2x_2 \leq 11$$

$$x_2 + 3x_4 \leq 13$$

$$x_1, x_4 \geq 0, x_2 \text{ unrestricted in sign}, \ x_3 \leq 0$$

8

LINEAR PROGRAMMING: GEOMETRIC CONCEPTS

This chapter introduces basic geometric concepts and terminology relevant to various simplex-based algorithms (Chapter 9) for solving linear programs and helpful to comprehending the various cutting plane methods embedded in the branch-and-cut method (Chapter 12) for solving integer programs. The geometry of the LP objective function, solution space, and requirement space is described. The geometry of convex sets in general, and polyhedra specifically, must be understood to motivate the linear algebra-based algorithms and methods to follow in later chapters.

8.1 GEOMETRIC SOLUTION

Recall that the feasible region of any LP is a *polyhedron* or *polyhedral set* and a polyhedron is the set of all points in E^n that simultaneously satisfy a set of m linear constraints:

$$P = \left\{ (x_1, x_2, \ldots, x_n) : \sum_{j=1}^{n} a_{ij}x_j \le b_i, \ i = 1, \ldots, m \right\}$$

In matrix notation, the system of constraints is $\mathbf{Ax} \le \mathbf{b}$ and $P = \{\mathbf{x}: \mathbf{Ax} \le \mathbf{b}\}$. It is understood that any lower or upper bound constraints on $\mathbf{x}$, including $\mathbf{x} \ge \mathbf{0}$, can be represented as a special form of $\sum_{j=1}^{n} a_{ij}x_j \le b_i$. We now examine the geometry of the LP problem in detail.

Applied Integer Programming: Modeling and Solution, By Der-San Chen, Robert G. Batson, and Yu Dang
Copyright © 2010 John Wiley & Sons, Inc.

8.1.1   Objective Function

Consider the objective function $z = \mathbf{c}^T\mathbf{x}$ subject to the above constraints. Variable z can take on any real value k. Hence, as k varies, we may consider the set $\{\mathbf{x}: \mathbf{c}^T\mathbf{x} = k\}$ to be an infinite number of parallel lines in E^2, parallel planes in E^3, or parallel hyperplanes in a higher dimension, each corresponding to a different value of k. For example, if $z = x_1 + 3x_2$, we know from calculus that the gradient of z

$$\nabla z = \begin{pmatrix} \dfrac{\partial z}{\partial x_1} \\[2mm] \dfrac{\partial z}{\partial x_2} \end{pmatrix} = \begin{pmatrix} 1 \\ 3 \end{pmatrix}$$

is the steepest ascent direction and the "equi-profit" contour for say $k = 0, 3, 6$, as shown in Figure 8.1. Note that the contours are parallel lines, perpendicular to the gradient vector $\mathbf{c} = (1, 3)^T$. Moving $\mathbf{x}$ in the $\mathbf{c}^T$ direction yields the greatest increase per unit change in the constant k. If the objective function is minimized, then the direction of steepest descent is $-\mathbf{c}^T$.

8.1.2   Solution Space

Consider the solution space for the following problem:

$$\begin{aligned} \text{Maximize} \quad & z = x_1 + 3x_2 \\ \text{subject to} \quad & x_1 + x_2 \le 3 \\ & x_1 - x_2 \ge 1 \\ & x_1, x \ge 0 \end{aligned}$$

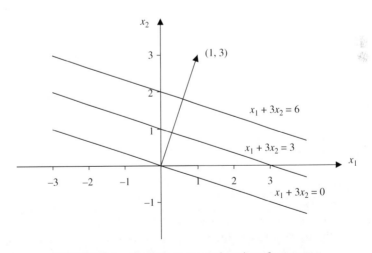

FIGURE 8.1   Gradient vector and equi-profit contours.

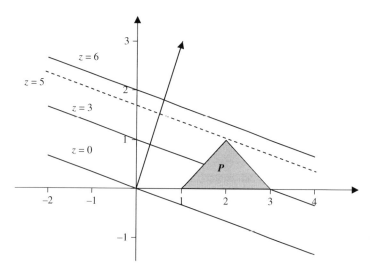

FIGURE 8.2 Bounded feasible region with various objective values.

The geometry of this problem is shown in Figure 8.2, where the shaded area is the solution space. Note that the contour $z = 6$ does not touch the feasible region, but $z = 5$ just does at the point $(2, 1)$. The example illustrates several points:

- An LP with a *bounded* feasible region *always* has a finite optimal solution.
- The optimal solution of a bounded LP, if unique, will occur at one and only one extreme point of P.
- If a bounded LP has two extreme points optimal (hence, alternative optima), then there are an infinite number of optimal points expressed by the line segment between them.

By bounded feasible region or set P, we mean there exists a nonnegative constant ε such that $P \subseteq \{x: |x| \leq \varepsilon\}$, a spheroid in E^n of diameter ε. An LP with an *unbounded* feasible region may or may not have a finite optimal value, depending on the objective function. The following objective functions are plotted in Figure 8.3:

(a) Maximize $z = -x_1 - x_2$
(b) Maximize $z = -x_1 + x_2$
(c) Maximize $z = x_1 + x_2$
(d) Maximize $z = -0.5x_1 + 4x_2$

Note that objective function (a) has a unique, finite optimum at $(0, 0)$, (b) has alternative optima expressed by a line segment, (c) has an unbounded solution, and (d) has a finite optimal ray.

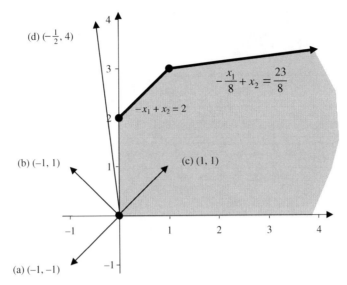

FIGURE 8.3 Optimal solutions of various objective functions.

8.1.3 Requirements Space

There is another geometric interpretation of the system of constraints in the space of the requirements and resource vectors, that is, in E^m. We shall first discuss case (A), a system of equality constraints, and then (B), a system of inequalities.

Definition 8.1 A *convex cone* C is a convex set with the additional property that $\lambda \mathbf{x} \in C$ for each $\mathbf{x} \in C$ and $\lambda \geq 0$.

The origin is always an element of a convex cone, and if $\mathbf{x} \in C$, the ray $\{\lambda \mathbf{x}: \lambda \geq 0\}$ belongs to C.

Case A: Equality Constraints $(\mathbf{Ax} = \mathbf{b}, \mathbf{x} \geq 0)$ Let $\mathbf{A} = (\mathbf{a}_1, \mathbf{a}_2, \ldots, \mathbf{a}_n)$, then the LP is feasible if $\mathbf{b}$ is within the convex cone generated by $\{\mathbf{a}_1, \mathbf{a}_2, \ldots, \mathbf{a}_n\}$.

Example 8.1

$$3x_1 + 2x_2 + x_3 = 1$$
$$-x_1 + x_2 + 2x_4 = 3$$
$$x_1, x_2, x_3, x_4 \geq 0$$

As shown in Figure 8.4, the system has a feasible solution because the vector $\mathbf{b} = (1, 3)^T$ falls within the convex cone generated by $\mathbf{a}_1 = (3, -1)^T$, $\mathbf{a}_2 = (2, 1)^T$,

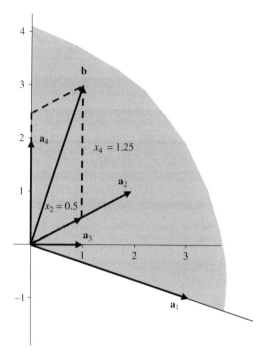

FIGURE 8.4   Geometric detection for feasibility.

$\mathbf{a}_3 = (1, 0)^{\mathrm{T}}$, and $\mathbf{a}_4 = (0, 2)^{\mathrm{T}}$. In fact, $\mathbf{b}$ can be generated by $(\mathbf{a}_2, \mathbf{a}_4)$, $(\mathbf{a}_3, \mathbf{a}_4)$, or $(\mathbf{a}_1, \mathbf{a}_4)$. For example,

$$\mathbf{b} = x_2\mathbf{a}_2 + x_4\mathbf{a}_4$$

$$\begin{pmatrix} 1 \\ 3 \end{pmatrix} = x_2 \begin{pmatrix} 2 \\ 1 \end{pmatrix} + x_4 \begin{pmatrix} 0 \\ 2 \end{pmatrix} \rightarrow x_2 = \frac{1}{2}, x_4 = \frac{5}{4}$$

Note that all variables are nonnegative: $x_2 > 0$, $x_4 > 0$, $x_1 = x_3 = 0$. Thus, this is a feasible solution. Likewise, we may verify that $\mathbf{b}$ can be generated by $(\mathbf{a}_3, \mathbf{a}_4)$ or $(\mathbf{a}_1, \mathbf{a}_4)$ by solving a system of equations for their associated variables.

Conversely, if $\mathbf{b}$ does not fall within the convex cone of the columns of $\mathbf{A}$, the LP is infeasible (all basic solutions of $\mathbf{Ax} = \mathbf{b}$ are infeasible). Consider the following example.

Example 8.2

$$2x_1 + 2x_2 + x_3 = -1$$

$$-x_1 + x_2 + 2x_4 = 2$$

$$x_1, x_2, x_3, x_4 \geq 0$$

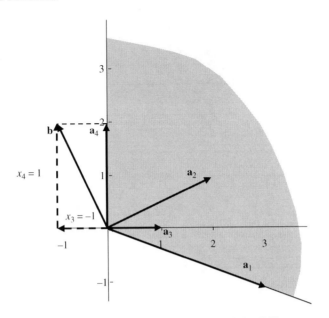

FIGURE 8.5   Geometric detection for infeasibility.

As can be seen in Figure 8.5, the vector $\mathbf{b} = (-1, 2)^{\mathrm{T}}$ falls outside the convex cone generated by the columns of $\mathbf{A}$. This situation implies that at least one variable takes on a negative value. For example,

$$\mathbf{b} = x_3\mathbf{a}_3 + x_4\mathbf{a}_4$$

$$\begin{pmatrix} -1 \\ 2 \end{pmatrix} = x_3 \begin{pmatrix} 1 \\ 0 \end{pmatrix} + x_4 \begin{pmatrix} 0 \\ 1 \end{pmatrix} \rightarrow x_3 = -1 \text{ and } x_4 = 2$$

Note that $x_3 < 0$, and thus $\mathbf{b}$ cannot be generated by vectors $\mathbf{a}_3$ and $\mathbf{a}_4$. Similarly, we can verify that $\mathbf{b}$ cannot be generated by any other pair of vectors, and hence there is no feasible solution for this system. The reader should verify that every basic solution of $\mathbf{Ax} = \mathbf{b}$ is infeasible (has at least one negative component).

Given a feasible LP, there is a geometric explanation of how a bounded optimal objective value arises. Let us illustrate this condition with an example.

Example 8.3

$$\text{Minimize} \quad z = -x_1 - 2x_2$$

$$\text{subject to} \quad x_1 + 3x_2 + 2x_3 = 3$$

$$x_1, x_2, x_3 \geq 0$$

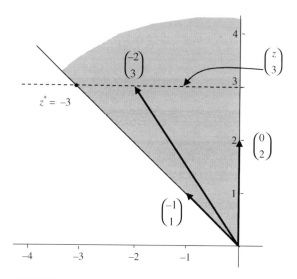

FIGURE 8.6 Bounded objective in requirement space.

The problem is to choose x_1, x_2, $x_3 \geq 0$ such that z is minimized in

$$\begin{pmatrix} -1 \\ 1 \end{pmatrix} x_1 + \begin{pmatrix} -2 \\ 3 \end{pmatrix} x_2 + \begin{pmatrix} 0 \\ 2 \end{pmatrix} x_3 = \begin{pmatrix} z \\ 3 \end{pmatrix}$$

In Figure 8.6, we first draw the vectors for the coefficient columns and then draw a horizontal line for the right-hand side column toward the direction of decreasing values (for minimization). When the line (dotted) hits the leftmost vector or its extension, then a minimum is found at $z = -3$.

To illustrate the geometric condition for an *unbounded* solution, we use Example 8.4 and Figure 8.7.

Example 8.4

$$\text{Minimize} \quad z = -x_1 - 2x_2$$

$$\text{subject to} \quad x_1 + 3x_2 - 2x_3 = 3$$

$$x_1, x_2, x_3 \geq 0$$

Case B: Inequality Constraints   As noted above, the *requirement space* $\{\mathbf{Ax}: \mathbf{x} \geq \mathbf{0}\}$ is the convex cone generated by $\{\mathbf{a}_1, \mathbf{a}_2, \ldots, \mathbf{a}_n\}$. If a feasible solution exists, this requirement space in E^m must overlap the collection of vectors that are less than or equal to the requirement vector $\mathbf{b}$ (another convex cone). Figure 8.8 shows (a) a feasible system and (b) an infeasible system.

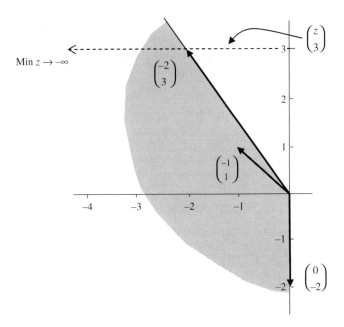

FIGURE 8.7 Unbounded objective in requirement space.

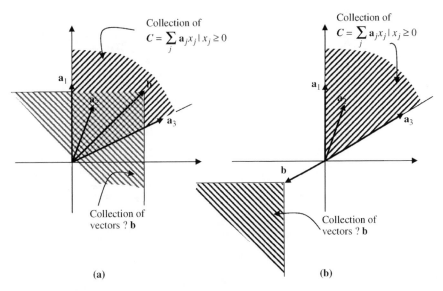

FIGURE 8.8 Geometry of (a) feasibility versus (b) infeasibility for an LP.

8.2   CONVEX SETS

This section introduces concepts and properties of convex sets in E^n, including polyhedra, and both convex and concave functions defined for vectors in E^n.

8.2.1   Convex Sets and Polyhedra

Definition 8.2   A set X in E^n is *convex* if given any two points $\mathbf{x}^1$ and $\mathbf{x}^2$ in X, and then $\alpha\mathbf{x}^1 + (1 - \alpha)\mathbf{x}^2$ is also in X for each $\alpha, 0 \le \alpha \le 1$. Each point along this line segment from $\mathbf{x}^1$ to $\mathbf{x}^2$ is called a *convex combination* of $\mathbf{x}^1$ and $\mathbf{x}^2$.

It can be shown that the solution of *every linear* equation or inequality forms a convex set such as the sets defined below:

$$A = \{\mathbf{x} : x_1 + 2x_2 = 5\}$$
$$B = \{\mathbf{x} : x_1 + 2x_2 \ge 5\}$$
$$C = \{\mathbf{x} : x_1 + 2x_2 \le 5\}$$

Moreover, the *intersection* of two or more convex sets forms a convex set. This implies that the set of feasible solutions of an LP forms a convex set. For example, Figure 8.9a is a convex set while the set depicted in Figure 8.9b is not convex because at least one point on the line segment between points x^1 and x^2 falls outside the set.

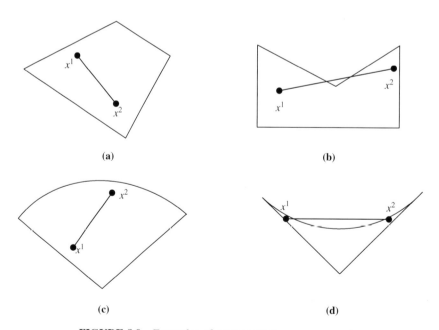

FIGURE 8.9   Examples of convex versus nonconvex sets.

However, a set containing nonlinear constraints may or may not be convex. For example, $D = \{\mathbf{x} : x_1^2 + x_2^2 \leq 1\}$ is convex, while $U = \{\mathbf{x} : x_1^2 + x_2^2 = 1\}$ and V $\{\mathbf{x}\ \text{integer}: x_1 + 2x_2 \leq 4\}$ are not convex. Figure 8.9c is convex but Figure 8.9d is not convex.

A hyperplane in E^n generalizes the concept of straight line in E^2 and plane in E^3.

Definition 8.3 A *hyperplane* in E^n is a set of the form $X = \{\mathbf{x}: \mathbf{p}^T\mathbf{x} = k\}$ where nonzero $\mathbf{p} \in E^n$ and k is a constant. Hence, a hyperplane is the solution set to a linear equation in E^n. A hyperplane clearly separates E^n into two *half-spaces*, each is a convex set containing the hyperplane, $H_1 = \{\mathbf{x}: \mathbf{p}^T\mathbf{x} \leq k\}$ and $H_2 = \{\mathbf{x}: \mathbf{p}^T\mathbf{x} \geq k\}$.

Example 8.5 Consider the linear equation $2x_1 + 3x_2 = 6$ in E^2. This equation may be written as

$$(2 \quad 3)\begin{pmatrix} x_1 \\ x_2 \end{pmatrix} = 6 \quad \text{or} \quad \mathbf{p}^T\mathbf{x} = k$$

and as Figure 8.10 illustrates, each $\mathbf{x} \in X$ is perpendicular to $\mathbf{p}$.

Definition 8.4 A point $\mathbf{x}$ in a convex set X is called an *extreme point* of X if it cannot be represented as a strict $(0 < \alpha < 1)$ convex combination of two distinct points in X. In a polyhedron, the line segment formed by all convex combinations of two *adjacent* extreme points is called an *edge* of X.

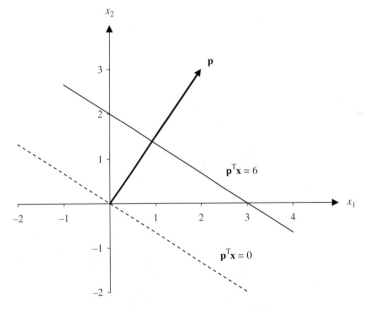

FIGURE 8.10 A hyperplane in E^2.

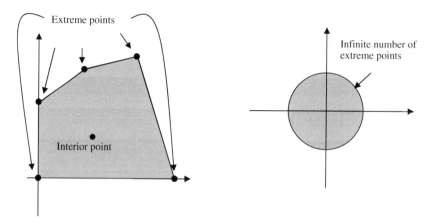

FIGURE 8.11   Extreme points of polyhedral and nonlinear convex sets.

A polyhedron P always has a finite number of extreme points and the line segment connecting two extreme points may include interior points of P. If a convex set contains any nonlinear constraint, then a convex set may have an infinite number of extreme points as illustrated in Figure 8.11.

Other examples of convex sets are rays and line segments in E^n.

Definition 8.5   A *ray* is a set in E^n of the form $X = \{\mathbf{x}: \mathbf{x} = \mathbf{x}^0 + \lambda\mathbf{d}, \mathbf{d} \neq 0, \lambda \geq 0\}$. The point $\mathbf{x}^0$ is called the *vertex* of the ray, and the vector $\mathbf{d}$ is the *direction of the ray*.

There is of course a ray in the "opposite direction" by using $-\mathbf{d}$ as the direction. A line segment emanating from $\mathbf{x}^0$ in a particular direction $\mathbf{d}$ and of length l may be produced by limiting λ to $0 \leq \lambda \leq u$ (where $u = l/\|\mathbf{d}\|$) in the definition above. More generally, the *line segment from* $\mathbf{x}^1$ *to* $\mathbf{x}^2$ is given by the set of all convex combinations of $\mathbf{x}^1$ and $\mathbf{x}^2$

$$L = \{\mathbf{x} : \mathbf{x} = \alpha\mathbf{x}^1 + (1-\alpha)\mathbf{x}^2, 0 \leq \alpha \leq 1\}$$

A more general concept than line segment is that of convex hull of any set of points.

Definition 8.6   Given a set $P \subseteq E^n$, it is possible to find a "minimal" convex set containing P. The *convex hull* of P is the intersection of all convex sets containing P, denoted by $\text{Conv}(P)$. The convex hull of a finite number of points is called a *convex polytope*, a special type of bounded polyhedron.

It is clear that a convex polytope X may be generated by all possible convex combinations of its n extreme points, say set E:

$$X = \left\{\mathbf{x} : \mathbf{x} = \sum_{i=1}^{n} \alpha_i\mathbf{x}^i, \text{ where } \mathbf{x}^i \in E, 0 \leq \alpha \leq 1, \text{and } \sum_{i=1}^{n} \alpha_i = 1\right\}$$

That is, $X = \text{Conv}(E)$.

8.2.2 Directions of Unbounded Convex Sets

The feasible region of a linear program may be unbounded. Unbounded convex sets have at least one "direction" in which the set recedes to infinity, whereas bounded convex sets has no such direction. We define and extend this concept formally.

Definition 8.7 Given an unbounded convex set X, a nonzero vector $\mathbf{d}$ is called a *direction* (of recession) of X if for each $\mathbf{x}^0 \in X$, the ray $\{\mathbf{x}: \mathbf{x} = \mathbf{x}^0 + \lambda\mathbf{d}, \mathbf{d} \neq 0, \lambda \geq 0\}$ is contained within X.

Note that a convex set may have multiple directions. For example, the first quadrant in E^2, $\{\mathbf{x}: x_1 \geq 0, x_2 \geq 0\}$, has any nonzero $\mathbf{d} = (d_1, d_2)$ with $d_1 \geq 0, d_2 \geq 0$ as a direction. However, we are most interested in the "extreme directions" associated with the two rays formed by the positive x_1-axis and positive x_2-axis.

Definition 8.8 A direction of an unbounded convex set X is called an *extreme direction* if it cannot be represented as a *positive* combination of two *distinct* directions of X. Two directions $\mathbf{d}^1$ and $\mathbf{d}^2$ are *distinct* if $\mathbf{d}^1$ cannot be expressed as $k\mathbf{d}^2$ for some positive scalar k.

For example, $\mathbf{d}^1 = (1, 0)^T$ and $\mathbf{d}^2 = (2, 1)^T$ are distinct because we cannot find a scalar k such that $\mathbf{d}^1 = k\mathbf{d}^2$, since the system of equations, $1 = 2k$ and $0 = k$, has no solution.

A property of extreme directions of X is that any other direction $\mathbf{d}$ of X can be expressed as a positive combination of extreme directions of X: $\mathbf{d} = \sum \alpha_i \mathbf{d}^i$, $\alpha_i \geq 0$ and $\mathbf{d}^i$ extreme direction for every i. Therefore, there is an obvious analogy between extreme points of a convex set and extreme directions of an unbounded convex set.

Definition 8.9 An *extreme ray* of an unbounded convex set is a ray whose direction is an extreme direction. For example, the positive x_1-axis and positive x_2-axis are extreme rays of the first quadrant in E^2. Obviously, the set of extreme rays of X has the form $\{\mathbf{x}: \mathbf{x} = \mathbf{x}^0 + \lambda\mathbf{d}, \mathbf{x}^0 \in X, \mathbf{d} \text{ an extreme direction of } X, \lambda \geq 0\}$.

8.2.3 Convex and Polyhedral Cones

Definition 8.10 A *convex cone* is a convex set C that consists of rays emanating from the origin, that is, C is a convex set with the additional property that $\lambda\mathbf{x} \in C$ for each $\mathbf{x} \in C$ and $\lambda \geq 0$.

Definition 8.11 A *polyhedral cone* C is a convex cone of the form, $C = \{\mathbf{x}: A\mathbf{x} \leq 0\}$. That is, C is the intersection of a finite number of half-spaces whose hyperplanes pass through the origin.

Example 8.6 Figure 8.12 depicts three cases: (a) a polyhedron P that is bounded, hence not a cone; (b) a polyhedron Q that is a cone; and (c) an unbounded polyhedron R that is not a cone.

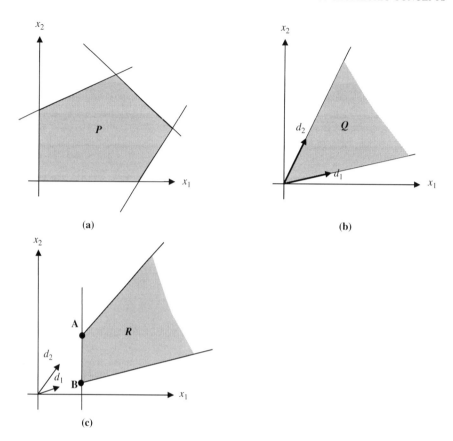

FIGURE 8.12   Example polyhedra.

8.2.4   Convex and Concave Functions

The reader probably encountered the concept of convex and concave functions in calculus. Convexity (concavity) is a strong property that often replaces differentiability as a desirable property in constrained optimization problems.

Definition 8.12   A real-valued function $f(\mathbf{x})$, $\mathbf{x} \in E^n$, is *convex* on E^n if the following inequality holds for any two points $\mathbf{x}^1$ and $\mathbf{x}^2$ in E^n: $f[\lambda\mathbf{x}^1 + (1 - \lambda)\mathbf{x}^2] \leq \lambda f(\mathbf{x}^1) + (1 - \lambda)f(\mathbf{x}^2)$ for all $0 \leq \lambda \leq 1$.

See Figure 8.13a and note the geometric interpretation of convexity of f. Any $\mathbf{x}$ between $\mathbf{x}^1$ and $\mathbf{x}^2$ has its function value $f(\mathbf{x})$ below the correspond point on the line segment joining $(\mathbf{x}^1, f(\mathbf{x}^1))$ and $(\mathbf{x}^2, f(\mathbf{x}^2))$.

Definition 8.13   A real-valued function $f(\mathbf{x})$, $\mathbf{x} \in E^n$, is *concave* on E^n if the following inequality holds for any two points $\mathbf{x}^1$ and $\mathbf{x}^2$ in E^n: $f[\lambda\mathbf{x}^1 + (1 - \lambda)\mathbf{x}^2] \geq \lambda f(\mathbf{x}^1) + (1 - \lambda)f(\mathbf{x}^2)$ for all $0 \leq \lambda \leq 1$.

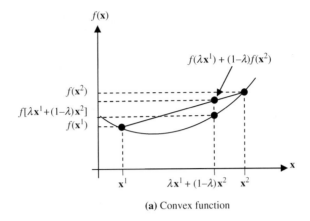

(a) Convex function

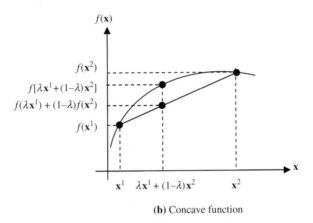

(b) Concave function

FIGURE 8.13   Example convex and concave functions on E^1.

See Figure 8.13b and note the geometric interpretation of concavity of f. Also, note the following obvious properties:

1. A function f is concave (convex) if and only if the function $g = -f$ is convex (concave).
2. A function f is linear if and only if f is both concave and convex.
3. The definition of convex and concave f can be reduced to a specific subset $X \in E^n$; some functions may be convex on a certain subset(s) of E^n, but not the entire space.

An interesting relationship between convex and concave functions on E^n and convex sets in E^{n+1} exists.

Definition 8.14   The *epigraph* of a function $f(\mathbf{x})$, $\mathbf{x} \in \boldsymbol{E}^n$, is the set in $\boldsymbol{E}^{n+1}$ defined by $\{(\mathbf{x}, y): \mathbf{x} \in \boldsymbol{E}^n, y \in \boldsymbol{E}^1, y \geq f(\mathbf{x})\}$. The *hypergraph* of f is defined similarly to be the set in $\boldsymbol{E}^{n+1}$: $\{(\mathbf{x}, y): \mathbf{x} \in \boldsymbol{E}^n, y \in \boldsymbol{E}^1, y \leq f(\mathbf{x})\}$.

It can be shown that a function $f(\mathbf{x})$, $\mathbf{x} \in \boldsymbol{E}^n$, is convex if and only if its epigraph is a convex set in $\boldsymbol{E}^{n+1}$. Similarly, a function $f(\mathbf{x})$, $\mathbf{x} \in \boldsymbol{E}^n$, is concave if and only if its hypergraph is a convex set in $\boldsymbol{E}^{n+1}$.

The reader can envision the epigraph of the function in Figure 8.13a by shading in all of $\mathbf{E}^2$ on or above the points of the graph of f. Similarly, the hypergraph of the function f in Figure 8.13b is generated by shading on or below the graph of f.

8.3   DESCRIBING A BOUNDED POLYHEDRON

8.3.1   Representation by Extreme Points

It can be shown that given a nonempty *bounded polyhedron* (or *polytope*) $\boldsymbol{P} = \{\mathbf{x}: \mathbf{A}\mathbf{x} \leq \mathbf{b}, \mathbf{x} \geq \mathbf{0}\}$ with extreme points $\mathbf{x}^1$, $\mathbf{x}^2$, ..., $\mathbf{x}^p$, any point $\mathbf{x} \in \boldsymbol{P}$ can be represented as a convex combination of extreme points; that is, $\mathbf{x} = \sum_{j=1}^{p} \alpha_j \mathbf{x}^j$ for some particular values of $\alpha_j \geq 0$, where $\sum_{j=1}^{p} \alpha_j = 1$. This property is very important in the simplex method of linear programming, so we elaborate on it here.

8.3.2   Example Application of Representation Theorem

Consider the polytope in $\mathbf{E}^2$ depicted in Figure 8.14. The point $\mathbf{x}^*$ is an interior point that happens to fall on the line segment connecting $\mathbf{x}^5$ and $\mathbf{y}$ on the edge between $\mathbf{x}^2$ and $\mathbf{x}^3$.

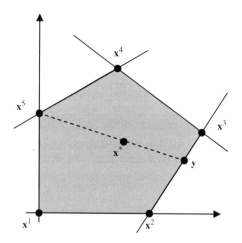

FIGURE 8.14   A polytope with five extreme points.

To illustrate the representation property, let us show that $\mathbf{x}^*$ is a convex combination of the five extreme points:

$$
\begin{aligned}
\mathbf{x}^* &= \alpha_5\mathbf{x}_5 + (1-\alpha_5)\mathbf{y} \\
&= \alpha 5\mathbf{x}^5 + (1-\alpha_5)[\alpha_2\mathbf{x}^2 + (1-\alpha_2)\mathbf{x}^3] \\
&= \alpha_5\mathbf{x}^5 + (1-\alpha_5)\alpha_2\mathbf{x}^2 + (1-\alpha_5)(1-\alpha_2)\mathbf{x}^3
\end{aligned}
$$

Now since $0 \le \alpha_5 \le 1$ and $0 \le \alpha_2 \le 1$, it follows that $0 \le (1-\alpha_5)\alpha_2 \le 1$ and $0 \le (1-\alpha_5)(1-\alpha_2) \le 1$. It is clear that $\alpha_5 + (1-\alpha_5)\alpha_2 + (1-\alpha_5)(1-\alpha_2) = 1$. Therefore, $\mathbf{x}^*$ is a convex combination of the five extreme points, with coefficients of $\mathbf{x}^1$ and $\mathbf{x}^4$ set to zero.

8.4 DESCRIBING UNBOUNDED POLYHEDRON

An unbounded polyhedron (or polytope) can be described by the set of all extreme points and the set of all extreme directions. First, we will show how to find all extreme directions algebraically. Then we will provide a precise mathematical expression that describes an unbounded polyhedron by the extreme points and extreme directions.

8.4.1 Finding Extreme Direction Algebraically

Theorem 8.1   The directions of an unbounded polyhedron $X = \{\mathbf{x}: A\mathbf{x} \le \mathbf{b}, \mathbf{x} \ge \mathbf{0}\}$ are nonzero vectors $\mathbf{d}$ in the set $\{\mathbf{d}: A\mathbf{d} \le \mathbf{0}, \mathbf{d} \ge \mathbf{0}, \mathbf{d} \ne \mathbf{0}\}$, known as the *recession cone* of X.

Recall that an extreme direction of X is a direction that cannot be represented as a positive combination of two distinct directions of X.

Definition 8.15   The set of *recession directions* of X is obtained from the recession cone by adjoining a normalization constraint to the recession cone definition:

$$
D = \{\mathbf{d}: A\mathbf{d} \le \mathbf{0}, \mathbf{d} \ge \mathbf{0}, \mathbf{1}^T\mathbf{d} = 1\}
$$

The set D is illustrated in Figure 8.15 for a three-constraint feasible region X. Note that D is always a bounded polyhedron because it is bounded by $\mathbf{1}^T\mathbf{d} = 1$.

Theorem 8.2   The vector $\mathbf{d}$ is an extreme point of D if and only if $\mathbf{d}$ is an extreme direction of X.

Example 8.7   To illustrate the algebraic process of finding the extreme direction of an LP feasible region, consider the polyhedral set X given by the inequalities

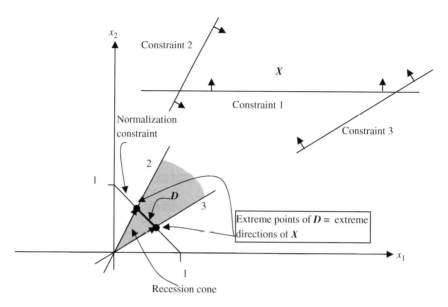

FIGURE 8.15   Recession cone and its normalized set of directions.

$$-x_1 - 2x_2 \leq 1$$
$$-5x_1 + x_2 \leq 6$$
$$-x_1 + x_2 \leq 4$$
$$-x_1 + 3x_2 \leq 12$$
$$x_1, x_2 \text{ unrestricted in sign}$$

The set is illustrated in Figure 8.16. Its extreme points are given as

$$\mathbf{x}^1 = \left(-\frac{13}{11}, \frac{1}{11}\right)^T, \mathbf{x}^2 = \left(-\frac{1}{2}, \frac{7}{2}\right)^T, \text{ and } \mathbf{x}^3 = (0, 4)^T$$

The set D above is given by all (d_1, d_2) that satisfy

$$d_1 + d_2 = 1 \tag{8.1}$$

$$-d_1 - 2d_2 \leq 0 \tag{8.2}$$

$$-5d_1 + d_2 \leq 0 \tag{8.3}$$

$$-d_1 + d_2 \leq 0 \tag{8.4}$$

$$-d_1 + 3d_2 \leq 0 \tag{8.5}$$

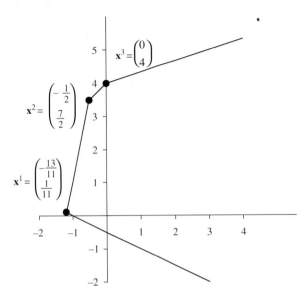

FIGURE 8.16 Boundary of feasible region for Example 8.8.

Adjoining slack variables to constraints (8.2)–(8.5) and solving the resulting system, leads to two extreme points of D (or extreme directions of X):

$$\mathbf{d}^1 = \begin{pmatrix} \frac{3}{4} \\ \frac{1}{4} \end{pmatrix}, \mathbf{d}^2 = \begin{pmatrix} 2 \\ -1 \end{pmatrix}$$

Figure 8.17 illustrates graphically the determination of the set D. The reader should also verify that if $x_1 \geq 0$, $x_2 \geq 0$ are adjoined to the original model, then with $d_1 \geq 0$, $d_2 \geq 0$ in the solution process, $\mathbf{d}^1 = \begin{pmatrix} 3/4 \\ 1/4 \end{pmatrix}$ as before, and $\mathbf{d}^2 = \begin{pmatrix} 1 \\ 0 \end{pmatrix}$.

Example 8.8 Using the normalizing equation, find all extreme directions of the LP feasible region defined by

$$x_1 - x_2 + x_3 \leq 10$$
$$2x_1 - x_2 + 2x_3 \leq 40$$
$$x_1, x_2, x_3 \geq 0$$

Create the system $\mathbf{Ad} \leq \mathbf{0}$, $\mathbf{d} \geq \mathbf{0}$, $\mathbf{d} \neq \mathbf{0}$, $\mathbf{1d} = 1$

$$d_1 - d_2 + d_3 \leq 0$$
$$2d_1 - d_2 + 2d_3 \leq 0$$
$$d_1 + d_2 + d_3 = 1$$

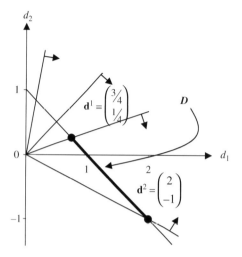

FIGURE 8.17   Extreme points of set D for Example 8.8.

Adjoin slack variables d_4 and d_5 yielding the system

$$d_1 - d_2 + d_3 + d_4 = 0$$

$$2d_1 - d_2 + 2d_3 + d_5 = 0$$

$$d_1 + d_2 + d_3 = 1$$

$$d_1, d_2, d_3, d_4, d_5 \geq 0$$

This system potentially has $\binom{5}{3} = 10$ basic solutions. It may be shown that only three of these are basic feasible solutions:

$$\mathbf{x_{B_1}} = (d_2, d_4, d_5) = (1, 1, 1)$$

$$\mathbf{x_{B_2}} = (d_2, d_3, d_4) = \left(\frac{2}{3}, \frac{1}{3}, \frac{1}{3}\right)$$

$$\mathbf{x_{B_3}} = (d_1, d_2, d_4) = \left(\frac{1}{3}, \frac{2}{3}, \frac{1}{3}\right)$$

We conclude that there are three extreme directions of X (extreme points of D):

$$(0, 1, 0, 1, 1)^{\mathrm{T}}, \left(0, \frac{2}{3}, \frac{1}{3}, \frac{1}{3}, 0\right)^{\mathrm{T}}, \left(\frac{1}{3}, \frac{2}{3}, 0, \frac{1}{3}, 0\right)^{\mathrm{T}}$$

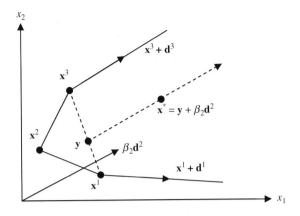

FIGURE 8.18   Representation of a feasible point using extreme points and directions.

8.4.2   Representing by Extreme Points and Extreme Directions

We now present the representation theorem for an unbounded polyhedron.

Theorem 8.3   (Representation Theorem) Given a nonempty polyhedron $X =$ $\{\mathbf{x}: \mathbf{A}\mathbf{x} \leq \mathbf{b}, \mathbf{x} \geq \mathbf{0}\}$, its set of extreme points $S_P = \{\mathbf{x}^1, \mathbf{x}^2, \ldots, \mathbf{x}^p\}$, and its set of extreme directions $S_\mathbf{d} = \{\mathbf{d}^1, \mathbf{d}^2, \ldots, \mathbf{d}^q\}$, any point $\mathbf{x}$ in X can be expressed as the sum of a convex combination of points in S_p and a positive linear combination of directions in $S_\mathbf{d}$:

$$\mathbf{x} = \sum_{i=1}^{p} \alpha_i \mathbf{x}^i + \sum_{j=1}^{q} \beta_j \mathbf{d}^j$$

where $\sum_{i=1}^{p} \alpha_i = 1$, $\alpha_i \geq 0$, $i = 1, \ldots, p$; $\beta_j \geq 0$, $j = 1, \ldots, q$.

8.4.3   Example of Representation Theorem

As an example of an application in E^2 of the above theorem, consider the unbounded polyhedron X depicted in Figure 8.18. Note that there are three extreme points and two extreme directions of X. A point along the upper extreme ray would have a unique representation as $\mathbf{x}^3 + \lambda \mathbf{d}^2$ for a particular $\lambda \geq 0$, whereas $\mathbf{x}^*$ has multiple possible representations, one (as depicted) being $\mathbf{x}^* = \mathbf{y} + \beta_2 \mathbf{d}^2 = \alpha \mathbf{x}^1 + (1 - \alpha)\mathbf{x}^3 + \beta_2 \mathbf{d}^2$ for particular values of $0 < \alpha < 1$ and $\beta_2 > 0$.

8.5   FACES, FACETS, AND DIMENSION OF A POLYHEDRON

In this section, we provide some additional geometrically motivated definitions and insights into the nature of extreme points and higher dimensional faces of a

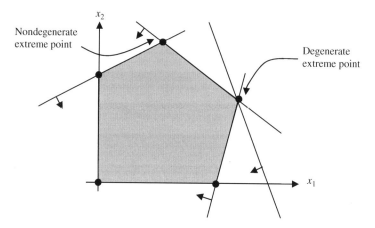

FIGURE 8.19 Degenerate polyhedron.

polyhedron. We will assume a polyhedron in E^n defined by $X = \{\mathbf{x}: \mathbf{A}\mathbf{x} \le \mathbf{b}, \mathbf{x} \ge \mathbf{0}\}$ where $\mathbf{A}$ is $m \times n$ matrix, $\mathbf{x}$ is $n \times 1$ matrix, and $\mathbf{b}$ is $m \times 1$ matrix.

Hence, there are $(m + n)$ inequality constraints, corresponding to $(m + n)$ *defining half-spaces* whose intersection is X. We will call the $(m + n)$ hyperplanes formed by the boundary of each of these half-spaces as the *defining hyperplanes* of X. Each hyperplane corresponds to the solution of an equation in E^n; a set of n defining hyperplanes are linearly independent if the coefficient matrix associated with this set has full row rank $(=n)$. An *extreme point* $\mathbf{x}$ of the polyhedron X in E^n is the (unique) solution of n linearly independent defining hyperplanes of X. If more than n defining hyperplanes of X pass through an extreme point $\mathbf{x}$, then such an extreme point is called a *degenerate* extreme point (see Figure 8.19). A polyhedron that contains at least one degenerate extreme point is called a *degenerate polyhedron*, and the corresponding LP has *degeneracy*.

Definition 8.16 A constraint $\alpha^T\mathbf{x} \le \beta$ is said to be *binding* (*active*) at a point $\mathbf{x}^* \in X$ if $\alpha^T\mathbf{x}^* = \beta$.

Above, we defined an extreme point $\mathbf{x}$ of X to be the unique solution of some n linearly independent defining hyperplanes binding at $\mathbf{x}$. A more general concept is that of proper face of X.

Definition 8.17 A *proper face* F of X is a nonempty set of points in X formed by the intersection of some set of binding defining hyperplanes of X. The *dimension of a face* of X is $\dim(F) = n - \mathrm{rank}(F)$, where $\mathrm{rank}(F) = $ maximum number of linearly independent defining hyperplanes binding at all points of F. *Note*: $0 \le \dim(F) \le n - 1$.

For example, in E^3 a face can have 1, 2, or 3 binding hyperplanes, so a face can be of dimension 2, 1, or 0. Of course, in the case of a degenerate extreme point $\mathbf{x}$, there would be four or more binding hyperplanes at $\mathbf{x}$. The extreme points of X are the zero-dimensional faces; the *edges* of X are the one-dimensional faces; and the planes,

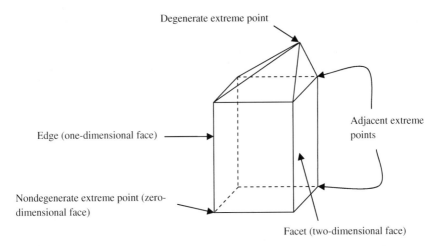

FIGURE 8.20   Proper faces of a polyhedron of full dimension in $\mathbf{E}^3$.

two-dimensional faces are called *facets*, a term reserved for the highest dimensional proper face of X. So if X is *full dimensional* (n), the dimension of a facet is dim $(X) - 1 = n - 1$, or 2 in the case of $X \subseteq \mathbf{E}^3$. A face loses one dimension (or degree of freedom) for every additional linearly independent binding hyperplane associated with it. In E^n, a face can have one of $1, 2, 3, \ldots, n$ binding defining hyperplanes, except for degenerate extreme points.

Figure 8.20 shows a full-dimensional polyhedron in $\mathbf{E}^3$ with nine defining hyperplanes. It has nine extreme points, five of which are degenerate, and nine two-dimensional faces (facets). As shown in Figure 8.20, two extreme points are adjacent if the line segment joining them is an edge of X. Hence, adjacent extreme points have $(n - 1)$ binding linearly independent defining hyperplanes in common.

8.6   DESCRIBING A POLYHEDRON BY FACETS

In cutting plane methods and branch-and-cut methods for solving mixed integer programs, it is necessary to generate a sequence of linear inequalities, each of which is used to form a new facet of the "updated" LP feasible region, enclosing the MIP feasible region, by means of intersecting its half-space with a "current" LP feasible region of interest. Thus, the knowledge of "minimal" hyperplane representation of polyhedra is useful background for Chapters 11 and 12.

Definition 8.18   A polyhedron $P \subseteq E^n$ is called *full dimensional* if it contains n linearly independent directions. By this we mean that at any interior point $\mathbf{x}^0$ of P, there exists a set of directions $\{\mathbf{d}^1, \mathbf{d}^2, \ldots, \mathbf{d}^n\}$ and an $\varepsilon_0 > 0$ such that $\mathbf{x}^0 + \varepsilon \mathbf{d}^i \in P$, for all $0 < \varepsilon < \varepsilon_0$. Equivalently, the spheroid $\{\mathbf{x}: \|\mathbf{x}^0 - \mathbf{x}\| < \varepsilon_0\} \subseteq P$. Hence, a full-dimensional polyhedron P has the property that there is no hyperplane $H = \{\mathbf{x} \in E^n | \alpha^\mathsf{T}\mathbf{x} = \beta\}$ such that $P \subseteq H$.

Theorem 8.4 Any full-dimensional polyhedron P can be uniquely represented by a set of inequalities $P = \{x \in \mathbf{R}^n: a_i x \leq b_i, i = 1, \ldots, m\}$ where each inequality is unique within a positive multiple, and each of which defines a facet of P.

The set of m inequalities in Theorem 8.4 is minimal in the sense that if one is removed, the resulting polyhedron is no longer P.

Definition 8.19 An inequality $a^T x \leq b$ is a *valid inequality* for $X \subseteq E^n$ if $a^T x \leq b$ for all $x \in X$. In other words, X is contained within the half-space defined by $a^T x \leq b$.

Theorem 8.5 If P is full dimensional, a valid inequality $a^T x \leq b$ is necessary in the description of P if and only if it defines a facet of P.

Wolsey (1998) provides the following example in E^2. Of the seven inequalities listed, only (8.6) and (8.9)–(8.11) are necessary. Inequalities (8.7), (8.8), and (8.12) although valid for P are redundant and would not be included among the "minimal" set of valid inequalities described in Theorems 8.4 and 8.5.

Example 8.9 (Wolsey 1998[1]) The reader is encouraged to verify that inequalities (8.7), (8.8), and (8.12) are not necessary in the minimal (facet) description of P:

$$x_1 \leq 2 \tag{8.6}$$

$$x_1 + x_2 \leq 4 \tag{8.7}$$

$$x_1 + 2x_2 \leq 10 \tag{8.8}$$

$$x_1 + 2x_2 \leq 6 \tag{8.9}$$

$$x_1 + x_2 \leq 2 \tag{8.10}$$

$$x_1 \geq 0 \tag{8.11}$$

$$x_2 \geq 0 \tag{8.12}$$

8.7 CORRESPONDENCE BETWEEN ALGEBRAIC AND GEOMETRIC TERMS

To summarize this chapter, Table 8.1 is provided to show the correspondence between the algebraic expression of a set related to an LP feasible region and its geometric concept in E^n. In this table, we assume that P is a full-dimensional polyhedron represented (Theorem 8.5) by $P = \{x : \sum_{j=1}^n a_{ij} x_j \leq b_i, i = 1, \ldots, m; x_j \geq 0, j = 1, 2, \ldots, n\}$.

TABLE 8.1 Correspondence between Algebraic and Geometric Concepts in LP

Algebraic Description	Geometric Term
$\mathbf{Ax} < \mathbf{b}$ and $\mathbf{x} > \mathbf{0}$	Interior point $\mathbf{x}$ of P
$\mathbf{Ax} \le \mathbf{b}, \mathbf{x} \ge \mathbf{0}$, and $\sum_{j=1}^{n} a_{ij}x_j = b_i$ for at least one i, or $x_j = 0$ for at least one j	Boundary point $\mathbf{x}$ of P
All *feasible* $\mathbf{x}$ satisfying $\sum_{j=1}^{n} a_{ij}x_j \le b_i$ for a specific subset of the indices $i = 1, \ldots, m$	Face of P
All *feasible* $\mathbf{x}$ satisfying $\sum_{j=1}^{n} a_{ij}x_j = b_i$ for exactly one I	Facet of P (or $(n-1)$-dimensional face)
All *feasible* $\mathbf{x}$ satisfying $\sum_{j=1}^{n} a_{ij}x_j = b_i$ for exactly $n-1$ of the indices $i = 1, \ldots, m$	Edge of P (or one-dimensional face)
A *feasible* $\mathbf{x}$ satisfying for exactly n of the indices $i = 1, \ldots, m$	Extreme point $\mathbf{x}$ of P (zero-dimensional face)
$\{\mathbf{x} : \sum_{j=1}^{n} a_{ij}x_j = b_i\}$ for a given i among $i = 1, \ldots, m$	Defining hyperplane of P
$\{\mathbf{x} : \sum_{j=1}^{n} a_{ij}x_j \le b_i\}$ for a given i among $i = 1, \ldots, m$	Defining half-space of P
$D = \{\mathbf{d} : \mathbf{Ad} \le \mathbf{0}, \mathbf{d} > \mathbf{0}, \mathbf{1d} = 1\}$	Directions of recession of an unbounded P
$\text{Cone}(D) = \{\lambda\mathbf{d} : \mathbf{d} \in D, \lambda \ge 0\}$	Recession cone of unbounded P
$D = \phi$	P is a bounded polyhedron (polytope)

8.8 NOTES

Sections 8.2 and 8.3

A standard reference on convex sets and functions is Rockafellar (1970). Minkowski published his "summation" theorem in 1911, the origin of the Representation Theorem (Theorem 8.3). Stability theory (Batson, 1979) uses this theorem.

Section 8.5

Much more detail on faces and facets of convex polytopes may be found in Grünbraum (1967).

8.9 EXERCISES

8.1 Consider the polyhedron P shown in the Figure 8.21. Is it possible that it is the feasible region of some LP problem? Why or why not?

8.2 Sketch the feasible region of the following LP problem. Do you think whether it has optimal solution or not without solving the problem? If yes, it is finite and unique?

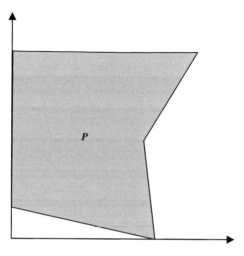

FIGURE 8.21   A polyhedron.

$$\text{Maximize} \quad z = x_1 + 8x_2$$
$$\text{subject to} \quad x_1 - x_2 \leq 0$$
$$x_1 + x_2 \geq 2$$
$$-5x_1 + x_2 \leq 5$$
$$x_1, x_2 \geq 0$$

8.3   Sketch the feasible region of the following LP problem. Try to tell if it has unbounded solution using the concept of convex cone.

$$\text{Minimize} \quad z = -2x_1 + x_2$$
$$\text{subject to} \quad x_1 + 2x_2 \leq 11$$
$$x_1 + 3x_2 \leq 21$$
$$4x_1 - x_2 \leq 3$$
$$x_1, x_2 \text{ unrestricted in sign}$$

8.4   Give examples of LP problems in E^2 that satisfy

(1) The feasible region is bounded with objective function optimized at a unique extreme point.

(2) The feasible region is bounded, and objective function is optimized at an edge.

(3) The feasible region is unbounded, but the objective function is optimized at a unique extreme point.

(4) The feasible region is unbounded, and the objective function is unbounded.

8.5   Consider the following sets: Are they convex? If not, explain why using two specific points in S.

      (1) $S = \{\mathbf{x} \text{ continuous: } x_1 x_2 \geq 4,\ x_1 \leq 4,\ x_2 \leq 4\}$

      (2) $S = \{\mathbf{x} \geq \mathbf{0}:\ x_1 x_2 \leq 25,\ x_1 - x_2 \leq 10,\ x_2 \leq 7\}$

      (3) $S = \{\mathbf{x} \text{ continuous: } |x_1 - x_2| \geq 4,\ x_1 + x_2 \leq 10,\ x_1 \geq 2\}$

8.6 Decide if each of the following functions is convex, concave, or neither. Justify your conclusion (by numerical proof, by using the epigraph or hypergraph, or by providing points in conflict with the definition).

      (1) $f(x) = -2x^3$

      (2) $f(x) = x^{-1},\ x \neq 0$

      (3) $f(\mathbf{x}) = \frac{x_1^2}{a^2} + \frac{x_2^2}{b^2},\ a,\ b > 0$ and constant

      (4) $f(\mathbf{x}) = \frac{x_1^2}{a^2} - \frac{x_2^2}{b^2},\ a,\ b > 0$ and constant

8.7 Show that $\text{Conv}(\mathbf{x}^1, \mathbf{x}^2, \ldots, \mathbf{x}^n) = \text{Conv}[\mathbf{x}^1, \mathbf{x}^2, \ldots, \mathbf{x}^t, \text{Conv}(\mathbf{x}^{t+1}, \mathbf{x}^{t+2}, \ldots, \mathbf{x}^n)]$.

8.8 Given two distinct sets S_1 and S_2, where $S_1 \subset S_2$, show that $\text{Conv}(S_1) \subset \text{Conv}(S_2)$.

8.9 Show that the convex hull of a convex set is itself.

8.10 Consider the feasible region of an IP problem. If the corresponding LP convex hull is unbounded, is it possible that the feasible region of the IP is bounded? Why or why not?

8.11 Prove that (a) a function is linear if and only if it is both concave and convex; (b) a function f is convex (concave) if and only if the function $g = -f$ is convex (concave).

8.12 Find all the extreme points of the following polyhedron formed by the feasible region of an LP problem with three decision variables. List all possible simplexes.

$$x_1 + x_2 + x_3 \leq 1$$
$$x_1, x_2, x_3 \geq 0$$

8.13 Following the procedure described in Example 8.8, find all extreme directions (if any) of the feasible region of the LP problem in Exercise 8.1.

8.14 What do you think is the relationship between degeneracy and the necessity of constraints in defining the facets of the corresponding polyhedron?

8.15 Show another way to represent the point $\mathbf{x}^*$ in Figure 8.18.

8.16 Find all extreme points and extreme rays of the polyhedron defined as follows: $P = \{\mathbf{x} \geq \mathbf{0}:\ -3x_1 + x_2 \leq 3,\ x_1 - x_2 \leq 5,\ -2x_1 - x_2 \leq -7\}$. Represent the point (10, 25) and (5, 5) using the extreme points and extreme rays you found.

8.17 Plot the feasible region of the polyhedron described in Exercise 8.12. Find all its faces and facets.

8.18 Consider the following set of constraints for an LP problem. Which one(s) is necessary in the description of the facets? Which not? Why?

$$-x_1 + x_2 \le 2$$
$$x_1 - 2x_2 \le 6$$
$$x_1 + 4x_2 \ge 5$$
$$3x_1 + x_2 \le 18$$
$$x_1 \ge 0$$
$$x_2 \ge 0$$

8.19 Consider the following constraint set for an LP problem. Sketch a graph in E^2 showing $\mathbf{a}_1, \mathbf{a}_2, \mathbf{a}_3$, the cone generated by these three vectors, and then add b to the graph. Show that $\mathbf{Ax} = \mathbf{b}, \mathbf{x} \ge \mathbf{0}$ has no solution but that $\mathbf{Ax} \le \mathbf{b}, \mathbf{x} \ge \mathbf{0}$ has feasible solutions:

$$x_2 + 2x_3 = 2$$
$$3x_1 + 2x_2 + 2x_3 = 1$$
$$x_1, x_2, x_3 \ge 0$$

9

LINEAR PROGRAMMING: SOLUTION METHODS

The modern methods for solving a large-scale integer program require the *optimization* and *reoptimization* of a usually long sequence of LP relaxation problems, which in turn are often solved by a variety of simplex-based methods. This chapter reviews three simplex-based methods that are the building blocks for solving integer programs. *The simplex method* provides the foundation for *optimizing* a long sequence of LP relaxations. *The simplex method for upper bounded variables* is used for reducing the problem size by *implicitly* handling the upper and lower bounds on variables (or *single-variable constraints*, more generally). *The dual simplex method* is most effective for *reoptimizing* the current optimum, after additional constraints are added, without resolving the augmented LP problem from scratch. *The revised simplex method* produces the same sequence of bases as the simplex method, but depends on updating $\mathbf{B}^{-1}$ (m columns) rather than on the entire simplex tableau (n columns).

9.1 LINEAR PROGRAMS IN CANONICAL FORM

Recall the following LP problem in standard form:

$$\text{Maximize } z = \sum_j c_j x_j \tag{9.1}$$

Applied Integer Programming: Modeling and Solution, By Der-San Chen, Robert G. Batson, and Yu Dang
Copyright © 2010 John Wiley & Sons, Inc.

$$\text{subject to} \sum_j a_{ij} x_j \leq b_i \quad (i = 1, 2, \ldots, m) \tag{9.2}$$

$$x_j \geq 0 \quad (j = 1, 2, \ldots, n) \tag{9.3}$$

where b_i ($i = 1, 2, \ldots, m$) can be a positive or negative number.

Because the simplex-based methods work on systems of linear equations rather than inequalities, the standard LP problem must be first converted to a system of equations. This can be accomplished by adding a nonnegative *slack variable* s_i to each inequality constraint (9.2) and transferring all the variable terms in the objective function (9.1) to the left-hand side of an equation and leaving the constant term on the right-hand side:

$$\text{Maximize} \quad z - \sum_j c_j x_j = 0 \tag{9.4}$$

$$\text{subject to} \quad \sum_j a_{ij} x_j + s_i = b_i \quad (i = 1, 2, \ldots, m) \tag{9.5}$$

$$x_j \geq 0 \quad (j = 1, 2, \ldots, n) \tag{9.6}$$

$$s_i \geq 0 \quad (i = 1, 2, \ldots, m) \tag{9.7}$$

Based on this new formulation, therefore, solving a linear program can be viewed as performing the following three tasks:

1. Find solutions to the augmented system of linear equations in (9.4) and (9.5).
2. Use the nonnegative conditions (9.6) and (9.7) to indicate and maintain the feasibility of a solution.
3. Maximize the objective function, which is rewritten as equation (9.4).

Note that (9.4) can also be viewed as a combination of two parts:

$$\text{maximize } z$$

$$\text{subject to } z - \sum_j c_j x_j = 0$$

Also note that the augmented system of equations (9.4)–(9.5) has a particular form of coefficient matrix called *canonical form*. In this form, a solution can be read immediately from the right-hand side of all equations: all $x_j = 0$, all $s_i = b_i$, and $z = 0$. Clearly, a *feasible* solution is readily available if *all* $b_i \geq 0$. Moreover, this LP canonical form, after detaching the coefficients from the variables, appears to be the so-called *simplex tableau*.

If there exists any $b_i < 0$, the system has *an infeasible solution* because it violates at least one of (9.6) and (9.7). In this case therefore, to obtain a starting basic *feasible* solution, a Phase I problem must be constructed and solved. The details of this procedure will be given in Section 9.3.

9.2 BASIC FEASIBLE SOLUTIONS AND REDUCED COSTS

9.2.1 Basic Feasible Solution

Definition 9.1 Given that a system $\mathbf{Ax} = \mathbf{b}$, where the number of solutions are infinite, and rank $(\mathbf{A}) = m$ $(m < n)$, a unique solution can be obtained by setting any $n - m$ variables to 0 and solving for the remaining system of m variables in m equations. Such a solution, if it exists, is called a *basic solution*. The variables that are set to 0 are called *nonbasic variables*, denoted by $\mathbf{x_N}$. The variables that are solved are called *basic variables*, denoted by $\mathbf{x_B}$. A basic solution that contains all nonnegative values is called a *basic feasible solution*. A basic solution that contains any *negative* component is called a *basic infeasible solution*. The $m \times m$ coefficient matrix associated with a given set of basic variables is called a *basis*, or *basis matrix*, and is denoted as $\mathbf{B}$.

Let $\mathbf{x} = (\mathbf{x_B}, \mathbf{x_N})^T$, $\mathbf{c} = (\mathbf{c_B}, \mathbf{c_N})^T$, and $\mathbf{A} = (\mathbf{B}, \mathbf{N})$. Then, the LP can be expressed by the following partitioned form:

$$\text{Maximize} \quad z = \mathbf{c_B^T x_B} + \mathbf{c_N^T x_N}$$
$$\text{subject to} \quad \mathbf{B x_B} + \mathbf{N x_N} = \mathbf{b}$$
$$\mathbf{x_B}, \mathbf{x_N} \geq \mathbf{0}$$

Example 9.1 Consider the following system of two equations in four unknowns (or variables):

$$x_1 + x_2 + x_3 = 6$$
$$2x_1 + x_2 + x_4 = 8$$

A basic solution to this system can be obtained by assigning 0 to *any* two variables and solving the remaining system of two equations in two variables. This system has a maximum of six basic solutions:

$$C_2^4 = \frac{4!}{2!(4-2)!} = 6$$

These six basic solutions are listed in Table 9.1. Note that basic solutions 3 and 4 are *infeasible* because one of their basic variables has a *negative* value while the remaining basic solutions are *feasible*.

For $\mathbf{x_B} = \begin{pmatrix} x_3 \\ x_4 \end{pmatrix}$, the basis $\mathbf{B} = \begin{pmatrix} 1 & 0 \\ 0 & 1 \end{pmatrix}$ and for $\mathbf{x_B} = \begin{pmatrix} x_1 \\ x_2 \end{pmatrix}$, the basis $\mathbf{B} = \begin{pmatrix} 1 & 1 \\ 2 & 1 \end{pmatrix}$

TABLE 9.1 Basic Solutions in Example 9.1

	Basic Solution					
	1	2	3	4	5	6
Nonbasic variables $\mathbf{x_N}$	$x_1 = 0,$ $x_2 = 0$	$x_1 = 0,$ $x_3 = 0$	$x_1 = 0,$ $x_4 = 0$	$x_2 = 0,$ $x_3 = 0$	$x_2 = 0,$ $x_4 = 0$	$x_3 = 0,$ $x_4 = 0$
Basic variables $\mathbf{x_B}$	$x_3 = 6,$ $x_4 = 8$	$x_2 = 6,$ $x_4 = 2$	$x_2 = 8,$ $x_3 = -2$	$x_1 = 6,$ $x_4 = -4$	$x_1 = 4,$ $x_3 = 2$	$x_1 = 2,$ $x_2 = 4$

In general, the number of basic solutions possible in a system of m equations in n variables is calculated by

$$C_m^n = \frac{n!}{m!(n-m)!}$$

We now formally define a canonical system of linear equations.

Definition 9.2   A system of linear equations is said to be in a *canonical form* if each equation contains a *basic variable* whose coefficient is 1 in that equation and whose coefficient in all other equations is 0.

Therefore, in a canonical system, every equation contains only one basic variable in the current basis whose value equals to the right-hand-side constant, and the rest of the variables are nonbasic with a value of 0. Thus, a basic solution can be obtained by letting each basic variable equal to the right-hand side of its respective equation and setting the nonbasic variables equal to zero.

Example 9.2   Consider the following LP in *standard form*:

$$\text{Max } z = 4x_1 + 3x_2$$

$$x_1 + x_2 \le 6$$

$$2x_1 + x_2 \le 8$$

$$x_1, x_2 \ge 0$$

After transferring the objective function and adding nonnegative slack variables s_1 and s_2 to equalize the inequality constraints, we obtain a *canonical* system:

$$\text{Max} \quad z - 4x_1 - 3x_2 = 0$$

$$x_1 + x_2 + s_1 = 6$$

$$2x_1 + x_2 + s_2 = 8$$

$$x_1, x_2 \ge 0$$

$$s_1, s_2 \ge 0$$

Let x_1 and x_2 be nonbasic variables, then the remaining variables s_1 and s_2 are basic. A basic solution is $x_1 = x_2 = 0$, $s_1 = 6$, $s_2 = 8$, and $z = 0$, which forms a solution vector (including z-value component) equal to the right-hand side of the equations. For simplicity, we detach the coefficients from the variables resulting in a simplex tableau given in Table 9.2.

Note that z can also be viewed as a basic variable for the *objective equation (row)*. Just like *the constraint equations* (or *rows*), this objective row is updated during

TABLE 9.2 Simplex Tableau for Example 9.2

Basic Variable	z	x_1	x_2	s_1	s_2	RHS
z	1	-4	-3	0	0	0
s_1	0	1	1	1	0	6
s_2	0	2	1	0	1	8

elementary row operations embedded in a simplex pivot. We shall refer to this equation as the *objective row* or *row 0* of the simplex tableau. The constraint rows are rows 1 through m.

9.2.2 Adjacent Basic Feasible Solution

Given a basic feasible solution in Table 9.2, one can generate an *adjacent* basic feasible solution by exchanging only one nonbasic variable for a basic variable while keeping all other variables unchanged (of course, their values typically change). This can be accomplished by the following steps: (a) determine a current nonbasic variable to become basic, (b) determine a current basic variable to become nonbasic, and (c) perform necessary row operations for exchanging the two variables determined in (a) and (b) and updating the values of the basic variables and z.

Definition 9.3 A nonbasic variable is called an *entering variable* if it is selected to become basic in the next basis. Its associated coefficient column is called a *pivot column*. A basic variable is called a *leaving variable* if it is selected to become nonbasic in the next basis. Its associated coefficient row is called a *pivot row*. The element that intersects a pivot column and a pivot row is called a *pivot or pivot element*. A *pivoting operation* is a sequence of elementary row operations that makes the pivot element "1" and all other elements "0" in the pivot column. Two basic feasible solutions are said to be *adjacent* if the set of their basic variables differ by only one basic variable. Geometrically, these two basic feasible solutions will correspond to two extreme points except in the case of degeneracy, in which case two or more bases correspond to the same extreme point.

In Table 9.2, suppose we select x_1 as the entering variable. This will make either s_1 or s_2 leave the basis. If s_1 is to leave, the coefficient of x_1 must be "1" in row 1 and "0" in both rows 2 and 0. To achieve this, we apply the following row operations: multiply row 1 by -2 and add to row 2, and then multiply row 1 by 4 and add to row 0, resulting in Table 9.3. Because s_2 is negative, the basic solution $(6, 0, 0, -4)$ is *infeasible*. So, selecting s_1 to exit was an improper choice.

Suppose x_1 is still the entering variable. If we let s_2, instead of s_1, leave the basis, then the coefficient of x_1 must be "1" in row 2 and "0" in rows 1 and 0. After pivoting, Table 9.4 is generated. Because all RHS values are nonnegative, the basic solution $(4, 0, 2, 0)$ is feasible with $z = 16$.

Note that the different results in Tables 9.3 and 9.4 indicate that a certain choice of leaving variable may cause the next basic solution *infeasible*. The question is how to

TABLE 9.3   Updated Simplex Tableau After Pivot 1

Basic Variable	z	x_1	x_2	s_1	s_2	RHS
z	1	0	1	4	0	24
x_1	0	1	1	1	0	6
s_2	0	0	-1	-2	1	-4

TABLE 9.4   Updated Simplex Tableau After Pivot 2

Basic Variable	z	x_1	x_2	s_1	s_2	RHS
z	1	0	-1	0	2	16
s_1	0	0	0.5	1	-0.5	2
x_1	0	1	0.5	0	0.5	4

ensure that the next basic solution will remain *feasible* if the current basic solution is *feasible*. To achieve this, we must choose a leaving variable such that its ratio of the right-hand side to the corresponding positive component in the pivot column is minimal. That is, from Table 9.1, we calculate the minimum ratio,

$$\min\{6/1, 8/2\} = 4$$

Because the minimum ratio 4 corresponds to s_2 row, s_2 must be the leaving variable to ensure the next basic solution is *feasible*.

Because *row* 0 of Table 9.4 still contains a negative value, which implies the objective value can be increased further, another simplex iteration is needed. Choosing x_2 as the entering variable and s_1 as the leaving variable, we obtain Table 9.5, which yields a *nonnegative* objective row and an optimum solution is found.

9.2.3   Reduced Costs

Examining Tables 9.2–9.5, we see that each row of a simplex tableau represents a basic variable written in terms of nonbasic variables. In Table 9.5 for instance, $x_2 + 2s_1 + s_2 = 4$ or $x_2 = 4 - 2s_1 + s_2$. In effect, the dimension of the original solution space is reduced from $(n + m)$, the number of basic and nonbasic variables,

TABLE 9.5   Optimal Tableau After Pivot 3

Basic Variable	z	x_1	x_2	s_1	s_2	RHS
z	1	0	0	2	1	20
x_2	0	0	1	2	-1	4
x_1	0	1	0	-1	1	2

to a *subspace* of dimension n equal to the number of nonbasic variables. Premultiplying B^{-1} on the equation $Bx_B + Nx_N = b$, we obtain

$$Ix_B + B^{-1}Nx_N = B^{-1}b$$

or $x_B = B^{-1}b - B^{-1}Nx_N$. Substituting it into $z = c_B^T x_B + c_N^T x_N$, we have

$$z = (c_B^T B^{-1} N - c_N^T) x_N$$

$$= c_B^T B^{-1} b - \sum_j (z_j - c_j) x_j, \ j \text{ nonbasic}$$

Definition 9.4 The subspace that contains only the nonbasic variables is referred to as a *reduced space*. The components of the objective row in a reduced space are called *reduced costs*, denoted by $\bar{c}$:

$$\bar{c}^T = (\bar{c}_B^T, \ \bar{c}_N^T) = (0^T, \ c_B^T B^{-1} N - c_N^T)$$

Note that the cost vector associated with the set of *basic variables* is a *null vector* 0.

9.3   THE SIMPLEX METHOD

The simplex method is an iterative algorithm consisting of the following steps:

1. *Initialization:* Find an initial basic solution that is feasible.
2. *Iteration:* Find a basic solution that is better, adjacent, and feasible.
3. *Optimality test:* Test if the current solution is optimal. If not, repeat step 2.

First, we shall address the iteration step, the core of the simplex method.

9.3.1   Better and Feasible Solution

The *iteration step* is aimed at finding a new basic solution that is *better, feasible,* and *adjacent* than a given feasible basic solution. When no better solution can be found, then an optimum solution has to be obtained. This iteration contains three basic steps: (1) determining the entering variable, (2) determining the leaving variable, and (3) pivoting on the pivot element for exchange of variables and updating the data in the tableau. A new basic solution will be *better* if an *entering variable* is properly chosen. A new basic solution will be *feasible* if a *leaving variable* is properly chosen. A new basic solution will be *adjacent* to the current one if only one basic variable from the old basic solution is exchanged with the old basic solution, which can be accomplished by a *pivot operation*. Consider the given simplex tableau in Table 9.6.

TABLE 9.6   The Simplex Tableau Immediately Before Pivoting

Basic Variable		$\mathbf{x_B}$				$\mathbf{x_N}$			RHS Solution
	z	$x_{B_1} \dots$	$x_{B_r} \dots$	x_{B_m}	$\dots$	$x_j \dots$	x_k	$\dots$	
z	1	0 $\dots$	0 $\dots$	0	$\dots$	$\bar{c}_j \dots$	$\bar{c}_k$	$\dots$	$\bar{b}_o$
x_{B_1} $\vdots$	0	1 $\dots$	0 $\dots$	0	$\dots$	$\bar{a}_{1j} \dots$	$\bar{a}_{1k}$	$\dots$	$\bar{b}_1$ $\vdots$
x_{B_r} $\vdots$	0	0 $\dots$	1 $\dots$	0	$\dots$	$\bar{a}_{rj} \dots$	$\bar{a}_{rk}$	$\dots$	$\bar{b}_r$ $\vdots$
x_{B_m}	0	0 $\dots$	0 $\dots$	1	$\dots$	$\bar{a}_{mj} \dots$	$\bar{a}_{mk}$	$\dots$	$\bar{b}_m$

The entering variable, denoted by x_k, is chosen among the current nonbasic variables, denoted by $x_j \in \mathbf{x_N}$ ($j = 1, 2, \dots, n$), such that it will improve the current objective value. This can be accomplished by selecting the x_k with the most (or any) negative *reduced cost*. Mathematically,

$$x_k = \{x_j \in \mathbf{x_N} : \min_j \bar{c}_j, \bar{c}_j < 0\} \qquad (9.8)$$

The coefficients column k associated with the entering variable x_k is the *pivot column*. The leaving variable, denoted by x_{B_r}, is chosen among the current basic variables, denoted by $x_{B_i} \in \mathbf{x_B}$ ($i = 1, 2, \dots, m$), such that it has a minimum (positive) ratio θ defined by

$$\theta = \frac{\bar{b}_r}{\bar{a}_{rk}} = \min_i \left\{ \frac{\bar{b}_i}{\bar{a}_{ik}}, \bar{a}_{ik} > 0 \right\} \qquad (9.9)$$

The coefficient row r associated with the leaving variable x_{B_r} is the *pivot row*. The rationale for selecting the minimum ratio is justified below.

Consider the simplex tableau given in Table 9.6. While holding ($n - 1$) nonbasic variables fixed at zero and increasing the nonbasic variable x_k from zero to positive, we will have the following system of equations for the objective function and constraints:

$$z + \bar{c}_k x_k = \bar{b}_o \quad \text{or} \quad z = \bar{b}_o - \bar{c}_k x_k \qquad (9.10)$$

$$\text{and} \quad x_{B_i} + \bar{a}_{ik} x_k = \bar{b}_i \quad \text{or} \quad x_{B_i} = \bar{b}_i - \bar{a}_{ik} x_k \quad (i = 1, 2, \dots, m) \qquad (9.11)$$

Because we want a *new* solution to remain *feasible*, meaning that the new x_{B_i} must be ≥ 0 for all i,

$$x_{B_i} = \bar{b}_i - \bar{a}_{ik} x_k \geq 0 \quad \text{for} \quad i = 1, 2, \dots, m \qquad (9.12)$$

If $\bar{a}_{ik} < 0$, then x_{B_i} increases as x_k increases and so x_{B_i} continues to be nonnegative without bound. If $\bar{a}_{ik} = 0$, then there is no change in x_{B_i} as x_k increases. Clearly, if $\bar{a}_{ik} \leq 0$ for all $i = 1, 2, \ldots, m$, then the problem has an *unbounded solution*. Moreover, if there exists *any nonpositive column j*, not necessarily the pivot column, with a negative component in row 0 (or $\bar{c}_j < 0$), then x_j can increase to infinity without making the new x_{B_i} negative.

If $\bar{a}_{ik} > 0$, then x_{B_i} decreases as x_k increases. To satisfy nonnegativity, x_k is increased until the first basic variable x_{B_i} drops to zero. Examining the system of inequalities in (9.12), it is clear that the first basic variable dropping to zero corresponds to the minimum of $\bar{b}_i / \bar{a}_{ik}$ for positive $\bar{a}_{ik}$. Mathematically, we can increase x_k until equal to the amount of θ determined by (9.9). From (9.10), the new objective value will be $(\bar{b}_o - \bar{c}_k \theta)$.

9.3.2 Updating Simplex Tableau by Pivoting

Now we address how to find an *adjacent* basic feasible solution. Given the entering variable x_k (pivot column k) and the leaving variable x_{B_r} (pivot row r), the *pivot element $\bar{a}_{rk}$* can be determined by the intersection of row r and column k. To update the simplex tableau in Table 9.6, the following *pivoting operation* is performed.

1. Divide row r by $\bar{a}_{rk}$.
2. For all $i \neq r$, update the ith row by adding to it $(-\bar{a}_{ik})$ times the new rth row.
3. Update row 0 by adding to it $\bar{c}_k$ times the new rth row.

After pivoting operation, we obtain Table 9.7. Note that the positions of x_{B_r} and x_k are exchanged. That is, x_{B_r} appears in the rows of basic variables and x_k in the columns of nonbasic variables.

TABLE 9.7 The Simplex Tableau After Pivoting

Basic Variable	z	x_{B_1} ...	x_{B_r}	... x_{B_m}	...	x_j	... x_k ...	RHS Solution
z	1	0 ...	$\dfrac{\bar{b}_r}{\bar{a}_{rk}}$	... 0	...	$\bar{c}_j - \dfrac{\bar{a}_{rj}}{\bar{a}_{rk}}\bar{c}_k$	... 0 ...	$\bar{b}_o - \dfrac{\bar{b}_r}{\bar{a}_{rk}}\bar{c}_k$
x_{B_1}	0	1 ...	$\dfrac{\bar{a}_{1k}}{\bar{a}_{rk}}$	... 0	...	$\bar{a}_{1j} - \dfrac{\bar{a}_{rj}}{\bar{a}_{rk}}\bar{a}_{1k}$	... 0 ...	$\bar{b}_1 - \dfrac{\bar{b}_r}{\bar{a}_{rk}}\bar{a}_{1k}$
$\vdots$	$\vdots$ $\vdots$	$\vdots$	$\vdots$		$\vdots$		$\vdots$	$\vdots$
x_k	0	0 ...	$\dfrac{1}{\bar{a}_{rk}}$	... 0	...	$\dfrac{\bar{a}_{rj}}{\bar{a}_{rk}}$	... 1 ...	$\dfrac{\bar{b}_r}{\bar{a}_{rk}}$
$\vdots$	$\vdots$ $\vdots$	$\vdots$	$\vdots$		$\vdots$		$\vdots$	$\vdots$
x_{B_m}	0	0 ...	$\dfrac{\bar{a}_{mk}}{\bar{a}_{rk}}$	... 1	...	$\bar{a}_{mj} - \dfrac{\bar{a}_{rj}}{\bar{a}_{rk}}\bar{a}_{mk}$	... 0 ...	$\bar{b}_m - \dfrac{\bar{b}_r}{\bar{a}_{rk}}\bar{a}_{mk}$

9.3.3   Optimality Test

An optimum solution is found if there exists no adjacent basic feasible solution that can improve the objective value. In a maximization problem, the optimality condition is satisfied if $\bar{c}_j \geq 0$ for all $j = 1, 2, \ldots, m$. If at optimality, there exists a nonbasic variable, say x_p, with $\bar{c}_p = 0$, then this variable can enter the basis to obtain an alternate optimum solution with the same objective value.

9.3.4   Initial Basic Feasible Solution

In the preceding section, we assume that an LP problem has all constraints in $\leq$ form and all $b_i \geq 0$. In this case, a basic feasible solution is naturally obtained after adding a nonnegative slack variables s_i to each constraint. However, if there is any $b_i < 0$ or any constraint in $\geq$ or $=$ form, then *artificial variables* are added to become *basic variables for a starting basis*. Unfortunately, this basic solution is infeasible to the original problem because of the presence of artificial variables with positive values. To obtain a *feasible* basic solution to the original problem, a phase I problem of the two-phase method is constructed to drive all artificial variables out of basis (and hence equal 0). The construction procedure is given below.

1. Convert each constraint so that the right-hand side is nonnegative. This requires that any constraint with a negative right-hand side be multiplied by -1. The resultant constraint has one of the three forms: $\leq, =,$ or $\geq$. If it is in $\leq$ form, then add a nonnegative *slack variable*; in $=$ form, add a nonnegative artificial variable; in $\geq$ form, subtract a nonnegative slack variable and add a nonnegative *artificial variable*.

2. Solve a phase I problem whose objective function is minimizing the sum of artificial variables subject to the same set of constraints. The sum of artificial variables is obtained by assigning a cost of 1 to each artificial variable and 0 to each of nonartificial variables.

Example 9.3   This example is extended by adding the following additional constraint to Example 9.2.

$$-2x_1 + x_2 \geq 2$$

Applying step 1, we obtain

$$-2x_1 + x_2 - s_3 + x^a = 2$$
$$s_3, xa \geq 0$$

Applying step 2, we minimize $z^a = x^a$,

$$\text{or}\quad \text{maximize} -z^a = -x^a$$
$$\text{or}\quad \text{maximize} -z^a + x^a = 0$$

Setting up a tableau format for phase I problem, we have the following tableau:

Basic Variable	$-z^a$	x_1	x_2	s_1	s_2	s_3	x^a	RHS
$-z^a$	1	0	0	0	0	0	1	0
s_1	0	1	1	1	0	0	0	6
s_2	0	2	1	0	1	0	0	8
x^a	0	-2	1	0	0	-1	1	2

Note that the above tableau is not yet in *canonical form* because the coefficient of x^a in row 0 is nonzero. To zero it out, we multiply row x^a by -1 and add the resultant row to row 0, resulting in the following tableau. Now the artificial variable x^a becomes a *basic* variable to the *transformed* problem.

Basic Variable	$-z^a$	x_1	x_2	s_1	s_2	s_3	x^a	RHS
$-z^a$	1	2	-1	0	0	1	0	-2
s_1	0	1	1	1	0	0	0	6
s_2	0	2	1	0	1	0	0	8
x^a	0	-2	1	0	0	-1	1	2

Let x_2 be the entering variable and x_a be the leaving variable. After pivoting, we have the following tableau.

Basic Variable	$-z^a$	x_1	x_2	s_1	s_2	s_3	x^a	RHS
$-z^a$	1	0	0	0	0	0	1	0
s_1	0	3	0	1	0	1	-1	4
s_2	0	4	0	0	1	1	-1	6
x_2	0	-2	1	0	0	-1	1	2

Because the artificial variable is driven out of basis and hence has a value of 0, we obtain a basic *feasible* solution for the *original* problem: $x_1 = 0$ and $x_2 = 2$.

Once a starting basic feasible solution is obtained, we proceed to the phase II problem to find an optimum solution using the original objective function and the last tableau of the phase I problem. To begin with, we must drop the columns associated with all the artificial variables, drop the objective row of phase I, and replace it with the original objective row. We obtain the following tableau.

Basic Variable	z	x_1	x_2	s_1	s_2	s_3	RHS
z	1	-4	-3	0	0	0	0
s_1	0	3	0	1	0	1	4
s_2	0	4	0	0	1	1	6
x_2	0	-2	1	0	0	-1	2

Note that this tableau is not yet in canonical form because in row 0 the coefficient of the basic variable x_2 is nonzero. To obtain a canonical form, we zero it out by multiplying row x_2 by 3 and adding the resultant row to the objective row.

Basic Variable	z	x_1	x_2	s_1	s_2	s_3	RHS
z	1	−10	0	0	0	−3	6
s_1	0	3	0	1	0	1	4
s_2	0	4	0	0	1	1	6
x_2	0	−2	1	0	0	−1	2

This tableau has a basic feasible solution $(0, 2, 4, 6, 0)$ but it is not optimal because the objective row contains negative components. Letting the entering variable be x_1 and the leaving variable be s_1, we have the following tableau.

Basic Variable	z	x_1	x_2	s_1	s_2	s_3	RHS
z	1	0	0	10/3	0	1/3	58/3
x_1	0	1	0	1/3	0	1/3	4/3
s_2	0	0	0	−4/3	1	−1/3	2/3
x_2	0	0	1	2/3	0	−1/3	14/3

Since all the components in the objective row are nonnegative, an optimum solution is found: $(4/3, 14/3, 0, 2/3, 0)$ with an objective value $z = 58/3$.

9.4 INTERPRETING THE SIMPLEX TABLEAU

9.4.1 Entire Simplex Tableau

Every simplex tableau provides information about the *current* basic feasible solution and its *n adjacent* basic feasible solutions. Geometrically, a basic feasible solution corresponds to an extreme point of the feasible region. Recall that a convex hull of $n + 1$ points is called a *simplex*, hence the name of the simplex method.

9.4.2 Rows of Simplex Tableau

Every *row* of a simplex tableau represents an *equation* with all variable terms on the left-hand side and a constant term on the right-hand side of the equality sign. Moreover, the coefficients of all but one basic variables are zero. Note that the objective function can also be expressed as an equation, $z - \sum \bar{c}_j x_j = 0$ (where x_j is nonbasic) with a new variable z being treated as a basic variable. This objective row is also referred to as row 0 and the remaining rows are rows 1 through m. The coefficient $\bar{c}_j = z_j - c_j$ is referred to as a *reduced cost* because it is a cost coefficient expressed in the *reduced space* of n nonbasic variables.

Note that each equation contains exactly one basic variable with coefficient equal to 1 and one or more nonbasic variables with coefficients of any values. Moreover, different equations have distinct basic variables.

9.4.3 Columns of Simplex Tableau

The right-hand side column of a simplex tableau contains the objective value and the m values of basic variables for the current basic feasible solution. Note that the values of all nonbasic variables are always 0 and do not appear in the tableau.

The left-hand side column associated with each *basic variable* always contains a unit column vector with a "1" corresponding to the basic variable and a "0" to each of the nonbasic variables, including row 0. The left-hand side column associated with each *nonbasic variable* provides information about the basic feasible solutions adjacent to the current one. The objective component of the left-hand side column predicts *the negative* rate of the change in objective function value if the corresponding nonbasic variable is increased by one unit. The remaining components of the same left-hand side column predict the amount of each resource to be consumed if a nonbasic variable is increased by one. In case of a negative component, the resource is added rather than consumed. Therefore, in the calculation of minimum ratio, the negative and zero components are excluded.

9.4.4 Pivot Column and Pivot Row

The negative of row-0 component in the pivot column represents the unit improvement in the objective value if the entering variable is increased by one unit. The ratio of a right-hand side to a positive component of the pivot column represents the maximum amount that the corresponding nonbasic variable can be increased without exceeding the resource availability on the right-hand side of the equation. To satisfy the limits of all resources, a minimum ratio must be used. Otherwise, the new solution will be infeasible, indicated by negative values on the right-hand side.

9.4.5 Predicting the New Objective Value Before Updating

Prior to updating a simplex tableau, the new objective value of the next tableau can be predicted by the following formula:

$$\text{New} \quad z = \text{current } z + \text{total improvement in the objective value}$$

$$= \text{current } z + (\text{unit improvement})(\text{amount of improvement})$$

$$\text{or,} \quad \hat{b}_o = \bar{b}_o - \bar{c}_k \theta$$

where $\hat{b}_o$ denotes the new objective value.

9.5 GEOMETRIC INTERPRETATION OF THE SIMPLEX METHOD

9.5.1 Basic Feasible Solution Versus Extreme Point

Recall that in Chapter 8, a point $\mathbf{x}$ in a polyhedron P (feasible region) is called an extreme point of P if it cannot be represented as a strict $(0 < \alpha < 1)$ convex combination of two distinct points in P. Here we will show that every extreme point corresponds one-for-one to a basic feasible solution in the absence of degeneracy. To illustrate this, we compare the extreme points in Figure 9.1 with the basic feasible solutions in Table 9.1 for Example 9.2. Figure 9.1 shows a feasible region (polyhedron) with four extreme points and two infeasible points outside the feasible region: $(6, 0)$ and $(0, 8)$.

Table 9.1 lists four *basic feasible solutions* and two *basic infeasible* solutions. Comparing Table 9.1 with Figure 9.1, we see that every basic feasible solution corresponds to an extreme point. Furthermore, every basic infeasible solution corresponds to a point *outside* the feasible region. The two-dimensional figure does not show the values of slack variables s_1 and s_2, whose values can be obtained by substituting the values of x_1 and x_2 into the respective equations. For example, extreme point $(6, 0)$ has $s_1 = 0$ and $s_2 = -4$ and extreme point $(0, 8)$ has $s_1 = -2$ and $s_2 = 0$. Both points are outside the feasible region and correspond to basic infeasible solutions.

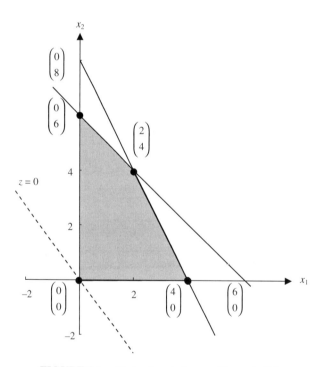

FIGURE 9.1 Six basic solutions to Example 9.2.

TABLE 9.8 Basic Solutions Associated with Figure 9.2

					Basic Solution					
	1	2	3	4	5	6	7	8	9	10
$\mathbf{x}_N$	$x_1=0,$ $x_2=0$	$x_1=0,$ $s_1=0$	$x_1=0,$ $s_2=0$	$x_1=0,$ $s_3=0$	$x_2=0,$ $s_1=0$	$x_2=0,$ $s_2=0$	$x_2=0,$ $s_3=0$	$s_1=0,$ $s_2=0$	$s_1=0,$ $s_3=0$	$s_2=0,$ $s_3=0$
$\mathbf{x}_B$	$s_1=6,$ $s_2=8,$ $s_3=4$	$x_2=6,$ $s_2=2,$ $s_3=4$	$x_2=8,$ $s_1=-2,$ $s_3=4$	No solution	$x_1=6,$ $s_2=-4,$ $s_3=-2$	$x_1=2,$ $s_1=4,$ $s_3=-2$	$x_1=4,$ $s_1=2,$ $s_2=0$	$x_1=2,$ $x_2=4,$ $s_3=2$	$x_1=4,$ $x_2=2,$ $s_2=-2$	$x_1=4,$ $x_2=0,$ $s_1=2$

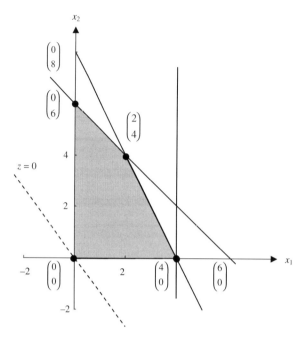

FIGURE 9.2   A degenerate solution created by $x_1 \leq 4$.

A degenerate solution has more than one basic feasible solution corresponding to an extreme point. To show this, we add the constraint $x_1 \leq 4$ to the problem in Example 9.2. Figure 9.2 shows the same feasible region with four extreme points, while Table 9.8 shows the five basic feasible solutions. Note that every basic feasible solution corresponds to an extreme point, except that the two *degenerate solutions* 7 and 10 correspond to the same extreme point $(4, 0)$. The only difference between the two is that they belong to different *bases*. In other words, they have two different sets of basic variables: the *zero-valued* variable x_2 is *basic* in solution 10, but *nonbasic* in solution 7.

9.5.2   Explanation of "Simplex Method" Nomenclature

Corresponding to *adjacent* basic feasible solutions, *adjacent* extreme points are hereby defined.

Definition 9.5   Two extreme points of a polyhedron X are said to be *adjacent* if they are joined by a line segment forming an edge of X.

Consider Figure 9.1. To the extreme point $(0, 0)$, for example, points $(4, 0)$ and $(0, 6)$ are its *adjacent extreme points*. To the extreme point $(4, 0)$, points $(2, 4)$ and $(0, 0)$ are its adjacent extreme points.

Recall that in Chapter 8, a specific class of the *bounded* polyhedron in $\mathbf{E}^n$, formed by all convex combinations of $n + 1$ linear independent vectors, is called a *simplex*.

The *simplex method* searches the feasible region for an optimum extreme point by sequentially examining a subset of simplexes comprising the boundary of the polyhedron. Each simplex is formed by the convex combination of the current extreme point, a basic feasible solution, and n adjacent extreme points. Each simplex iteration geometrically moves from the current extreme point to an adjacent extreme point along an edge of one of these simplexes.

Take Figure 9.3a for an example. There are four extreme points and hence four simplexes. The four extreme points are denoted by $x^1 = (0, 0)$, $x^2 = (0, 6)$, $x^3 = (2, 4)$, and $x^4 = (4, 0)$. The four simplexes are $S_1 = \{x^1, x^2, x^4\}$, $S_2 = \{x^2, x^1, x^3\}$, $S_3 = \{x^3, x^2, x^4\}$, and $S_4 = \{x^4, x^3, x^1\}$. Simplex S_1 is indicated by a shaded triangle in Figure 9.3a, and simplex S_2 by a shaded triangle in Figure 9.3b. Solving the LP problem in Example 9.2, the simplex method begins with S_1 and moves to S_4, and then to S_3 when x^3 is the optimum. The corresponding edges traveled are $[x^1, x^4]$ and $[x^4, x^3]$. An alternate sequence is $S_1 \rightarrow S_2 \rightarrow S_3$. Note that each simplex tableau contains the information about the current extreme point (basic feasible solution) and its adjacent extreme points. The pivot procedure decides whether to exchange the current extreme point for an adjacent extreme point and determines the coordinates (values) of the next extreme point.

9.5.3 Identifying an Extreme Ray in a Simplex Tableau

Recalling from Chapter 8, we know that an *unbounded* polyhedron P of an LP problem can be described in terms of *extreme points* and *extreme rays*. An extreme ray of P is defined as $x = x_0 + d\lambda$, $\lambda \geq 0$, where x_0 is the *root* or *vertex* of the extreme ray, d is the *extreme direction*, and λ is a nonnegative scalar, unbounded above. Note that x_0 is an extreme point of P. In Chapter 8, we learned how to calculate algebraically an extreme direction. Here we will show how to identify an extreme ray and extreme direction from a given simplex tableau. First, we use a simple graphical example and then derive the algebraic relationship.

Example 9.4 (Extreme Ray) Consider the following LP problem:

$$
\begin{aligned}
\text{Maximize} \quad & z = 4x_1 + 3x_2 \\
\text{subject to} \quad & -x_1 + x_2 \leq 4 \\
& x_1 - 2x_2 \leq 2 \\
& x_1, x_2 \geq 0
\end{aligned}
$$

Solving it by the graphical method, we obtain Figure 9.4 in which the extreme ray is expressed by

$$
\begin{pmatrix} 2 \\ 0 \end{pmatrix} + \lambda \begin{pmatrix} 2 \\ 1 \end{pmatrix}, \quad \lambda = 0
$$

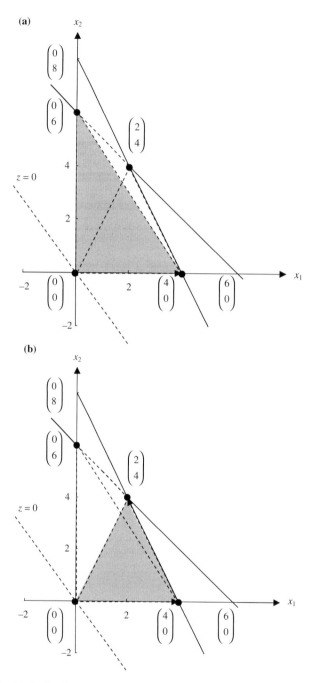

FIGURE 9.3 (a) A simplex associated with Example 9.2. (b) Another simplex associated with Example 9.2.

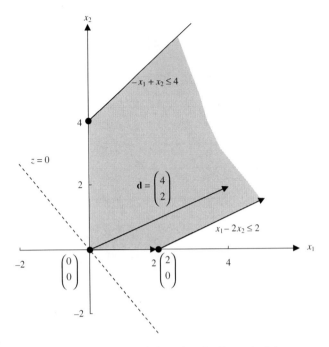

FIGURE 9.4 Feasible region for Example 9.4.

Alternatively, solving the problem by the simplex method, we obtain the following two simplex tableaus:

Basic Variable	z	x_1	x_2	x_3	x_4	RHS
z	1	−4	−3	0	0	0
x_3	0	−1	1	1	0	4
x_4	0	1	−2	0	1	2

Basic Variable	z	x_1	x_2	x_3	x_4	RHS
z	1	0	−11	0	4	8
x_3	0	0	−1	1	1	6
x_1	0	1	−2	0	1	2

The last simplex tableau indicates that the problem has an unbounded solution because x_2 column contains all nonpositive values. This condition implies that the variable x_2 can increase its value to ∞ without making the new RHS negative (or violating any constraints). This simplex tableau indicates that the current basic feasible solution is

$$\mathbf{x_B} = (x_3, x_1)^T = (6, 2)^T$$

$$\text{and} \quad \mathbf{x_N} = (x_2, x_4)^T = (0, 0)^T$$

Rearranging in order of variables, we have the following *root* of an extreme ray

$$\mathbf{x_0} = (x_1, x_2, x_3, x_4)^T = (2, 0, 6, 0)^T$$

Also from this simplex tableau, we have the pivot column,

$$\bar{\mathbf{a}}_2 = \begin{pmatrix} -1 \\ -2 \end{pmatrix}$$

Given that x_2 is the entering variable, the following condition must be satisfied to ensure that the next solution is feasible:

$$\begin{pmatrix} 2 \\ 0 \\ 6 \\ 0 \end{pmatrix} - \begin{pmatrix} -2 \\ 0 \\ -1 \\ 0 \end{pmatrix} x_2 \geq \begin{pmatrix} 0 \\ 0 \\ 0 \\ 0 \end{pmatrix}$$

where $x_2 \geq 0$. The extreme direction of this ray is $\mathbf{d}^T = (2, 1, 1, 0)$, as can be seen in (9.16).

Now we show the general relationship algebraically. Recall that for a maximization problem, if we have a basic feasible solution with $\bar{c}_k < 0$ for some nonbasic variable x_k and $\bar{a}_{ik} \leq 0$ for all $i = 1, 2, \ldots, m$, then the problem has an unbounded solution. This has been shown in Section 9.4.1 using the system of equations in (9.13), which is restated here:

$$x_{B_i} = \bar{b}_i - \bar{a}_{ik} x_k \quad (i = 1, 2, \ldots, m)$$

$$\text{or in matrix form,} \quad \mathbf{x_B} = \bar{\mathbf{b}} - \bar{\mathbf{a}}_k x_k$$

Because the coefficient of the entering variable x_k is 1, the vector of the next nonbasic variables must be

$$\mathbf{x_N} = \begin{pmatrix} 0 \\ \vdots \\ 1 \\ \vdots \\ 0 \end{pmatrix} = \mathbf{e}_k$$

Putting $\mathbf{x_B}$ and x_N together, we obtain

$$\mathbf{x} = \begin{pmatrix} \mathbf{x_B} \\ x_N \end{pmatrix} = \begin{pmatrix} \bar{\mathbf{b}} - \bar{\mathbf{a}}_k x_k \\ \mathbf{e}_k \end{pmatrix} = \begin{pmatrix} \bar{\mathbf{b}} \\ \mathbf{0} \end{pmatrix} + \begin{pmatrix} -\bar{\mathbf{a}}_k \\ \mathbf{e}_k \end{pmatrix} x_k \qquad (9.13)$$

Comparing (9.13) with the definition of *extreme ray* given below,

$$\mathbf{x} = \mathbf{x}_0 + \lambda \mathbf{d}, \quad \lambda \geq 0 \qquad (9.14)$$

with the *root*, the current extreme point,

$$\mathbf{x}_0 = \begin{pmatrix} \bar{\mathbf{b}} \\ \mathbf{0} \end{pmatrix} \qquad (9.15)$$

the *extreme (ray) direction*,

$$\mathbf{d} = \begin{pmatrix} -\bar{\mathbf{a}}_k \\ \mathbf{e}_k \end{pmatrix} \qquad (9.16)$$

and the unbounded step size

$$\lambda = x_k$$

9.6 THE SIMPLEX METHOD FOR UPPER BOUNDED VARIABLES

Quite often in practice, a variable has a lower bound other than 0 and has a finite upper bound other than infinity. Let l_j and u_j denote the lower and upper bounds of variable x_j, respectively. Then, we have the following lower bound and upper bound constraints as follows:

$$x_j \geq l_j \qquad (9.17)$$

$$x_j \leq u_j \qquad (9.18)$$

The lower bound constraint (9.17) can be easily handled by a variable substitution. Let a new variable $x'_j = x_j - l_j$ for any lower bound constraint. We can then obtain a new problem containing new variables x'_j for all lower bound constraints, each with standard lower bound zero.

However, the upper bound constraint (9.18) cannot be handled similarly because despite substituting new variables $x''_j = u_j - x_j$, we still require upper bound constraints for the new variables. Nevertheless, the upper bound constraints can be handled *implicitly* by a modification of the ordinary simplex method without explicitly treating them as ordinary constraints. As a result, the number of constraints in an upper bounded linear program can be greatly reduced.

Now we are ready to describe the *simplex method for upper bounded variables*, also known as the *upper bound technique*. The basic concept is to allow any upper bound variable x_j to be nonbasic if $x_j = 0$ (as usual) *or* if $x_j = u_j$. To attain this, we use the following rules: For each upper bounded variable with $x_j = u_j$, we define a new variable $\bar{x}_j$ by the relationship $x_j + \bar{x}_j = u_j$ or $\bar{x}_j = u_j - x_j$. Note that if $x_j = 0$, then $\bar{x}_j = u_j$, whereas if $x_j = u_j$, then $\bar{x}_j = 0$. Whenever we want x_j to equal its upper bound u_j, we simply replace it with $u_j - \bar{x}_j$.

Suppose a basic feasible solution is available and we are solving a maximization problem. At each simplex iteration, we choose the entering variable x_k as in the ordinary simplex algorithm. There are three possible cases that limit the amount by which x_k can increase:

Case 1: x_k cannot exceed the minimum ratio θ as usual. Otherwise, it will cause one or more current basic variables to become negative.

Case 2: x_k cannot exceed the amount by which it will cause one or more current basic variables to exceed its upper bound. We shall denote this amount as θ'.

Case 3: x_k cannot exceed its upper bound u_k.

Any increase of x_k must be within these three limits (i.e., the minimum of θ, θ', and u_k). The simplex algorithm for upper bounded variables for a maximization problem is as follows:

Step 0 (Initialization). Find a starting basic feasible solution as in the ordinary simplex method. Introduce a new variable $\bar{x}_j$ for each upper bound constraint, $x_j \leq u_j$, such that $x_j + \bar{x}_j = u_j$, where u_j is a constant.

Step 1 (Optimality Test). Check if the usual optimality condition is satisfied. If yes, an optimum solution is found; otherwise, go to next step.

Step 2 (Entering Variable). Select the entering variable x_k as in the ordinary simplex method.

Step 3 (Leaving Variable and Pivoting). Compute θ, θ', and Δ as follows:

$$\theta = \min_i \left\{ \frac{\bar{b}_i}{\bar{a}_{ik}}, \bar{a}_{ik} > 0 \right\}$$

$$\theta' = \min_i \left\{ \frac{u_i - \bar{b}_i}{-\bar{a}_{ik}}, \bar{a}_{ik} < 0 \right\}$$

$$\Delta = \min\{\theta, \theta', u_k\}$$

There are three cases of Δ: (1) If $\Delta = \theta$, then determine the leaving variable x_{B_r} and perform the ordinary pivoting. (2) If $\Delta = \theta'$, then replace the leaving variable x_{B_r} with $u_{B_r} - \bar{x}_{B_r}$ in row r and the "label" for x_{B_r} with $\bar{x}_{B_r}$ and perform the ordinary pivoting. (3) If $\Delta = u_k$, then replace the entering variable x_k with $u_k - \bar{x}_k$ in each row of the tableau, and x_k with $\bar{x}_k$ in the "label" row. In any case, go to step 1 for an optimality test.

Example 9.5 (Bounded Variables) Consider the following LP problem

$$\text{Maximize} \quad z = 4x_1 + 3x_2$$

$$\text{subject to} \quad x_1 + x_2 \leq 6$$

$$2x_1 + x_2 \leq 8$$

$$x_1 \geq 1$$

$$1 \leq x_2 \leq 3$$

Assuming $x_1' = x_1 - 1$, $x_2' = x_2 - 1$, the problem can be transformed into

$$\text{Maximize} \quad z = 4x_1' + 3x_2' + 7$$

$$\text{subject to} \quad x_1' + x_2' + s_1 = 4$$

$$2x_1' + x_2' + s_2 = 5$$

$$x_2' \leq 2$$

$$x_1', x_2' \geq 0$$

Let $x_2' + \bar{x}'_2 = 2$. The starting basis consists of (s_1, s_2) with the initial tableau as follows:

Basic Variable	z	x_1'	x_2'	s_1	s_2	RHS
z	1	-4	-3	0	0	7
s_1	0	1	1	1	0	4
s_2	0	2	1	0	1	5

Clearly, the optimality condition is not satisfied. Select x_1' as the entering variable. Then, we have

$$\theta = \min\left\{\frac{4}{1}, \frac{5}{2}\right\} = 2.5$$

Note that θ' does not exist in this case since both $\bar{a}_{11}$ and $\bar{a}_{21}$ are nonnegative and that x_1' has no upper bound. Thus, $\Delta = \theta = 2.5$, which makes s_2 the leaving variable. After pivoting, we obtain the following updated tableau.

Basic Variable	z	x_1'	x_2'	s_1	s_2	RHS
z	1	0	-1	0	2	17
s_1	0	0	0.5	1	-0.5	1.5
x_1'	0	1	0.5	0	0.5	2.5

The optimality condition still does not hold, so choose x_2' to be the entering variable. Compute

$$\theta = \min\left\{\frac{1.5}{0.5}, \frac{2.5}{0.5}\right\} = 3$$

Note that θ' does not exist. Since x_2' has upper bound 2, $\Delta = \min\{\theta = 3, u_2' = 2\} = 2$. Replacing x_2' with $2 - \bar{x}'_2$, we obtain the following tableau:

Basic Variable	z	x_1'	$\bar{x}'_2$	s_1	s_2	RHS
z	1	0	1	0	2	19
s_1	0	0	-0.5	1	-0.5	0.5
x_1'	0	1	-0.5	0	0.5	1.5

Now the optimality condition is satisfied; hence, the optimal solution to the transformed problem is $x_1' = 1.5$, $\bar{x}'_2 = 0$ (or $x_2' = 2-0 = 2$), and $z = 19$. Transforming back to the original problem using the relations $x_1' = x_1 - 1$ and $x_2' = x_2 - 1$, we have an optimal solution to the original problem: $x_1 = 2.5$, $x_2 = 3$, and $z = 19$.

Handling lower bounded variables by substitution and upper bounded variables by this method greatly increase the efficiencies for solving LP problems. To illustrate this, suppose we are solving an LP with 100 bounded variables with 10 other constraints. If we use the ordinary simplex method, the size of basis for each tableau would be $210 \times 210 \, (=44,100)$. If we use these two handling techniques, the size would be only 10×10 or 100. Moreover, solving an integer program by the branch-and-bound method (to be covered in Chapter 11) mainly contains two branches using lower and upper bounded variables. The savings in computation are evident.

9.7 THE DUAL SIMPLEX METHOD

There are three uses of the dual simplex method: (1) finding a new LP optimum *after* one or more constraints are added to the current LP optimum, (2) finding a new LP optimum after changing the right-hand side of constraints, and (3) solving an ordinary linear program.

For cases 1 and 2, the addition of constraints or change of the right-hand side may cause the current basic solution to become *infeasible*. In other words, the augmented simplex tableau may contain *negative* values on the right-hand side while the objective row remains *nonnegative* (dual feasible) in a maximization problem. These are the typical *starting conditions* for the dual simplex method. For case 3 where the objective row contains some negative values, we can augment a *big-M artificial constraint* to the original simplex tableau and perform row operations to obtain a canonical form for a starting basis. See note 9.7 for details.

For solving an integer program, the first use is the most important for efficient *reoptimization* because the dual simplex method is applied within the IP algorithms such as the branch-and-bound, cutting plane, and branch-and-cut. The dual simplex algorithm for a maximization problem is described below.

Step 0 (Initialization). Obtain a starting dual feasible solution. In the ordinary simplex tableau, this implies that *all* components of the objective row are *nonnegative*, or the updated values $\bar{\mathbf{c}} \geq \mathbf{0}$. Initially, we construct a *basic* solution with only *slack variables* as basic variables (no artificial variables are ever needed). This may cause some right-hand side values to become negative. In the initial simplex tableau, $\bar{\mathbf{b}}$ may be equal to $\mathbf{b}$ where some components are negative.

Step 1 (Optimality Test). Check if $\bar{\mathbf{b}} \geq \mathbf{0}$. If yes, the current solution is optimal. Otherwise, go to next step.

Step 2 (Leaving Variable). Determine the leaving variable x_r by selecting a pivot row r with the most negative value on the right-hand side, that is, $\bar{b}_r = \min_i \{\bar{b}_i : \bar{b}_i < 0\}$.

Step 3 (Infeasibility Test). If $\bar{a}_{rj} \geq 0$ for all j, the given problem has no feasible solution. Otherwise, go to next step.

Step 4 (Entering Variable). Determine the entering variable by selecting the pivot column k based on the minimum ratio test:

$$\frac{\bar{c}_k}{-\bar{a}_{rk}} = \min_j\left\{\frac{\bar{c}_j}{-\bar{a}_{rj}}, \bar{a}_{rj} < 0\right\}$$

where $\bar{c}_j$ is the jth component of $\bar{c}_N$. Note that ties in entering variable are broken arbitrarily and that if the entering variable rule cannot be applied (e.g., $\bar{a}_{rj} \leq 0$ for all $j = 1, \ldots, n$), then the dual is unbounded and the primal is infeasible.

Step 5 (Pivoting). Update the current simplex tableau by pivoting on the pivot element $\bar{a}_{rk}$. Return to step 1.

Example 9.6 (Dual Simplex)   Consider the following constraint that is added after an optimum solution is found for Example 9.2: $3x_1 + 2x_2 \leq 12$. We wanted to find a new optimum using the dual simplex method. Recall the current optimum tableau (Table 9.5) below:

Basic Variable	z	x_1	x_2	s_1	s_2	RHS
z	1	0	0	2	1	20
x_2	0	0	1	2	−1	4
x_1	0	1	0	−1	1	2

Appending the additional constraint in equation form after introducing a slack variable s_3, we obtain the following tableau:

Basic Variable	z	x_1	x_2	s_1	s_2	s_3	RHS
z	1	0	0	2	1	0	20
x_2	0	0	1	2	−1	0	4
x_1	0	1	0	−1	1	0	2
s_3	0	3	2	0	0	1	12

Note that this tableau does not have a canonical form and hence is not a simplex tableau. To obtain a canonical form, we add to s_3 row (-2) multiple of x_2 row and (-3) multiple of x_1 row.

Basic Variable	z	x_1	x_2	s_1	s_2	s_3	RHS
z	1	0	0	2	0	0	20
x_2	0	0	1	2	−1	0	4
x_1	0	1	0	−1	1	0	2
s_3	0	0	0	−1	−1	1	−2

Because $\bar{\mathbf{c}} \geq \mathbf{0}$ and $\bar{\mathbf{b}}$ has a negative component, we have a starting condition for the dual simplex method. Let s_3 be the leaving variable. Compute $\min\{2/1, 0/1\}$, choose s_2 as the entering variable, and perform pivoting.

Basic Variable	z	x_1	x_2	s_1	s_2	s_3	RHS
z	1	0	0	2	0	0	20
x_2	0	0	1	3	0	0	6
x_1	0	1	0	-2	0	0	0
s_2	0	0	0	1	1	1	2

Because all $\bar{b}_i \geq 0$, we obtain an optimum solution $(0, 6, 0, 2)$ with $z = 20$.

9.8 THE REVISED SIMPLEX METHOD

Recall the LP problem in partitioned form

$$\text{Maximize} \quad z = \mathbf{c}_B^T \mathbf{x}_B + \mathbf{c}_N^T \mathbf{x}_N$$

$$\text{subject to} \quad \mathbf{B}\mathbf{x}_B + \mathbf{N}\mathbf{x}_N = \mathbf{b}$$

$$\mathbf{x}_B, \mathbf{x}_N \geq \mathbf{0}$$

where $\mathbf{x}_B$ and $\mathbf{x}_N$, respectively, denote vectors of basic and nonbasic variables; $\mathbf{c}_B$ and $\mathbf{c}_N$, respectively, are associated objective coefficients; and $\mathbf{B}$ and $\mathbf{N}$, respectively, are coefficient matrices associated with the constraints.

Multiplying the equality constraints by $\mathbf{B}^{-1}$, we obtain

$$\mathbf{I}\mathbf{x}_B + \mathbf{B}^{-1}\mathbf{N}\mathbf{x}_N = \mathbf{B}^{-1}\mathbf{b}$$

Writing the objective function in terms of $\mathbf{x}_N$, we obtain

$$\text{Maximize } z = (-\mathbf{c}_B^T \mathbf{B}^{-1}\mathbf{N} + \mathbf{c}_N^T)\mathbf{x}_N + \mathbf{c}_B^T \mathbf{B}^{-1}\mathbf{b}$$

Transferring the variable term to the left-hand side of the objective row and combining it with the constraints, we obtain the following "ordinary" simplex tableau:

z	$\mathbf{x}_B$	$\mathbf{x}_N$	RHS
1	$\mathbf{0}^T$	$\mathbf{c}_B^T \mathbf{B}^{-1}\mathbf{N} - \mathbf{c}_N^T$	$\mathbf{c}_B^T \mathbf{B}^{-1}\mathbf{b}$
0	$\mathbf{I}$	$\mathbf{B}^{-1}\mathbf{N}$	$\mathbf{B}^{-1}\mathbf{b}$

From the foregoing ordinary simplex tableau, we observe the following notes:

1. Once the set of basic variables x_B is specified, the corresponding simplex tableau can be calculated directly from the original data.
2. There is no need to update and store the coefficient matrix associated with the basic variables because that matrix is always identity $\mathbf{I}$, and the reduced cost associated with each basic variables is always zero.
3. All other entries of the ordinary simplex tableau are characterized by pre-multiplying the original data by $\mathbf{B}^{-1}$. Therefore, only $\mathbf{B}^{-1}$ is required to be updated.
4. The reduced cost vector $\bar{\mathbf{c}}_N^T$ associated with the nonbasic variables is calculated by $\bar{\mathbf{c}}_N^T = \mathbf{c}_B^T \mathbf{B}^{-1} \mathbf{N} - \mathbf{c}_N^T$, whose jth entry $\bar{c}_j = \mathbf{c}_B^T \mathbf{B}^{-1} \mathbf{a}_j - c_j$, where column $\mathbf{a}_j \in \mathbf{N}$.
5. The dual solution $\mathbf{u}^T$ is updated by $\mathbf{u}^T = \mathbf{c}_B^T \mathbf{B}^{-1}$, which requires to be computed only once for each simplex iteration.
6. The updated coefficient columns associated with nonbasic variables are $\mathbf{B}^{-1}\mathbf{N}$, whose jth column is $\bar{\mathbf{a}}_j = \mathbf{B}^{-1}\mathbf{a}_j$.
7. The primal solution is computed by $\mathbf{x}_B = \mathbf{B}^{-1}\mathbf{b} = \bar{\mathbf{b}}$, $\mathbf{x}_N = \mathbf{0}$, and $z = \mathbf{c}_B^T \mathbf{B}^{-1} \mathbf{b} = \mathbf{c}_B^T \bar{\mathbf{b}}$.

Based on the foregoing results, the ordinary simplex tableau can be further simplified to the *revised simplex tableau* as shown below.

$\mathbf{c}_B^T \mathbf{B}^{-1} = \mathbf{u}^T$	$\mathbf{c}_B^T \mathbf{B}^{-1} \mathbf{b} = z$
$\mathbf{B}^{-1}$	$\mathbf{B}^{-1}\mathbf{b} = \bar{\mathbf{b}}$

Using the above revised simplex tableau, the revised simplex algorithm for the maximization problem may now be described as follows:

Step 0 (Initialization). Find an initial revised simplex tableau using slack variables (and/or artificial variables, if needed). In the presence of artificial variables, the two-phase method is applied as usual to obtain a starting basic feasible solution. Initially, $\mathbf{B} = \mathbf{B}^{-1} = \mathbf{I}$, $\mathbf{u}^T = \mathbf{c}_B^T \mathbf{B}^{-1} = \mathbf{c}_B^T$, $\mathbf{B}^{-1}\mathbf{b} = \mathbf{b}$, and $z = \mathbf{c}_B^T \mathbf{B}^{-1} \mathbf{b} = \mathbf{c}_B^T \mathbf{b}$.

Step 1 (Pivot Column). For all nonbasic j and $\mathbf{a}_j \in \mathbf{N} = \mathbf{A} \backslash \mathbf{B}$, compute

$$\bar{\mathbf{a}}_j = \mathbf{B}^{-1}\mathbf{a}_j$$

$$\bar{c}_j = \mathbf{u}^T \bar{\mathbf{a}}_j - c_j$$

Determine the entering variable x_k by

$$k = \{j : \min_j(\bar{c}_j : \bar{c}_j < 0)\}$$

and the pivot column

$$\begin{pmatrix} \bar{c}_k \\ \bar{\mathbf{a}}_k \end{pmatrix}$$

Step 2 (Optimality). A finite maximum solution has been found if $\bar{c}_k \geq 0$ and at least one entry of $\bar{\mathbf{a}}_k$ is positive. An unbounded maximum exists if $\bar{c}_k \geq 0$ and $\bar{\mathbf{a}}_k \leq \mathbf{0}$. Otherwise, go to step 3.

Step 3 (Pivot Row). Append pivot column k and determine the leaving variable x_{B_r} or pivot row r by

$$r = \left\{ i : \min_i \left(\frac{\bar{b}_i}{\bar{a}_{ik}}, \; \bar{a}_{ik} > 0 \right) \right\}$$

Go to step 4.

Step 4 (Pivoting). Update the revised simplex tableau by pivoting at $\bar{a}_{rk}$. Go to step 1.

The revised simplex method is actually the version of the simplex algorithm most implemented in software. It is also particularly useful in the branch-and-price algorithm for solving MIPs (Chapter 13). In the standard simplex, the most computation time would be due to updating every column in the simplex tableau with each iteration. Even though this assures the availability of all the data needed for the next pivot (pivot position not yet determined), we actually end up using only one column to make the decision on which variable will exit the basis. An example will illustrate the efficiency of this approach.

Example 9.7 Solve the following problem by the revised simplex method.

$$
\begin{aligned}
\text{Maximize} \qquad & 4x_1 + 3x_2 + x_3 + 7x_4 + 6x_5 \\
\text{subject to} \qquad & x_1 + 2x_2 + 3x_3 + x_4 - 3x_5 \leq 9 \\
& 2x_1 - x_2 + 2x_3 + 2x_4 + x_5 \leq 10 \\
& -3x_1 + 2x_2 + x_3 - x_4 + 2x_5 \leq 11 \\
& \mathbf{x} \geq \mathbf{0}
\end{aligned}
$$

Adding slack variables, we have

$$\text{Maximize} \quad z = 4x_1 + 3x_2 + x_3 + 7x_4 + 6x_5$$

$$\text{subject to} \quad x_1 + 2x_2 + 3x_3 + x_4 - 3x_5 + s_1 = 9$$

$$2x_1 - x_2 + 2x_3 + 2x_4 + x_5 + s_2 = 10$$

$$-3x_1 + 2x_2 + x_3 - x_4 + 2x_5 + s_3 = 11$$

$$\mathbf{x} \geq \mathbf{0}, \mathbf{s} \geq \mathbf{0}$$

Step 0 (Initialization)

$$\mathbf{x_B} = (s_1, s_2, s_3)^T, \mathbf{x_N} = (x_1, x_2, x_3, x_4, x_5)^T, \mathbf{c_B^T} = (0, 0, 0), \mathbf{c_N^T} = (4, 3, 1, 7, 6),$$
$$\mathbf{b} = (9, 10, 11)^T, \mathbf{B} = \mathbf{B}^{-1} = \mathbf{I}, \mathbf{u}^T = \mathbf{c_B^T} \mathbf{B}^{-1} = (0, 0, 0),$$
$$\bar{\mathbf{b}} = \mathbf{B}^{-1} \mathbf{b} = (9, 10, 11)^T, z = 0$$

Iteration 1

Compute $\bar{\mathbf{c}}_N^T = \mathbf{c_B^T} \mathbf{B}^{-1} \mathbf{N} - \mathbf{c_N^T} = (-4, -3, -1, -7, -6)$. We select x_4 as the entering variable. Then the pivoting column 4 is calculated by

$$\mathbf{B}^{-1} \mathbf{a}_4 = \begin{bmatrix} 1 & 0 & 0 \\ 0 & 1 & 0 \\ 0 & 0 & 1 \end{bmatrix} \begin{bmatrix} 1 \\ 2 \\ -1 \end{bmatrix} = \begin{bmatrix} 1 \\ 2 \\ -1 \end{bmatrix}, \bar{c}_4 = \mathbf{u}^T \bar{\mathbf{a}}_4 - c_4 = -7$$

Add the pivot column to the right of the revised simplex tableau, and s_2 becomes the leaving variable (why?).

z	0	0	0	0		−7
s_1	1	0	0	9		1
s_2	0	1	0	10		2
s_3	0	0	1	11		−1

After pivoting, the new tableau becomes

z	0	3.5	0	35		0
s_1	1	−0.5	0	4		0
x_4	0	0.5	0	5		1
s_3	0	0.5	1	16		0

Iteration 2

$\mathbf{u}^T = (0, 3.5, 0)$. Therefore, $\bar{\mathbf{c}}_N^T = \mathbf{c}_B^T \mathbf{B}^{-1} \mathbf{N} - \mathbf{c}_N^T = (3, -6.5, 6, 0, -2.5, 0, 3.5, 0)$. Then, x_2 is selected as the entering variable.

$$\mathbf{B}^{-1}\mathbf{a}_2 = \begin{bmatrix} 1 & -0.5 & 0 \\ 0 & 0.5 & 0 \\ 0 & 0.5 & 1 \end{bmatrix} \begin{bmatrix} 2 \\ -1 \\ 2 \end{bmatrix} = \begin{bmatrix} 2.5 \\ -0.5 \\ 1.5 \end{bmatrix}, \quad \bar{c}_2 = \mathbf{u}^T\bar{\mathbf{a}}_2 - c_2 = -6.5$$

Add the pivot column to the right of the revised simplex tableau and s_1 becomes the leaving variable.

z	0	3.5	0	35	-6.5
s_1	1	-0.5	0	4	2.5
x_4	0	0.5	0	5	-0.5
s_3	0	0.5	1	16	1.5

After pivoting, the new tableau becomes

z	2.6	2.2	0	45.4	0
x_2	0.4	-0.2	0	1.6	1
x_4	0.2	0.4	0	5.8	0
s_3	-0.6	0.8	1	13.6	0

Iteration 3

$\mathbf{u}^T = (2.6, 2.2, 0)$. Therefore, $\bar{\mathbf{c}}_N^T = \mathbf{c}_B^T \mathbf{B}^{-1} \mathbf{N} - \mathbf{c}_N^T = (3, 0, 11.2, 0, -11.6, 2.6, 2.2, 0)$. Then x_5 is selected as the entering variable.

$$\mathbf{B}^{-1}\mathbf{a}_5 = \begin{bmatrix} 0.4 & -0.2 & 0 \\ 0.2 & 0.4 & 0 \\ -0.6 & 0.8 & 1 \end{bmatrix} \begin{bmatrix} -3 \\ 1 \\ 2 \end{bmatrix} = \begin{bmatrix} -1.4 \\ -0.2 \\ 4.6 \end{bmatrix}, \quad \bar{c}_j = -11.6$$

Add the column to the right of the revised simplex tableau and s_3 becomes the leaving variable.

z	2.6	2.2	0	45.4	-11.6
x_2	0.4	-0.2	0	1.6	-1.4
x_4	0.2	0.4	0	5.8	-0.2
s_3	-0.6	0.8	1	13.6	4.6

After pivoting, the new tableau becomes

z	1.09	4.22	2.52	79.70	0
x_2	0.22	−0.04	0.30	5.74	1
x_4	0.17	0.43	0.04	6.39	0
x_5	−0.13	0.17	0.22	2.96	0

Iteration 4

$\mathbf{u}^T = (1.09,\ 4.22,\ 2.52)$. Therefore, $\bar{\mathbf{c}}_N^T = \mathbf{c}_B^T \mathbf{B}^{-1} \mathbf{N} - \mathbf{c}_N^T = (-2.04,\ 0,\ 13.22,\ 0,\ 0,\ 1.09,\ 4.22,\ 2.52)$. Then x_1 is selected as the entering variable.

$$\mathbf{B}^{-1}\mathbf{a}_1 = \begin{bmatrix} 0.22 & -0.04 & 0.30 \\ 0.17 & 0.43 & 0.04 \\ -0.13 & 0.17 & 0.22 \end{bmatrix} \begin{bmatrix} 1 \\ 2 \\ -3 \end{bmatrix} = \begin{bmatrix} -0.61 \\ 0.913 \\ -0.43 \end{bmatrix},\ \bar{c}_1 = -2.04$$

Add the column to the right of the revised simplex tableau and x_4 becomes the leaving variable.

z	1.09	4.22	2.52	79.70	−2.04
x_2	0.22	−0.4	0.30	5.74	−0.61
x_4	0.17	0.43	0.04	6.39	0.913
x_5	−0.13	0.17	0.22	2.96	−0.43

After pivoting, the new tableau becomes

z	1.48	5.19	2.62	94	0
x_2	0.33	0.33	0.33	10	0
x_1	0.19	0.48	0.05	7	1
x_5	−0.05	0.38	0.24	6	0

Iteration 5

$\mathbf{u}^T = (1.48,\ 5.19,\ 2.62)$. Therefore, $\bar{\mathbf{c}}_N^T = \mathbf{c}_B^T \mathbf{B}^{-1} \mathbf{N} - \mathbf{c}_N^T = (0,\ 0,\ 16.42,\ 2.23,\ 0,\ 1.48,\ 5.19,\ 2.62)$. Since all reduced costs are nonnegative, an optimum has been found and the algorithm terminates.

9.9   NOTES

Section 9.2

Many LP solvers provide several options for the user to select the entering variable. The options may include random selection, the first index, the most unit improvement, and the most total improvement.

Section 9.3

For finding the starting basic feasible solution, the big M method seems easier to comprehend and to compute manually, but it is not implementable in practice because we cannot give an appropriate value precisely for the big M. No matter what value you choose, for some problems it could be either too large or too small. In either case, it could cause great truncation and/or rounding errors that would mislead the solution. The two-phase procedure is the only method utilized in practice.

Theoretically, the degeneracy can cause a cycling problem for simplex iterations (see the example given by Beale, 1955). But in practice the cycling problem is highly unlikely. Although there are cycling prevention rules available (e.g., Bland's rule (Bland, 1977)) that guarantee finite convergence of the simplex algorithm, usually they are not implemented in commercial software because of substantial extra computational efforts.

Section 9.7

The starting dual feasible solution corresponds to the condition when a cutting constraint is introduced after an LP optimum is obtained. To reoptimize the augmented LP problem, the dual simplex method is much efficient than solving a new LP problem from scratch.

In case the dual simplex method is used as an independent algorithm for solving an LP problem from scratch and if the dual solution is infeasible, then make it feasible by adding the following redundant constraint (also called *artificial constraint*) obtained by summing over all nonbasic variables:

$$\sum_{j\in \mathbf{N}} x_j \leq M$$

$$\text{or} \quad \sum_{j\in \mathbf{N}} x_j + x_{n+1} = M$$

where M is a big value that can be set equal to the sum of finite upper bounds of all *nonbasic* variables and x_{n+1} is the associated slack variable. Perform the elementary row operations such that a *standard form* of the simplex tableau is obtained.

9.10 EXERCISES

9.1 Transform the following LP problem to canonical form.

$$\text{Minimize} \quad z = 3x_1 + 5x_2$$
$$\text{subject to} \quad 2x_1 + x_2 \geq 13$$
$$-x_1 + x_2 \leq 10$$
$$x_1, x_2 \geq 0$$

9.2 Consider the LP problem in Exercise 9.1 again. List all its extreme points and determine if each pair of points are adjacent.

9.3 Consider the following LP problem. Plot the feasible region and identify all the extreme points. Find the optimal solution by evaluating the objective function at each extreme point.

$$\text{Maximize} \quad z = 5x_1 + 3x_2$$
$$\text{subject to} \quad x_1 + x_2 \leq 12$$
$$-2x_1 + x_2 \leq 7$$
$$x_1 - x_2 \geq 3$$
$$x_1 \leq 5$$
$$x_1, x_2 \geq 0$$

9.4 The point (1/3, 1/3, 1/3) is feasible for the following problem. Is it also a basic solution? Why or why not?

$$\text{Minimize} \quad z = 2x_1 + 3x_2 - 2x_3$$
$$\text{subject to} \quad x_1 + x_2 + x_3 \leq 1$$
$$x_1 - 2x_2 + 2x_3 \leq 2$$
$$x_1, x_2, x_3 \geq 0$$

9.5 Consider the following LP problem. Verify that $\mathbf{x} = (0, 2.5, 0)^{\mathrm{T}}$ is optimal and that the dual price of binding constraint 3 is 1.5.

$$\text{Minimize} \quad z = 2x_1 + 3x_2 + x_3$$
$$x_1 + x_2 + x_3 \leq 3$$
$$2x_1 + 2x_3 \leq 3$$
$$x_1 + 2x_2 + 3x_3 \geq 5$$
$$\mathbf{x} \geq \mathbf{0}$$

9.6   Consider the following LP problem:

$$\text{Minimize} \quad z = x_1 - 2x_2 - 2x_3$$
$$\text{subject to} \quad 2x_1 + 2x_2 + x_3 \leq 11$$
$$x_1 + x_2 - x_3 \geq 4$$
$$x_1, x_2, x_3 \geq 0$$

(i) Rewrite the problem in canonical form.

(ii) Find an initial tableau using two-phase method (without solving the whole problem).

9.7   Go through the phase I of the following LP problem and stop at the phase I optimum. Is the original problem feasible? Why?

$$\text{Minimize} \quad z = 2x_1 + 3x_2$$
$$\text{subject to} \quad 2x_1 + x_2 \leq 8$$
$$x_1 + 3x_2 \geq 29$$
$$x_1, x_2 \geq 0$$

9.8   Consider the following LP problem:

$$\text{Maximize} \quad z = x_1 + 2x_2$$
$$\text{subject to} \quad 2x_1 + 5x_2 = 21$$
$$x_1 - x_2 \leq 10$$
$$x_1, x_2 \geq 0$$

(i) Rewrite the problem in canonical form.

(ii) Solve the problem using two-phase method.

9.9   Solve the following LP problem using the primal simplex method.

$$\text{Maximize} \quad z = 2x_1 + 3x_2$$
$$\text{subject to} \quad x_1 - 3x_2 \leq 4$$
$$-x_1 + x_2 \leq 2$$
$$3x_1 + x_2 \leq 10$$
$$x_1, x_2 \geq 0$$

9.10   Suppose variable x_j leaves the basis at some iteration p. Is it possible that x_j enters the basis at the end of iteration $p + 1$? Why or why not?

9.11 Solve the following LP problem using the primal simplex method. Note that some variables are unrestricted in sign.

$$\text{Maximize} \qquad z = -3x_1 + x_2 - x_3 + x_4$$

$$\text{subject to} \qquad 2x_1 - x_2 - x_3 + x_4 \leq 8$$

$$-2x_1 + 2x_2 + 2x_3 + 3x_4 \leq 10$$

$$-x_1 + x_2 - 3x_3 + x_4 \leq 3$$

$$x_2, x_4 \geq 0$$

9.12 Consider LP problem in Exercise 9.11. Suppose at some iteration the basis consists of x_2, s_1, and s_3, where s_1 and s_3 are the slack variables corresponding to the first and third constraints, respectively.
(i) Decide the right-hand side values at this iteration without actually solving the problem.
(ii) Decide the objective value at this iteration.

9.13 Consider the following LP problem:

$$\text{Minimize} \quad z = 13x_1 - 7x_2 + x_3 + 3x_4 - 5x_5 + x_6$$

$$\text{subject to} \qquad -x_1 + 2x_2 + x_3 + x_4 \leq 10$$

$$2x_1 - x_3 + x_4 + x_5 \leq 13$$

$$x_2 + 3x_3 - x_4 - x_6 \leq 7$$

$$x_1, x_2, x_3, x_4, x_5, x_6 \geq 0$$

(i) Solve the problem to optimum using the primal simplex method. At each iteration, apply the rule of "least index" when picking the entering variable. That is, among all the candidates of entering variable, always choose the one with least index.
(ii) At optimum, adjoin to the model the constraint $x_2 \leq 4$. Reoptimize the problem using the dual simplex method.

9.14 Solve the LP problem in Exercise 9.4 using the dual simplex method.

9.15 The shaded area of the Figure 9.5 shows the feasible region of an LP problem.
(i) Show the inequalities of the constraints.
(ii) Start the primal simplex iteration with the basic feasible solution $(x_1, x_2) = (3, 0)$ and iterate until an optimal solution is obtained.

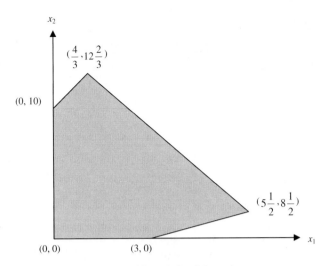

FIGURE 9.5   LP feasible region.

9.16   Consider the following LP problem:

$$\text{Minimize}\quad x_1 - 2x_2 + 2x_3 + x_4 - 5x_5 + 3x_6$$

$$\text{subject to}\qquad\qquad x_1 + x_2 + x_3 + x_5 \le 7$$

$$2x_2 - 2x_3 + 2x_5 + x_6 \le 13$$

$$x_2 \le 3$$

$$x_5 \le 5$$

$$x_1, x_2, x_3, x_4, x_5, x_6 \ge 0$$

Solve it using the upper bound technique.

9.17   Consider the problem in Exercise 9.13. Now add two constraints: $x_1 \le 5$, $x_2 \le 8$.

(**i**) Start from the optimal tableau you obtained in Exercise 9.15 and reoptimize by adding one constraint at a time.

(**ii**) What new extreme points are introduced into the problem, and which ones are gone?

(**iii**) Solve the whole problem from beginning by using upper bound technique.

9.18   Ben has three dogs, Uno, Dos, and Tres. Every day Ben feeds them with five types of food: beef, dog food, bread, bones, and chicken. Each type of food

TABLE 9.10   Food Prices

Food	Beef	Dog Food	Bread	Bones	Chicken
$/lb	$2.50	$1.00	$0.80	$1.20	$1.60

is bought in pounds. The price for each type of food is as follows (Table 9.10):

Ben's wife wants him to find the most cost-effective plan for feeding dogs, subject to the dogs' preferences. Table 9.11 shows the minimum amount of food consumed by each dog each day.

Despite this, Dos eats no less than 2.5 lb of chicken plus bread. Uno eats no less than 2.7 lb of meats (including dog food, chicken, and bones). The total amount of bread and beef fed to Tres cannot be less than 2.6 lb.

(i) Formulate the problem as an LP.

(ii) Solve the problem using LINGO®. Declare the variables and parameters by sets. Apply the domain defining functions as necessary. Show your output and interpret the solution.

(iii) When solving the model, manipulate the "linear solver options" by selecting primal simplex, dual simplex, and barrier, respectively. Compare the number of steps it took for LINGO® to solve the problem under each option.

9.19 Solve the following LP problem using revised simplex method.

$$\text{Maximize} \qquad -3x_1 + x_2 - x_3 + x_4$$
$$\text{subject to} \qquad 2x_1 - x_2 - x_3 + x_4 \leq 8$$
$$-2x_1 + 2x_2 + 2x_3 + 3x_4 \leq 10$$
$$-x_1 + x_2 - 3x_3 + x_4 \leq 3$$
$$\mathbf{x} \geq \mathbf{0}$$

TABLE 9.11   Minimum Daily Food Consumption

	Beef	Dog Food	Bread	Bones	Chicken
Uno	0	0	0.5	1.7	1.9
Dos	0	1.5	0.3	0.9	0.1
Tres	1.5	0.9	0.8	0.6	0.2

9.20   Solve the following LP problem using revised simplex method.

$$\text{Minimize} \qquad x_1 - 2x_2 - 2x_3$$
$$\text{subject to} \quad 2x_1 + 2x_2 + x_3 \leq 11$$
$$x_1 + x_2 - x_3 \geq 4$$
$$x_1, x_2, x_3 \geq 0$$

10

NETWORK OPTIMIZATION PROBLEMS AND SOLUTIONS

There are certain classes of integer programming problems whose special structures make them particularly easy to solve. Among them, the most noteworthy class is the network-structured problems whose LP solutions under certain conditions are *naturally integer*. This class includes the well-known transportation, assignment, transshipment, maximum flow, and shortest path problems, appearing in most introductory OR texts. These network and related models are widely formulated in real-world problems, as estimated (Taha, 2007) that "70% of real-world mathematical programming problems use network-related models."

The purposes of this chapter are (a) to formulate each of these problems as a special case of a larger class of problem called the *minimum cost network flow problem*, (b) to introduce a *unifying* solution algorithm (called the *network simplex*) that is much more efficient than the ordinary simplex algorithm, and (c) to provide the sufficient conditions (or model structures) that characterize such "easy" integer programs.

As a generalization of the transportation algorithm, the network simplex method performs the simplex operations *directly* on the network itself. Moreover, these simplex operations involve only additions and subtractions, unlike the ordinary simplex that requires multiplications/divisions. The empirical experience shows that this method enables one "to solve problems 200–300 times faster than a standard simplex method that ignores any inherent special structures other than sparsity" (Bazaraa et al., 2005) of the constraint matrix.

Before we define the class of minimum cost network flow problem and its individual problems, we need some basic knowledge of network (or graph) concepts.

Applied Integer Programming: Modeling and Solution, By Der-San Chen, Robert G. Batson, and Yu Dang
Copyright © 2010 John Wiley & Sons, Inc.

10.1   NETWORK FUNDAMENTALS

Definition 10.1   A network (or *graph*) **G** is a collection of nodes (or vertices) and a collection of arcs (or edges) joining pairs of nodes, denoted by $\mathbf{G} = (\mathbf{V}, \mathbf{E})$, $\mathbf{V} = \{1, 2, \ldots, m\}$, and $\mathbf{E} = \{(i, j): i, j \in \mathbf{V}\}$.

An arc in **G** may be either *directed* or *undirected*. A directed arc is an ordered pair of nodes (i, j) that allows the flow only going from node i to j. Nodes i and j, respectively, are called *initial* and *terminal nodes* of arc (i, j). An undirected arc allows the flow in either direction and may be replaced by two opposite directed arcs of the same capacity. In the context of this chapter, we will deal with *directed networks* (or *digraphs*) in which all arcs are directed.

Definition 10.2   A *path* (from node i_0 to i_p) is a sequence of arcs $\{(i_0, i_1), (i_1, i_2), \ldots, (i_{p-1}, i_p)\}$ in which the initial node of each arc is the same as the terminal node of the preceding node in the sequence and all nodes $i_0, i_1, \ldots, i_p$ are distinct. A *chain* is a sequence of arcs similar to a path, except that *not all arcs are necessarily* directed toward node i_p. Thus, every path is a chain but a chain may not be a path.

Consider, for example, Figure 10.1 in which the sequences $\{(1, 2), (2, 4)\}$ and $\{(1, 3), (3, 4)\}$ are paths from node 1 to node 4. These paths are also chains from node 1 to node 4. The sequences $\{(1, 2), (3, 2), (3, 4)\}$ is a chain from node 1 to 4, but not a path because arc $(3, 2)$ is not directed to node 4.

Definition 10.3   Given a network **G** $(\mathbf{V}, \mathbf{E})$ and a distance (cost) c_{ij} associated with each directed arc (i, j), the problem to determine a path from a specified node to another specified node with a minimal total distance is called a *shortest path* (*route*) *problem*.

Definition 10.4   A *circuit* is a path from some node i_0 to i_p plus the return arc (i_p, i_0). Thus, a circuit is a closed path. Similarly, a *cycle* is a closed chain and every circuit is a cycle but a cycle may not be a circuit.

Consider Figure 10.1, adding arc $(4, 1)$ to either of the above two paths will yield a circuit. Similarly, adding arc $(4, 1)$ to chain $\{(1, 2), (3, 2), (3, 4)\}$ will yield a cycle.

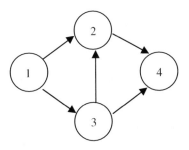

FIGURE 10.1   A directed network.

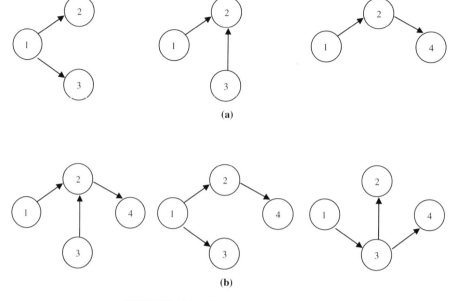

FIGURE 10.2   Trees and spanning trees.

Definition 10.5   Given a network with m nodes, a *tree* with k $(2 \leq k \leq m)$ nodes is a subnetwork that connects all k nodes with no cycles. A *spanning tree* is a tree that connects all m nodes in the given network with no cycles.

By definition, every arc $(k = 2)$ in a network is a tree (with two nodes). Examples of trees with $k = 3$ and $k = 4$ nodes are given in Figure 10.2a and b, respectively. The trees in Figure 10.2b are spanning trees of the network in Figure 10.1, but the trees in Figure 10.2a are not.

The following are some important properties of a tree:

1. Every *tree* (including spanning tree) with k nodes has exactly $(k - 1)$ arcs.
2. Adding any new arc from the original network to a spanning tree results in a unique cycle.
3. Every pair of nodes in a tree can be connected by a unique chain.

10.2   A CLASS OF EASY NETWORK PROBLEMS

In this section, we define the minimum cost network flow (MCNF) problem and then show how the transportation, assignment, transshipment, maximal flow, and shortest path problems can be viewed as special cases of an MCNF problem.

10.2.1   The Minimum Cost Network Flow Problem

The minimum cost network flow problem is defined as follows: Given a directed network $\mathbf{G}$ consisting of m nodes and n arcs joining pairs of nodes, let b_i be the net supply amount ($=$outflow $-$ inflow) at node i. There are three types of nodes: the supply or source node (if $b_i > 0$), demand or destination node (if $b_i < 0$), and transshipment or intermediate node (if $b_i - 0$). Associated with each arc (i, j) is a lower bound L_{ij} on flow through arc, a upper bound U_{ij} on flow through arc, and a cost c_{ij} of transporting a unit flow through arc. The problem is to determine the amount of flow x_{ij} through each arc (i, j) so that the total shipping cost is minimum.

We assume that the total supply (sum of all $b_i > 0$) equals the total demand (sum of all $b_i < 0$) in the network, that is, $\sum b_i = 0$, $i = 1, 2, \ldots, m$. If $\sum b_i < 0$, then the total supply cannot meet the total demand, and hence there is no feasible solution. If $\sum b_i > 0$, then we can make it equal to 0 by adding a *dummy* demand node, say $m + 1$, with $b_{m+1} = -\sum b_i$, and adding arcs with zero cost from each supply node to the dummy demand node.

In order to obtain a "uniform" flow balance equation for all nodes, the original network may be modified so that every node has both outgoing and incoming arc flows. To accomplish this, a "dummy" return arc is usually added. For example, if source node 1 has no incoming arc and sink node m has no outgoing arc, then a return arc $(m, 1)$ joining m and 1 with 0 cost is created. After modification, the MCNF problem becomes

$$\text{Minimize} \quad z = \sum_i \sum_j c_{ij} x_{ij}$$

$$\text{subject to} \quad \sum_{j=1}^{m} x_{ij} - \sum_{k=1}^{m} x_{ki} = b_i \quad \text{for each node } i \tag{10.1}$$

$$x_{ij} \leq U_{ij} \quad \text{for each arc } (i, j) \tag{10.2}$$

$$x_{ij} \geq L_{ij} \quad \text{for each arc } (i, j) \tag{10.3}$$

Constraints (10.1) stipulates that the net flow into and out of node i must equal b_i. These equations ensure that the flow may not be created or destroyed in the network. They are referred to as the *flow conservation equations* or *flow balance equations*. Constraints (10.2) and (10.3) ensure that the flow through each arc satisfies the upper and lower limits. For ease of representation, we will assume $L_{ij} = 0$. If any lower bound is other than 0, we can convert it to 0 by a simple variable substitution as described in Section 9.6.

10.2.2   Formulating the Transportation–Assignment Problem as an MCNF Problem

The "classical" transportation problem may be stated as follows. Given m_1 source nodes ($i = 1, 2, \ldots, m_1$), each with s_i units of supply, and m_2 destination nodes ($j = 1, 2,$

..., m_2), each with d_j units of demand. Let c_{ij} be the unit flow cost from node i to j, and x_{ij} be the amount of arc flow to be determined. The problem is to determine the amount of commodity to be shipped from each source i to each destination j so that the total transportation cost is minimized. Mathematically, the transportation problem becomes

$$\text{Minimize} \quad z = \sum_{i=1}^{m_1}\sum_{j=1}^{m_2} c_{ij}x_{ij} \tag{10.4}$$

$$\text{Subject to} \quad \sum_{j=1}^{m_2} x_{ij} = s_i \quad i = 1, 2, \ldots, m_1 \tag{10.5}$$

$$\sum_{i=1}^{m_1} x_{ij} = d_j \quad j = 1, 2, \ldots, m_2 \tag{10.6}$$

$$x_{ij} \geq 0 \qquad \text{all } i, j$$

Reformulate the transportation model as a special MCNFP using the following procedure and Figure 10.3.

1. Renumber m_1 source and m_2 destination nodes using a common index $i = 1, 2, \ldots, m_1, m_1 + 1, m_1 + 2, \ldots, m$, where $m = m_1 + m_2$. The unit transportation costs c_{ij} are also renumbered accordingly.
2. Set $b_i = s_i$ for $i = 1, 2, \ldots, m_1$ and set $b_{m_1+j} = -d_j$ for $j = 1, 2, \ldots, m_2$. Note that $\sum_i s_i = \sum_j d_j$ implies $\sum_{i=1}^{m_1} b_i + \sum_{i=1}^{m_2} b_{m_1+i} = \sum_{i=1}^{m} b_i = 0$.
3. Create a dummy source node (say node 0) with $b_0 = \sum_{i=1}^{m_1} b_i$ and connect arcs $(0, i)$ for $i = 1, 2, \ldots, m_1$ with unit cost $c_{0i} = 0$.

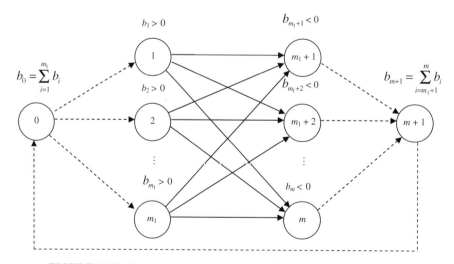

FIGURE 10.3   Formulating transportation problem as an MCNF problem.

4. Create a dummy sink node (say node $m + 1$) with $b_{m+1} = \sum_{i=m_1+1}^{m} b_i$ and connect arcs $(m_1 + i, m + 1)$ for $i = 1, 2, \ldots, m_2$ with unit cost $c_{i,m+1} = 0$.
5. Add a return arc $(m + 1, 0)$ with $c_{m+1,0} = 0$.
6. Ignore the upper bound constraints (10.2).

The "classical" assignment problem can be viewed as a special case of the transportation problem, in which the number of source nodes is equal to the number of destination nodes (i.e., $m_1 = m_2$). Demand d_j at each destination node is 1, and supply s_i at each source node is also 1. The objective is to "assign" each source to a unique destination so that the total cost associated with an assignment plan is minimized. Clearly, we can formulate the assignment problem as an MCNF problem as we did for the transportation problem with special values of $s_i = d_j = 1$ for all i, j.

10.2.3 Formulating the Transshipment Problem as an MCNF Problem

The transshipment problem is a generalization of transportation problem in which there are transshipment nodes in the network in addition to the sources and destinations. The transshipment node does not supply nor demand commodity (i.e., $b_i = 0$). This problem is a special MCNF problem in which there is no upper limit on each arc flow as stipulated in constraints (10.2).

10.2.4 Formulating the Maximum Flow Problem as an MCNF Problem

Given a directed network with a single source node 1 and a single sink node m. A commodity supplied at node 1 is to be shipped to node m using one or more paths from 1 to m. At each intermediate node, the sum of inflows must equal the sum of outflows. An arc (i, j) connecting nodes i and j is subject to a flow capacity U_{ij}. The maximum flow problem is to determine the maximum amount of flow that can be shipped from node 1 to node m. After adding a return arc $(m, 1)$, the problem can be formulated as follows:

$$\text{Minimize} \quad -z = \sum_j -x_{1j}$$

$$\text{subject to} \quad \sum_j x_{ij} - \sum_k x_{ki} = 0 \quad \text{for } i = 1, 2, \ldots, m$$

$$x_{ij} \leq U_{ij} \qquad\qquad \text{for all } i, j$$

$$x_{ij} \geq 0 \qquad\qquad \text{for all } i, j$$

The upper limit of the return arc $(m, 1)$ may be set to $U_{m1} = \min\{\sum_j U_{1j}, \sum_j U_{mj}\}$. Converting this problem to an MCNF problem, we set all $b_i = 0$, and let the objective coefficient $c_{ij} = -1$ for $i = 1$ and all j, and $c_{ij} = 0$ for $i \neq 1$ and all j.

10.2.5 Formulating the Shortest Path Problem as an MCNF Problem

Given a directed network with distance or cost c_{ij} on each arc (i, j). The shortest (longest) path problem is to find a path from the source node 1 to the sink node m at minimum (maximum) cost. It can be viewed as sending a unit of flow from a node to another. Converting to a MCNF problem, we set $U_{ij} = 1$ for all i, j, and $b_i = 0$ for all i.

10.3 TOTALLY UNIMODULAR MATRICES

An important property possessed by the MCNF problem is that every basic feasible solution is *naturally integer* provided that all b_i and U_{ij} are integer. In this section, we shall present a sufficient condition, called *total unimodularity*, that ensures this happens.

10.3.1 Definition

Definition 10.6   A square matrix whose determinant is 0, 1, or -1 is called *unimodular*. A matrix $\mathbf{M}$ is *totally unimodular* (TU) if the determinant of *every* square *submatrix* of $\mathbf{M}$ has value 0, 1, or -1.

Clearly, this definition implies that if matrix $\mathbf{M}$ is TU, then *all* of its elements must be 0, 1, or -1. This is because every element of a matrix is a 1×1 submatrix.

Example 10.1   Check each of the following matrices to see if it is totally unimodular.

Checking all nine possible square submatrices for matrix 1, we see that the determinant of every submatrix is 0, 1, or -1, and hence matrix 1 is TU. Matrix 2 is 5×6 and has numerous square submatrices. For instance, there are 75 4×4 submatrices and 200 3×3 submatrices (verify!). Therefore, the process of checking the determinant of each of these submatrices is arduous. It will turn out that the structure of matrix 2 guarantee that it is TU. Matrix 3 is clearly not TU because the matrix contains elements of value 2. Matrix 4 is not TU because its submatrix formed by the first three columns and three rows has the determinant -2.

Matrix 1   Matrix 2   Matrix 3   Matrix 4

$$
\begin{pmatrix} -1 & 0 & 0 \\ 1 & 1 & 0 \\ 0 & -1 & -1 \end{pmatrix}
\quad
\begin{pmatrix} 1 & 0 & 0 & 1 & 1 & 1 \\ 1 & 1 & 0 & 0 & 1 & 1 \\ 1 & 1 & 1 & 0 & 0 & 1 \\ 0 & 1 & 1 & 1 & 0 & 0 \\ 0 & 0 & 1 & 1 & 1 & 0 \end{pmatrix}
\quad
\begin{pmatrix} 1 & 1 & 0 & 1 & 0 \\ 1 & 2 & 0 & 1 & 0 \\ 0 & 0 & 0 & 1 & 1 \\ 0 & 0 & 2 & 1 & 1 \\ 1 & 0 & 1 & 0 & 0 \end{pmatrix}
\quad
\begin{pmatrix} 1 & 0 & 1 & -1 \\ 1 & 1 & 0 & -1 \\ 0 & -1 & -1 & 0 \\ 0 & 1 & 1 & 0 \end{pmatrix}
$$

10.3.2 Sufficient Condition for a Totally Unimodular Matrix

Theorem 10.1   (Sufficient Condition) An $m \times n$ matrix $\mathbf{M}$ is totally unimodular if the following conditions hold:

1. Every element of $\mathbf{M}$ is 0, 1, or -1.
2. Each column of $\mathbf{M}$ contains at most two nonzero elements.

3. The m rows of $\mathbf{M}$ can be partitioned into two mutually exclusive subsets $\mathbf{M}_1$ and $\mathbf{M}_2$ such that

(a) If any column contains two nonzero elements of the same sign, one element can be placed in $\mathbf{M}_1$ and the other in $\mathbf{M}_2$.

(b) If any column contains two nonzero elements of opposite signs, both elements can be placed in the same subset.

Example 10.2   Check if the following matrix satisfies the sufficient condition for TU.

$$
\begin{pmatrix}
-1 & 0 & 1 & 0 & 0 & 0 \\
0 & 0 & -1 & -1 & 0 & 0 \\
0 & 1 & 0 & -1 & 1 & 0 \\
1 & -1 & 0 & 0 & 0 & 1 \\
0 & 0 & 0 & 0 & 1 & -1
\end{pmatrix}
$$

1. Every element is 0, 1, or -1.
2. Each column contains two nonzero elements.
3. Begin the partitioning procedure with column 1. The two nonzero elements are of opposite sign, so both rows should be placed in the same subset, say, $\mathbf{M}_1 = \{R_1, R_4\}$.

Scanning column 2, again the two nonzero elements are of opposite sign, and because R_4 is already in $\mathbf{M}_1$, we have $\mathbf{M}_1 = \{R_1, R_4, R_3\}$. From column 3, rows 1 and 2 must be in the same set, and hence $\mathbf{M}_1 = \{R_1, R_4, R_3, R_2\}$. Column 5 has nonzero elements of the same sign, and so R_2 and R_3 should be in opposite sets. But, we already have R_2 and R_3 in the same set $\mathbf{M}_1$. This contradiction implies that this matrix does not satisfy condition 3 of Theorem 10.1.

Note that if one of two sets created in verification of condition 3 is empty, then $\mathbf{M}$ is TU. This means that any matrix satisfying 1 and 2, and having nonzero elements of opposite sign in every column, is automatically TU. Matrix 1 in Example 10.1 is an example of such a matrix. The sufficient condition for a matrix to be TU is not necessary—meaning that we *cannot claim that a matrix is not a TU just because it does not satisfy this sufficient condition.*

Definition 10.7   A network $G(V, E)$ is called *bipartite* if there exists two subsets of nodes (vertices), V_1 and V_2, such that $V_1 \cup V_2 = V$ and $V_1 \cap V_2 = \phi$, and every arc (edge) of G is incident to exactly one node of V_1 and one node of V_2.

The transportation and the assignment problems are examples of bipartite networks (graphs). The incidence matrix (matrix $\mathbf{A}$ in $\mathbf{A}\mathbf{x} = \mathbf{b}$) of a bipartite graph is totally unimodular.

10.3.3 Some Properties of Totally Unimodular Matrices

The following are some important properties of totally unimodular matrices:

1. The matrix obtained by adding (deleting) an identify matrix to (from) a TU matrix is also TU.
2. The transpose of a TU matrix is also TU.
3. The matrix obtained by pivoting on a TU matrix is also TU.
4. The matrix obtained by multiplying any row (column) of a TU matrix by -1 is also TU.
5. The matrix obtained by interchanging any two rows (columns) of a TU matrix is also TU.
6. The matrix obtained by deleting (adding) a unit row (column) of a TU matrix is also TU.

Definition 10.8 A 0–1 matrix is called an *interval matrix* if in each column (or row) the 1's appear *consecutively* (allowing wrapping around).

For example, the following matrix is an interval matrix with three consecutive 1's. In column 4, the third "1" wraps around in row 1. In column 5, the second and third "1" appear in rows 1 and 2, respectively.

$$\begin{pmatrix} 1 & 0 & 0 & 1 & 1 \\ 1 & 1 & 0 & 0 & 1 \\ 1 & 1 & 1 & 0 & 0 \\ 0 & 1 & 1 & 1 & 0 \\ 0 & 0 & 1 & 1 & 1 \end{pmatrix}$$

It can be shown that an interval matrix is TU. Recall that the IP formulation of the workforce/staff scheduling problem in Section 10.2.5 has an interval coefficient matrix. Hence, the solution to the workforce scheduling problem is always integer provided that all b_i are integer.

10.3.4 Matrix Structure of the MCNF Problem

To give an idea about the structure of the coefficient matrix associated with a system of *flow conservation equations* given in constraints (10.1), we illustrate an example for the transportation problem with two sources and three destinations. After renumbering and adding an artificial source node 0 and artificial sink node 6, we obtain the coefficient matrix and right-hand side in Table 10.1.

Note that each column contains exactly one "1" and one "-1". We can easily show that this matrix satisfies the sufficient condition of totally unimodularity. Likewise, it can be shown that the coefficient matrix of any MCNF problem is totally unimodular. The reader is encouraged to construct an example matrix for each network problem.

TABLE 10.1   Coefficient Matrix for a Transportation Problem

Node i	x_{01}	x_{02}	x_{13}	x_{14}	x_{15}	x_{23}	x_{24}	x_{25}	x_{36}	x_{46}	x_{56}	x_{60}	RHS
0	1	1										-1	b_0
1	-1		1	1	1								b_1
2		-1				1	1	1					b_2
3			-1			-1			1				b_3
4				-1			-1			1			b_4
5					-1			-1			1		b_5
6									-1	-1	-1	1	b_6

Each column (vector) of the coefficient matrix, denoted by $\mathbf{a}_{ij}$, may be represented by the difference of two unit vectors, $\mathbf{e}_i - \mathbf{e}_j$.

$$
\mathbf{a}_{ij} = \begin{pmatrix} 0 \\ \vdots \\ 1 \\ \vdots \\ -1 \\ \vdots \\ 0 \end{pmatrix} = \begin{pmatrix} 0 \\ \vdots \\ 1 \\ \vdots \\ 0 \\ \vdots \\ 0 \end{pmatrix} - \begin{pmatrix} 0 \\ \vdots \\ 0 \\ \vdots \\ 1 \\ \vdots \\ 0 \end{pmatrix} = \mathbf{e}_i - \mathbf{e}_j
$$

where i is the initial node and j is the terminal node of arc (i, j).

Note that the coefficient matrix corresponding to the upper bounded constraints (10.2) is an *identity matrix*. Thus, by property 1, the matrix obtained by combining (10.1) and (10.2) together is also TU.

10.3.5   Lower Triangular Matrix and Forward Substitution

Another important property about the coefficient matrix of the flow conservation equations is that it has a rank of $m - 1$ for an m-node network. After deleting any row, the remaining $(m - 1) \times (m - 1)$ submatrix will be nonsingular. From Section 7.2, we know that an equivalent lower (upper) triangular matrix with *nonzero diagonal elements* can be constructed and that a unique solution can be found by the forward (backward) substitution method. This property is utilized in the network simplex method to find efficiently the primal solution $\mathbf{x_B}$ for the system $\mathbf{Bx_B} = \mathbf{b}$ and the dual solution $\mathbf{u}$ for the system $\mathbf{u^T N} = \mathbf{c}$.

10.3.6   Naturally Integer Solution for the MCNF Problem

Now let us observe the structure of the coefficient matrix corresponding to the system of m flow conservation equations in (10.1). Because of the condition $\sum b_i = 0$, the system has at most $m - 1$ linearly independent equations, and in fact every equation is a linear combination of the remaining $m - 1$ equations. Therefore, any equation can be dropped from the system without affecting the feasible solution space.

Partitioning the set of variables into basic and nonbasic variables, we have $\mathbf{x} = (\mathbf{x_B}, \mathbf{x_N})^T$ and $\mathbf{A} = (\mathbf{B}, \mathbf{N})$. Thus, we may obtain a *basic* solution with $m - 1$ basic variables or less, depending on the degree of degeneracy. Assume that $\mathbf{B}$ is a $(m - 1) \times (m - 1)$ nonsingular matrix (or $\det \mathbf{B} \neq 0$) in the system $\mathbf{B}\mathbf{x_B} = \mathbf{b}$. To determine a basic solution, we let $\mathbf{x_N} = \mathbf{0}$ and compute the values of basic variables x_j $(j = 1, 2, \ldots, m - 1)$ by *Cramer's rule*:

$$x_j = \frac{\det \mathbf{B}_j}{\det \mathbf{B}}$$

where $\mathbf{B}_j$ is obtained from $\mathbf{B}$ by replacing the jth column of $\mathbf{B}$ by $\mathbf{b}$. Because $\det \mathbf{B} = 1$ or -1, we have $x_j = \pm \det \mathbf{B}_j$. Also, because $\mathbf{B}_j$ is a matrix containing all integer elements, the determinant of $\mathbf{B}_j$ must be integer, which in turn results in integer values for all x_j. We have just proved the following result.

Theorem 10.2 Every basic feasible (including basic optimal) solution to an LP problem $\mathbf{P} = \{\max \mathbf{c}^T\mathbf{x}: \mathbf{A}\mathbf{x} \leq \mathbf{b}, \mathbf{x} \geq \mathbf{0}\}$ is always an integer solution if $\mathbf{A}$ is TU, $\mathbf{b}$ is integer, and $\mathbf{P}$ has a finite optimal solution. In addition, the statement is also true if $\mathbf{I}\mathbf{x} \leq \mathbf{u}$ is adjoined to the constraint set of $\mathbf{P}$, provided $\mathbf{u}$ is integer.

Corollary 10.1 The MCNF problem has integer basic feasible solution and optimal solution if $\mathbf{b}$ and $\mathbf{u}$ are integer-valued vectors.

10.4 THE NETWORK SIMPLEX METHOD

The network simplex method is based on the simplex method for upper bounded variables while taking advantage of the special network structure of the MCNF problem. Without using the simplex tableaux, the method performs simplex iterations *directly* on the network itself and only additions and subtractions are required for calculations.

10.4.1 Feasible Spanning Trees Versus Basic Feasible Solutions

Suppose we are solving an MCNF problem with m nodes by the simplex method for bounded variables. The method in effect requires only carrying a $(m - 1) \times (m - 1)$ basis matrix obtained from constraints (10.1), while implicitly handling a maximum of $2n$ bounded constraints (10.2)–(10.3). To accomplish this, the variables in a basic feasible solution are classified into three types of variables:

1. Nonbasic variables at lower bounds ($x_{ij} = 0$).
2. Nonbasic variables at upper bounds ($x_{ij} = U_{ij}$).
3. Basic variables: In the absence of degeneracy, each variable x_{ij} will satisfy $0 < x_{ij} < U_{ij}$ and in degeneracy, some x_{ij} may be 0 or U_{ij}.

Recall that in the simplex tableau, the row-0 coefficients of all basic variables must be 0 or $\bar{c}_{ij} = 0$, where

$$\bar{c}_{ij} = \mathbf{c}_{\mathbf{B}}^{\mathrm{T}} \mathbf{B}^{-1} \mathbf{a}_{ij} - c_{ij} = (u_1, \dots, u_i, \dots, u_j, \dots, u_m) \mathbf{a}_{ij} - c_{ij}$$

But $\mathbf{a}_{ij} = \mathbf{e}_i - \mathbf{e}_j$. Hence, $\bar{c}_{ij} = u_i - u_j - c_{ij} = 0$, or $u_i - u_j = c_{ij}$, for every basic variable x_{ij}. The dual variables $u_1, \dots, u_m$ that corresponds to each of m nodes are also called *simplex multipliers*. Because there are only $m - 1$ linearly independent equations, we may set any u_i to 0 and solve for the remaining ones because c_{ij} is known, for all i, j.

After obtaining the dual solutions, we can calculate the reduced costs for all nonbasic variables by $\bar{c}_{ij} = u_i - u_j - c_{ij}$. From these reduced costs, we can determine whether the current solution is optimal.

Previously, we know that there are $m - 1$ linearly independent flow conservation equations in an MCNF model. This means that the rank of basis matrix is $m - 1$, and each basic feasible solution to an m-node MCNF problem will have $m - 1$ basic variables. Recall that, by definition, a spanning tree in an m-node network is a connected network containing exactly $m - 1$ arcs with no cycles. Therefore, a set of $m - 1$ variables will yield a basic feasible solution if and only if the arcs corresponding to the basic variables form a spanning tree for the network. These arcs are called *basic arcs*, and the remaining arcs in the given network are called *nonbasic arcs*. By property 2 of a tree, we know that adding a nonbasic arc to a spanning tree will form a unique *cycle*. This implies that a nonbasic arc/variable can be represented by the basic arcs/variables that form a spanning tree. Therefore, a feasible basic solution to an MCNF problem is a spanning tree that satisfies the bounds constraints on each arc. We call it a *feasible spanning tree*.

Recall that the simplex method always starts with a full-rank (m) constraint matrix. But earlier we concluded that the coefficient matrix of the MCNF problem is of rank $m - 1$. Therefore, an artificial variable is required to make up the difference so that the rank of the new matrix is m. The addition of an artificial variable is equivalent to creating an arc leading to node m (or any other node). This particular one-ended arc is called a *root arc* and the associated node (m) is called a *root node*. The feasible spanning tree with a root arc is called a *rooted spanning tree*.

10.4.2 The Network Algorithm

Now we are ready to describe the network simplex algorithm that works on the network directly. Assume that the given network has m nodes and n arcs, where $n > m$. Each node is associated with a given commodity amount b_i and each arc is associated with a given unit flow cost c_{ij} and given flow capacity U_{ij}. The lower bounds of all arc flows are assumed to be 0. The problem is to determine the amount of arc flow x_{ij}. The algorithm is as follows:

Step 0 (Initialization). Find a rooted feasible spanning tree comprising $m - 1$ basic arcs:

(a) Check if $\sum b_i = 0$. If not, add a dummy node and its associated dummy arcs.

(b) Designate a demand node ($b_i < 0$) as a root node and create a rooted arc incident to the root node.

(c) Create a feasible spanning tree in the following process. Begin with the end nodes and proceed toward the rooted node (say $m + 1$). Assign flows to arcs so that at each node the net flow (=outflows − inflows) equals b_i.

Step 1 (Node Potential). Compute the node potentials u_i (or dual variables or simplex multipliers) for all nodes as follows. Begin with the rooted node and set $u_{m+1} = 0$, and proceed toward the end nodes. Determine u_i iteratively by $u_i - u_j = c_{ij}$.

Step 2 (Entering Arc). Compute the reduced costs $\bar{c}_{ij}$ for all nonbasic arcs by $\bar{c}_{ij} = u_i - u_j - c_{ij}$ and then check the optimality (minimization) conditions for both types of nonbasic variables. If $x_{ij} = 0$, then the optimal condition is $\bar{c}_{ij} \leq 0$. If $x_{ij} = U_{ij}$, then the optimal condition is $\bar{c}_{ij} \geq 0$. If the current solution is not optimal, then choose an entering arc that most violates either optimal condition.

Step 3 (Leaving Arc). Form a unique cycle by adding the entering arc to the current spanning tree. Determine the amount of arc flow Δ that can increase without exceeding any arc capacity U_{ij} in the cycle and that can decrease without violating the lower bound $x_{ij} \geq 0$. The leaving arc will be the one that first hits either 0 or U_{ij}.

Step 4 (Updating). Find the new feasible spanning tree (feasible solution) by adjusting the arc flows in the cycle so that the flow conservation at each node is maintained. Go to step 1.

10.4.3 Numerical Example

We use the network in Figure 10.4 to show how the network simplex method works. Each node i is associated with a given b_i. Each arc is associated with a triplet ($\$c_{ij}$, U_{ij}, x_{ij}), where c_{ij} denotes the given unit flow cost, U_{ij} denotes the given upper bound, and x_{ij} denotes the amount of flow to be determined.

Step 0. We construct a starting feasible spanning tree beginning with the end node 1 toward the root node 5. We first assign $x_{13} = 10$ and $x_{14} = 30$, and then assign $x_{23} = 50$ and $x_{35} = 60$. These four basic arcs form a five-node spanning tree. As shown in Figure 10.5, we use the solid lines to represent the basic arcs (or variables) that form the spanning tree and the dashed lines to represent the nonbasic arcs (or variables). Initially, all nonbasic arcs are set at the lower bound L_{ij} (assumed to be 0 from now on). Note that arc $(1, 3)$ is a *basic* arc even though its arc flow reaches the arc capacity $U_{13} = 10$. The total cost is 800 (verify!). Add a root arc $x_5 = 0$ to the rooted node 5.

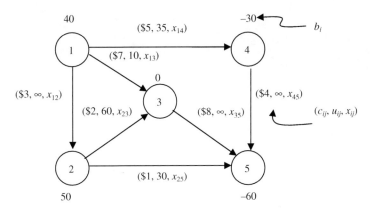

FIGURE 10.4 The given network (Taha, 2007).

Iteration 1.

Step 1. Consider the nodes in Figure 10.6. Compute the dual values u_i iteratively in the order of nodes 5, 3, 2, 1, and 4 via $u_i - u_j = c_{ij}$. These nodes are connected by the basic arcs in the spanning tree. Let $u_5 = 0$, and then compute $u_3 = u_5 + 8 = 8$, $u_2 = u_3 + 2 = 10$, $u_1 = u_3 + 7 = 15$, and $u_4 = u_1 - 5 = 10$.

Step 2. For all nonbasic arcs, compute the reduced costs $\bar{c}_{ij} = u_i - u_j - c_{ij}$:

$$\bar{c}_{12} = u_1 - u_2 - c_{12} = 15 - 10 - 3 = 2 > 0$$
$$\bar{c}_{25} = u_2 - u_5 - c_{25} = 10 - 0 - 1 = 9 > 0 \qquad (10.7)$$
$$\bar{c}_{45} = u_4 - u_5 - c_{45} = 10 - 0 - 4 = 6 > 0$$

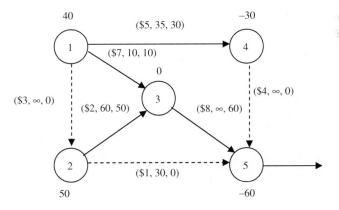

FIGURE 10.5 First feasible solution ($z = 800$).

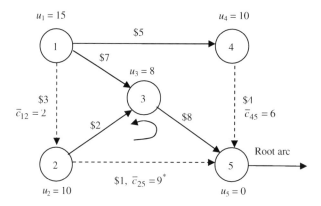

FIGURE 10.6 Steps 1–3 for the first feasible solution.

Because $x_{12} = x_{25} = x_{45} = 0$, all nonbasic arcs violate the optimality condition $\bar{c}_{ij} \leq 0$. Select arc $(2, 5)$ as the candidate entering arc because it violates this condition the most.

Step 3. Construct the cycle $\{(2, 5), (3, 5), (2, 3)\}$. If arc $(2, 5)$ increases the flow by Δ, then the flows of arcs $(3, 5)$ and $(2, 3)$ must decrease by Δ because of opposite direction of flow. The amount of Δ must satisfy the following conditions:

$$\text{For arc } (2, 5) : 0 + \Delta \leq 30$$
$$\text{For arc } (3, 5) : 60 - \Delta \geq 0$$
$$\text{For arc } (2, 3) : 50 - \Delta \geq 0$$

Therefore, $\Delta = \min\{30, 60, 50\} = 30$. Arc $(3, 5)$ is the leaving variable. Because arc $(2, 5)$ must be increased, we adjust the flow in the arcs of the cycle by an equal amount of Δ in order to maintain the feasibility of the new solution. To achieve this, we identify a positive $(+)$ flow in the cycle by the same flow direction of the entering arc and assign a negative $(-)$ flow in the cycle by the opposite flow direction of the entering arc. See Figure 10.6 for the assignment of $+\Delta$ or $-\Delta$ in the cycle. After adjusting the arc flow in the cycle, we obtain a new feasible solution in Figure 10.7 with new flow of 30 in arc $(2, 5)$, new flow of 30 in arc $(3, 5)$, and new flow of 20 in arc $(2, 3)$.

Note that because no current basic arcs $(2, 3)$ and $(3, 5)$ leave the basis at zero level, arc $(2, 5)$ remains nonbasic at level U_{ij}, switching from level 0. However, to maintain dealing with nonbasic arcs at level 0, we substitute the arc using its reverse arc by the relations:

$$x_{25} = U_{25} - x_{52} = 30 - x_{52} \quad \text{and} \quad 0 \leq x_{52} \leq 30$$

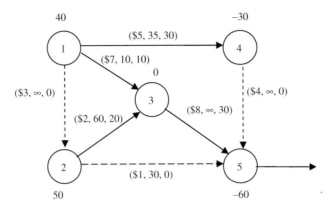

FIGURE 10.7   The second feasible solution ($z = 610$).

similar to what we did in the simplex method for upper bounded variables. This substitution causes changes in flow conservation equations at nodes 2 and 5 as well as on arcs. At node 2, the current flow equation is $x_{23} + x_{25} - x_{12} = 50$ and the new flow equation becomes $x_{23} - x_{12} - x_{52} = 20$. At node 5, the current flow equation is $0 - x_{25} - x_{35} - x_{45} = -60$ and the new flow equation becomes $x_{52} - x_{35} - x_{45} = -30$. The direction of flow in arc $(2, 5)$ is reversed to $(5, 2)$ with $x_{52} = 0$. The unit cost of flow on arc $(5, 2)$ is $-\$1$. These changes are shown in Figure 10.8.

Iteration 2.

Repeat steps 1–3 on the adjusted network. We obtain new u_i for nodes, $\bar{c}_{ij}$ for nonbasic arcs, flow increment Δ, and new cycle. All of these are shown in Figure 10.9.

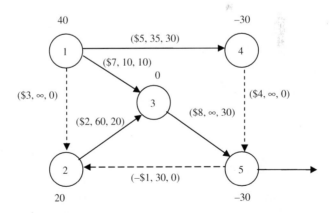

FIGURE 10.8   Adjusted network after the second feasible solution ($z' = 580$).

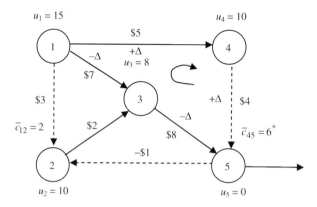

FIGURE 10.9   Steps 1–3 for the adjusted second feasible solution.

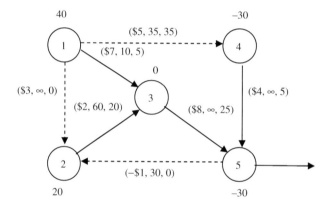

FIGURE 10.10   The third feasible solution ($z' = 550$).

Iteration 3.

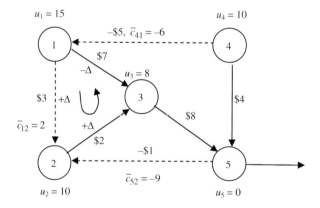

FIGURE 10.11   Steps 1–3 for the third feasible solution.

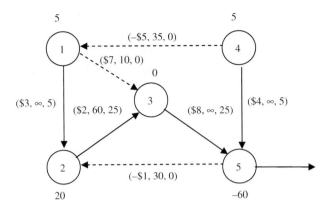

FIGURE 10.12   The fourth feasible solution ($z'' = 285$, $z' = 460$, $z = 490$).

Iteration 4.

Check the optimality following conditions:

$$\text{For } arc\ (1,3): \bar{c}_{13} = -2 \text{ and } x_{13} = 0\ (\text{satisfied})$$
$$\text{For } arc\ (4,1): \bar{c}_{41} = -4 \text{ and } x_{41} = 0\ (\text{satisfied})$$
$$\text{For } arc\ (5,2): \bar{c}_{52} = -9 \text{ and } x_{52} = 0\ (\text{satisfied})$$

Optimal solution:

$$\text{Basic variables}: x_{12} = 5, x_{23} = 25, x_{35} = 25, x_{45} = 5$$
$$\text{Nonbasic variables}: x_{13} = 0, x_{14} = 35 - x_{41} = 35 - 0 = 35,$$
$$x_{25} = 30 - x_{52} = 30 - 0 = 30$$
$$\text{Total cost} = \$490$$

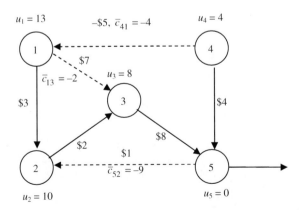

FIGURE 10.13   Steps 1–3 for the fourth feasible solution.

10.5   SOLUTION VIA LINGO®

The following LINGO® program can be used to find an optimum solution to the above example (five nodes, seven arcs), as well as any minimum cost network flow problem with appropriate network structure and input data (m, b_i, c_{ij}, U_{ij}).

```
MODEL:
SETS:
NODES/1..5/:DEMAND;
ARCS(NODES,NODES)/1,2 1,3 1,4 2,3 2,5 3,5 4,5/:CAP,FLOW,
COST;
ENDSETS
MINY=Y@SUM(ARCS:COST*FLOW);
@FOR(NODES(I):@SUM(ARCS(I,J):FLOW(I,J))
-@SUM(ARCS(K,I):FLOW(K,I))=DEMAND(I));
@FOR(ARCS:FLOW<=CAP);
DATA:
DEMAND = 40 50 0 -30 -60;
CAP = 100 10 35 60 30 100 100;
COST = 3 7 5 2 1 8 4;
ENDDATA
END
```

After running the program, we will obtain a standard (or default) output report that includes the optimum solution containing all decision variables (zero or nonzero), slack variables, reduced costs, dual prices, and even all input data. To avoid obtaining such a lengthy report, we may select an option in the following "Solution Report and Graph" menu to obtain specific set of information. For example, if we select the attributes "flow" and "nonzero," the report will contain an optimum solution containing only nonzero flow values as shown in Figure 10.14.

10.6   NOTES

In this chapter, we present only the "primal" network simplex algorithm for the MCNF problem. There are other algorithms such as the "dual" network simplex, the primal–dual, and the out-of-kilter. To know these algorithms, the interested reader may refer to Murty (1992), Ahuja et al. (1993), Bazaraa et al. (1990) and Phillips and Garcia-Diaz (1981).

We provide *how* the network simplex method can work directly on the network itself and give basic concepts and reasoning on its ties with the upper bound simplex method. For more details about the correspondence between the two methods, see Bazaraa et al. (2005).

Most introductory OR textbooks such as Hillier and Lieberman (2005), Taha (2007), and Winston (1994) ignore the procedure about how to find a starting feasible spanning tree. To learn how to construct it, read Bazaraa et al. (2005).

Variable	Value	Reduced Cost
FLOW(1, 2)	5.000000	0.000000
FLOW(1, 4)	35.00000	0.000000
FLOW(2, 3)	25.00000	0.000000
FLOW(2, 5)	30.00000	0.000000
FLOW(3, 5)	25.00000	0.000000
FLOW(4, 5)	5.000000	0.000000

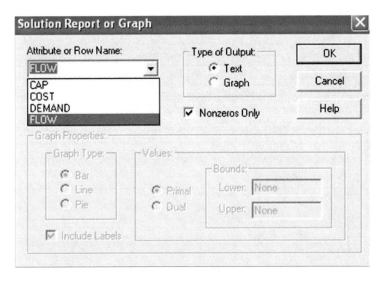

FIGURE 10.14   Solution to numerical example via LINGO.

The *implementation* of the network simplex method plays a very important role in making the method efficient. Knowing that the method is about finding a series of related spanning trees, the data (parameters and solutions) can be stored, accessed, and updated by using efficient tree or listed structures. The interested readers, especially computer programmers and analysts, should refer to Bazaraa et al. (2005), Murty (1992), and Ahuja et al. (1993).

10.7   EXERCISES

10.1   Given the graph shown in Figure 10.15, identify (a) a path from node 1 to node 6 and (b) a directed path from node 2 to node 7. If the numbers on each arc represent arc capacity, identify the capacity of the paths you find in (a) and (b).

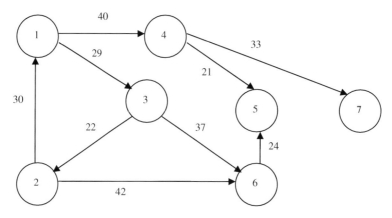

FIGURE 10.15 A directed graph.

10.2 Identify the vertex–edge incidence matrix of the graph in Exercise 10.1.

10.3 Given an $m \times n$ (0, 1, −1) matrix **M** where $m > n$, m, $n > 3$, how many submatrices do you need to evaluate so that the total unimodularity of matrix **M** can be identified? Show your reasoning.

10.4 Which of the following matrices are TU? Which are not? Why?

$$
\begin{pmatrix}
1 & 0 & 1 & 0 \\
-1 & 1 & 0 & 0 \\
0 & -1 & -1 & 0 \\
0 & -1 & -1 & 1 \\
0 & 0 & 0 & 1 \\
0 & 0 & 0 & 1
\end{pmatrix}
\qquad
\begin{pmatrix}
1 & 1 & 0 & 0 \\
1 & 0 & 1 & 1 \\
0 & 1 & 1 & 0 \\
1 & 1 & 0 & 1
\end{pmatrix}
\qquad
\begin{pmatrix}
-1 & 0 & 0 & 1 \\
1 & 0 & 1 & -1 \\
-1 & -1 & 0 & 1 \\
0 & -1 & -1 & 0 \\
0 & 1 & 1 & 0
\end{pmatrix}
$$
$$
\quad (1) \qquad\qquad\qquad (2) \qquad\qquad\qquad (3)
$$

10.5 Generate three more TU matrices using the TU matrix given below.

$$
\begin{pmatrix}
0 & 1 & 1 & -1 \\
-1 & 1 & 0 & 0 \\
0 & 0 & -1 & 1 \\
1 & -1 & 0 & 1
\end{pmatrix}
$$

10.6 If the TU matrix in Exercise 10.5 is the vertex–edge incidence matrix of some directed graph, draw the graph.

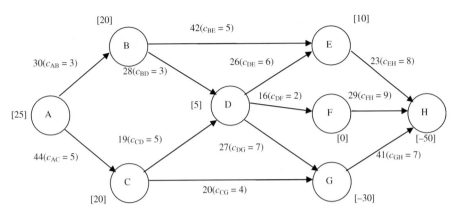

FIGURE 10.16 A maximum flow problem.

10.7 Consider the 0–1 matrix below. Is it possible that it is the vertex–edge incidence matrix of some undirected graph? If yes, draw the graph. Otherwise, why is it not?

$$\begin{pmatrix} 1 & 0 & 0 & 1 \\ 1 & 0 & 1 & 0 \\ 1 & 1 & 0 & 1 \\ 0 & 0 & 1 & 1 \\ 0 & 1 & 1 & 0 \end{pmatrix}$$

10.8 Check if the following matrices satisfy the sufficient condition of TU. If not, is it TU? Why?

$$\begin{pmatrix} 0 & -1 & 1 & 0 & 0 \\ 0 & 0 & -1 & -1 & 0 \\ 1 & 1 & 0 & -1 & -1 \\ 1 & 0 & 0 & 0 & 1 \end{pmatrix} \quad \begin{pmatrix} 1 & 0 & -1 & 0 \\ -1 & 1 & 0 & 0 \\ 0 & 1 & 0 & -1 \\ 0 & -1 & -1 & 1 \end{pmatrix} \quad \begin{pmatrix} 0 & -1 & 1 & 0 \\ 0 & 1 & 1 & -1 \\ -1 & 0 & 0 & 1 \end{pmatrix}$$
$$\quad\quad\quad (1) \quad\quad\quad\quad\quad\quad\quad (2) \quad\quad\quad\quad\quad\quad (3)$$

10.9 Consider the graph in Exercise 10.1 again. Identify (a) a tree and (b) a spanning tree (not minimal). Now assuming the numbers on each arc represent flow cost, identify a minimal spanning tree.

10.10 For the graph in Exercise 10.1, identify (a) a cycle and (b) a directed cycle.

10.11 Create two TU matrices using the sufficient conditions in Theorem 10.1. The two matrices must be of dimensions 5×5 and 6×4.

10.12 Consider a three-supplier, three-customer transportation problem. Show its coefficient matrix. Partition the columns of the matrix using the partitioning approach.

10.13 Solve the following maximum flow problem using the network simplex algorithm (Figure 10.16). Numbers on the arcs indicate the flow capacities.

PART III

SOLUTIONS

11

CLASSICAL SOLUTION APPROACHES

This chapter introduces three classical approaches for solving integer programs, namely, *branch-and-bound*, *cutting plane*, and *group theoretic*. Although all approaches are capable of solving integer programs, their degrees of success vary in software implementation. The cutting plane approach, when used as a stand-alone solver, has potential to solve IP programs of limited size, but may not work well in large-scale application. Similarly limited is the group theoretic approach, which has not been implemented as a stand-alone solver in practice. However, the valid inequality cuts generated by both cutting plane and group theoretic approaches can be useful when combined with branch-and-bound to yield a powerful branch-and-cut approach.

For over three decades, the branch-and-bound had been the prevailing solution method until the emergence of the *branch-and-cut* in early 1990s. Branch-and-cut combined branch-and-bound with the generated cutting planes into a much more efficient "hybrid" approach. Similarly, the group cuts generated from the group theoretical approach have also been incorporated, but at a lesser degree of integration. As a whole, extracting the strengths of these two approaches and injecting them into the branch-and-bound may greatly increase the modern solution power for integer programs. In what follows, we will introduce the concepts and background of these three solution approaches, and then exploit the potential strengths of each approach.

Applied Integer Programming: Modeling and Solution, By Der-San Chen, Robert G. Batson, and Yu Dang
Copyright © 2010 John Wiley & Sons, Inc.

11.1 BRANCH-AND-BOUND APPROACH

Branch-and-bound is a general-purpose approach capable of solving pure IP, mixed IP, and binary IP problems. For ease of exposition, sometimes we shall assume that the given problem is a pure IP problem because a similar algorithm can be applied to a mixed or binary IP problem. We also assume that the given problem is a maximization problem because modification of the algorithm for the minimization problem is straightforward.

11.1.1 Basic Concepts

Theoretically, any pure IP problem with *finite* bounds on integer variables can be solved by enumerating *all* possible combinations of integer values and determining a combination (solution) that satisfies all constraints and yields the maximal objective value—hence the name of *complete enumeration*. Unfortunately, the number of all possible combinations is prohibitively large to be evaluated even for a small problem. A problem of n integer variables with m values each has a total of m^n possible combinations (feasible and infeasible solutions). Therefore, complete enumeration is theoretically simple but practically intractable.

As a better alternative, *implicit enumeration* applies an intelligent enumeration scheme that can cover all possible solutions by explicitly evaluating only a small number of solutions while ignoring (or implicitly enumerating) a large number of *inferior* solutions. One such strategy is called *divide and conquer*. Basically, this strategy *divides* the given problem into a series of easier to solve subproblems that are systematically generated and solved (or *conquered*). The solutions of these generated subproblems are then put together to solve the original problem.

Branch-and-bound can be viewed as a divide-and-conquer approach to solving the IP problem, in which a branching process for dividing and a bounding process for conquering. Throughout the algorithm, a series of LP subproblems are systematically generated and solved. Then upper and lower bounds are progressively tightened on the objective value of the original IP problem.

A typical way to represent such a process is via a *branch-and-bound (B&B) tree*, which is a specialized enumeration tree for keeping track of how LP subproblems are generated and solved. The B&B tree by convention is drawn upside down with its root node at the top. The root node that represents the LP relaxation of the original IP problem (denoted by S_{LP}) is solved. If the LP optimum solution satisfies the integer requirement, the IP problem is solved. Otherwise, the LP objective value becomes the initial upper bound on the IP optimal objective value and the root node is partitioned into two successor nodes (subproblems) by two branches. These branches are valid cuts in terms of simple inequality constraints that have the following properties: (a) they cut off the current noninteger LP optimum point and other fractional region, and (b) the two successor nodes are mutually exclusive and their union contains the same integer feasible region as that of their predecessor (i.e., no integer points are eliminated). The solution of an LP relaxation on a node provides information about (a) whether a further branching from this node is needed (or whether the node can be *pruned*), and (b) a better lower bound (for maximization problem) on the objective of the original IP problem.

Note that in some texts, the term *pruned* may be replaced by *fathomed*, to indicate that no further exploration beyond that point is necessary.

There are three cases indicating that a node can be pruned: (1) the subproblem has no feasible LP solution, (2) the subproblem has an integer optimum solution, and (3) the upper bound of the subproblem optimum is less than or equal to the lower bound of the original problem. These three cases are, respectively, referred to as *pruned by infeasibility*, *pruned by optimality*, and *pruned by bound*. If a node is pruned by optimality, its optimum solution can be used to increase the lower bound on the objective value of the original IP problem.

Whenever an integer solution to a subproblem is obtained, it is a *candidate* optimum to the original IP problem. In the solution process of B&B, the best integer solution found so far is continuously updated. Such a solution is called an *incumbent solution*. To illustrate how the B&B algorithm works, we use the following two examples—one for pure and one for mixed IP problem.

Example 11.1   Solve the following pure IP problem by branch-and-bound approach.

$$\text{Maximize} \quad z = 5y_1 - 2y_2$$
$$\text{subject to} \quad -y_1 + 2y_2 \leq 5$$
$$3y_1 + 2y_2 \leq 19$$
$$y_1 + 3y_2 \geq 9$$
$$y_1, y_2 \geq 0 \text{ and integer}$$

We first solve the LP relaxation S_{LP}. As shown in Figure 11.1, the shaded area represents the LP feasible region and the solid lattice points the IP feasible solutions.

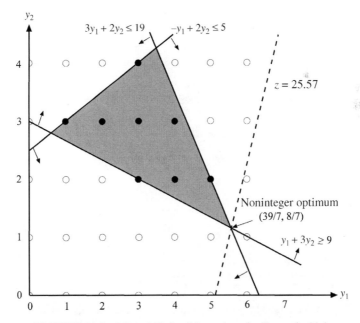

FIGURE 11.1   LP and IP feasible regions for Example 11.1.

We obtain the noninteger optimum $y_1 = 39/7$, $y_2 = 8/7$, and $z = 25.57$. Then the objective value 25.57 becomes an upper bound to the IP problem. We set the lower bound to $-\infty$. Since both variables are fractional, we need to branch on them in an attempt to obtain an integer optimum.

We arbitrarily select y_1 as the variable to be branched. Two subproblems are generated by adding the constraint of $y_1 \geq 6$ and $y_1 \leq 5$, respectively, to the LP relaxation. From Figure 11.2 we can see that the triangle area $\mathbf{S}'$ is cut off by $y_1 \leq 5$. Clearly, the branch with the added constraint $y_1 \geq 6$ is infeasible, so it is *pruned by infeasibility*. The other branch with the added constraint $y_1 \leq 5$ is optimized at $(y_1, y_2) = (5, 4/3)$, with objective value 22.33. So the new *upper bound* is updated to 22.33.

Again, the variable y_2 is fractional, so this time we branch on y_2. The two constraints $y_2 \geq 2$ and $y_2 \leq 1$ are then added. This time the area $\mathbf{S}''$ is cut off, as shown in Figure 11.3.

The branch with $y_2 \leq 1$ is infeasible, and hence is pruned by infeasibility. The branch with $y_2 \geq 3$ is optimized at $(y_1, y_2) = (5, 2)$, with objective value 21. Since this is a feasible solution to the IP problem, the value 21 becomes a new *lower bound* to the problem, replacing the initial lower bound $-\infty$, and (5, 2) is a *candidate solution*. Checking the tree, all branches are evaluated, so $(y_1, y_2) = (5, 2)$ is the optimal solution to the IP problem, and the optimal objective value is 21.

The branch-and-bound algorithm is usually depicted as an enumeration tree, in which the nodes denote the subproblems, and the branches correspond to constraints

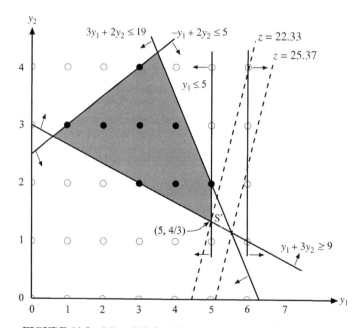

FIGURE 11.2   LP and IP feasible regions after the first branching.

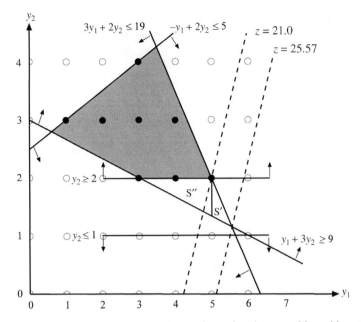

FIGURE 11.3 LP and IP solution regions after the second branching.

(cuts) that separate the subproblems from their parent subproblems. The number above each node is the optimal solution to the LP subproblem generated on that node (which is also the upper bound on that branch). The number below the node indicates the best lower bound on the original IP problem found so far. The previous procedure is depicted in Figure 11.4.

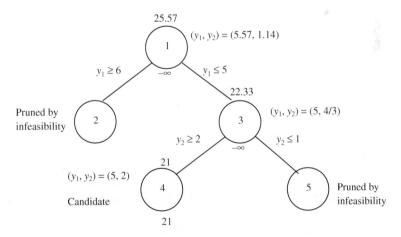

FIGURE 11.4 Branch-and-bound tree for Example 11.1.

When an LP solution contains several fractional integer variables, the decision of which integer variable to branch on next is needed. The following rules are commonly used for choosing a branching variable:

1. Variable with fractional value closest to 0.5
2. Variable with highest impact on objective function
3. Variable with the least index

A decision is also needed as to which unpruned node to explore first. The most commonly used search strategies include

1. Depth-first (last-in first-out; solve the most recently generated subproblem first)
2. Best-bound-first (best upper bound; branch on the active node with greatest z-value)

The goal of the depth-first strategy is to quickly obtain a primal feasible integer solution whose objective function value z^k is a lower bound on the given IP problem and can be used to prune nodes by optimality (rule 3). The best-bound-first strategy chooses the active node with the best upper bound (for maximization problem). The goal is to minimize the total number of nodes evaluated in the B&B tree. Performance of these branching rules depends on the problem structure. In practice, a compromise between the two is adopted. That is, apply the depth-first strategy first to get one feasible integer solution, followed by a mixture of both strategies.

Example 11.2   Solve the following mixed integer problem using branch-and-bound approach. At each step, apply the rule of best-bound-first, and at each node, select the variable with least index to branch first.

$$\text{Maximize} \quad z = -y_1 + 2y_2 + y_3 + 2x_1$$
$$\text{subject to} \quad y_1 + y_2 - y_3 + 3x_1 \leq 7$$
$$y_2 + 3y_3 - x_1 \leq 5$$
$$3y_1 + x_1 \geq 2$$
$$y_1, y_2, y_3 \geq 0 \text{ and integer}$$
$$x_1 \geq 0$$

After solving the LP relaxation, we obtain an LP optimum $y_1 = 6/11$, $y_2 = 59/11$, $y_3 = 0$, $x_1 = 4/11$, and $z = 120/11$. This solution violates the integer requirements of y_1 and y_2. We use this solution as the root node the branch-and-bound tree in Figure 11.5. The number of each node indicates the sequence of subproblems evaluated. Note

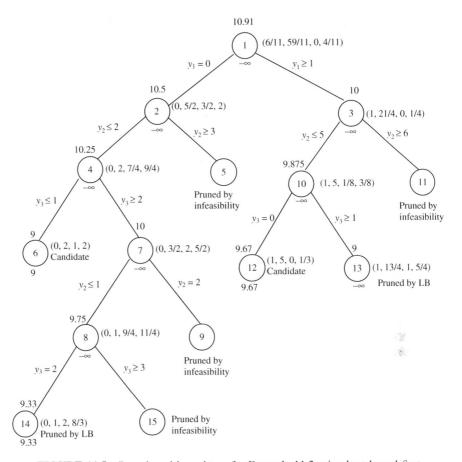

FIGURE 11.5 Branch-and-bound tree for Example 11.2 using best-bound-first.

that at node 1, the constraint $y_1 \leq 0$ was indicated on the left branch, but since $y_1 \geq 0$, y_1 has to be fixed at 0. At node 7, the constraint $y_2 \geq 2$ was intended to be added, but if we trace back along node 7, we would see that the constraint of $y_2 \leq 2$ was already added at node 2. Combining these two constraints, we have $y_2 = 2$. So is the constraint of $y_3 = 2$ at node 8. The problem is finally optimized at node 12, where $(1, 5, 0, 1/3)$ is the optimal solution, with objective value 9.67.

Figure 11.6 depicts the branch-and-bound tree for the same problem, where the "depth-first" rule is applied, and at each node, the variable (violating an integer constraint) with the largest absolute value cost coefficient is chosen to branch first. Ties are broken arbitrarily.

Depth-first is sometimes called last-in first-out (LIFO) because it solves the most recently generated subproblem first. It tends to pursue paths to the depths of the tree, then backtrack to where that path started, and finally plunge down into another

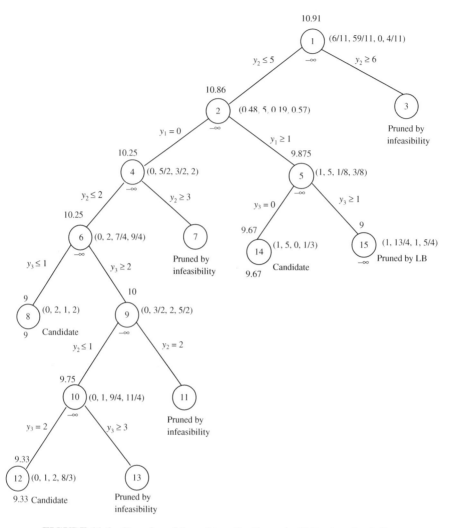

FIGURE 11.6 Branch-and-bound tree for Example 11.2 using depth-first.

depth search. Yet another name for depth-first is "backtracking." Best-bound-first is sometimes called "jumptracking" because it leads to searches that jump back and forth across the tree.

11.1.2 Branch-and-Bound Algorithm

Now we describe the general branch-and-bound algorithm using the following notation.

S = the given IP problem
S_{LP} = the LP relaxation of S

y_{LP} = the solution to the LP relaxation of the given IP

$\bar{z}$ = lowest (best) upper bound on z^* of the given IP problem

$\underline{z}$ = highest (best) lower bound on z^* of the given IP problem

These are *global* bounds that are periodically updated as the branching proceeds down the various paths in the tree, but are *not* shown on the tree. In Example 11.2, $\bar{z} = 10.91$ throughout and $\underline{z}$ was $-\infty$, then 9, 9.33, and 9.67 at nodes 1, 8, 12, and 14, respectively. Next comes more notation. Let

S^k = subproblem k of problem S

S_{LP}^k = the LP relaxation of subproblem k

z^k = the optimum objective value of S^k

$\bar{z}^k$ = best (lowest) upper bound of subproblem S^k (shown above node k)

$\underline{z}^k$ = best (highest) lower bound of subproblem S^k (shown below node k)

y_{LP}^k = the optimum solution of the LP subproblem S_{LP}^k

$\bar{y}_j$ = noninteger value of integer variable y_j (current numerical value of y_j)

$\lfloor a \rfloor$ = the largest integer $\leq a$ (or rounding down a)

$\lceil a \rceil$ = the smallest integer $\geq a$ (or rounding up a)

We now formally describe the B&B algorithm

Step 0 (Initialization). Solve the LP relaxation (S_{LP}) of the given IP problem (S). If it is infeasible, so is the IP problem—terminate. If the LP optimum solution satisfies the integer requirement, the IP problem is solved—terminate. Otherwise, initialize the best upper bound ($\bar{z}$) by the optimal objective value of problem S_{LP} and the best lower bound by $\underline{z} = -\infty$. Place S_{LP}^k on the active list of nodes (subproblems). Initially, there is no incumbent solution.

Step 1 (Choosing a Node). If the active list is empty, terminate. The incumbent solution y^* is optimal. Otherwise, choose a node (subproblem) S^k with S_{LP}^k by one of the rules (e.g., depth-first, best-bound-first, etc.)

Step 2 (Updating Upper Bound). Solve and set $\bar{z}^k$ equal to the LP optimum objective value. Keep the optimum LP solution y_{LP}^k.

Step 3 (Prune by Infeasibility). If S_{LP}^k has no feasible solution, prune the current node and go to step 1. Otherwise, go to step 4.

Step 4 (Prune by Bound). If $z^k \leq \underline{z}$, prune the current node and go to step 1. Otherwise, go to step 5.

Step 5 (Updating Lower Bound and Prune by Optimality).

(a) If the LP optimum y_{LP}^k is integer, a feasible solution to S is found, an incumbent solution to the given problem. Set $\underline{z}^k = y_{LP}^k$ and compare $\underline{z}^k$ with $\underline{z}$. If $\underline{z}_k > \underline{z}$, set $\underline{z} = \underline{z}_k$, otherwise $\underline{z}$ does not change. The current node is pruned because no better solution can be branched down from this node. Go to step 1.

(b) If the LP optimum y_{LP}^k is noninteger, go to step 6.

Step 6 (Branching). From the current node S^k choose a variable y_j with fractional value to generate two subproblems S_1^k and S_2^k defined by

$$S_1^k = S^k \cap \left\{ y : y_j \leq \lfloor \bar{y}_j \rfloor \right\}$$

$$S_2^k = S^k \cap \left\{ y : y_j \geq \lceil \bar{y}_j \rceil \right\}$$

Place both of these two nodes in the active list and go to step 1.

11.2 CUTTING PLANE APPROACH

In geometry, an equation in two variables is called a *plane* and an equation in n variables a *hyperplane*, strictly speaking. For simplicity, however, both in practice are often referred to as a plane, regardless of the number of variables. Strictly speaking, an inequality constraint in n variables is called a *half-space*, not a hyperplane. But an inequality constraint can always be converted to an equation by adding or subtracting a nonnegative slack variable. The term *cutting plane* is often used for an equality or inequality constraint that can cut off a fractional part of an LP feasible region, without excluding any *integer* feasible solution. In the cutting plane approach, one or more such cutting planes are added to the current LP simplex tableau, which in turn are resolved for a new LP optimum. This process is repeated until the prescribed integer requirements are satisfied. In this text, the collection of all such cutting plane methods will be called a *cutting plane approach* (more specifically, a *dual cutting plane approach*, due to the use of the dual simplex method for LP reoptimization).

11.2.1 Dual Cutting Plane Approach

A large variety of cutting plane methods were developed during the 1950s and 1960s. Among them, the most prominent ones belong to the class of *dual* cutting plane approach such as the fractional and mixed cutting plane methods developed by Gomory. This class shares a common solution algorithm when they are utilized as a *stand-alone* solver.

Step 1. Solve the integer program as if it were a linear program. If it is infeasible, so is the integer program and then stop. Else if an LP optimal solution satisfying the integer requirements is found, then the IP is solved. Otherwise, go to step 2.

Step 2. Select a row to be a *generating row* (or *source row*) from the LP optimum simplex tableau.

Step 3. Derive a cut constraint from the generating row and augment it to the current tableau, resulting in a primal infeasible solution.

Step 4. Apply the dual simplex method to reoptimize the augmented linear program. If the new LP optimum satisfies the integer requirements, the original MIP program is solved. Otherwise, go to step 2.

The main difference among various methods of cutting plane is *how* a cut constraint is generated. The main requirement is that a generated cut constraint must be *valid*, meaning that its addition will result in cutting off the current LP optimal point but will not eliminate any *integer feasible* solution. In other words, every valid cut has two properties:

- The current optimal solution to the LP relaxation problem will violate the cut constraints.
- Any feasible point to the corresponding IP or MIP problems will satisfy the cut constraint.

The class of dual cutting plane methods begins with an optimal LP solution and requires application of a series of dual simplex steps to reoptimize a series of new LP problems, each adding one or more constraints to the current simplex tableau (although some cuts may be dropped from later considerations).

There is another class of cutting plane methods known as the *primal cutting plane approach*. This approach commonly begins with a primal simplex tableau and creates a series of primal simplex tableaux, from which cuts are generated. As a result, all the subsequent simplex tableaux will remain primal feasibility and dual infeasibility. The primal simplex method is applied throughout the process until both primal and dual solutions are feasible, in which case an optimum solution if found. We shall not describe them in detail. The interested reader may refer the Section 11.5.

11.2.2   Fractional Cutting Plane Method

The fractional cutting plane method is capable of solving *pure* integer programs. This method requires that the starting IP problem must contain *all-integer coefficients* so that all slack variables, including those that are added after introduction of cuts, are ensured to be nonnegative integers. Note that the integer assumption of the starting IP problem does not limit the problem application because any coefficients consisting of *rational numbers* can always be made integers by multiplying an appropriate number.

The fractional cutting plane method begins with an *optimal* simplex tableau of the LP relaxation given below (recall that we use y_j to denote integer variables):

$$\text{Maximize} \quad z + \sum_k \bar{d}_k y_k = \bar{d}_0$$

$$\text{subject to} \quad y_{B_i} + \sum_k \bar{g}_{ik} y_k = \bar{b}_i \quad i = 1, 2, \ldots, m \tag{11.1}$$

$$y_{B_i} \geq 0 \qquad\qquad i = 1, 2, \ldots, m$$

$$y_k = 0 \qquad\qquad k = 1, 2, \ldots, p$$

where y_{B_i} and y_k denote basic variables and nonbasic variables, respectively. Note that the current LP optimum solution is $y_{B_i} = \bar{b}_i$ and $y_k = 0$, in which some $\bar{b}_i$ are assumed to be noninteger. Optimality conditions require that $\bar{d}_k \geq 0$ for all k.

To find an integer solution, we arbitrarily select a row with $\bar{b}_i$ noninteger. The selected row, say r, is called a *source row* or *generating row*, from which a fractional cut will be generated. Consider the source row

$$y_{B_r} + \sum_k \bar{g}_{rk} y_k = \bar{b}_r \quad (k = 1, \ldots, p)$$

which can be rewritten by separating fractional and integral parts of all data:

$$y_{B_r} + \sum_k \{(\bar{g}_{rk} - \lfloor \bar{g}_{rk} \rfloor) + \lfloor \bar{g}_{rk} \rfloor\} y_k = (\bar{b}_r - \lfloor \bar{b}_r \rfloor) + \lfloor \bar{b}_r \rfloor$$

where $\lfloor a \rfloor$ denotes the largest integer $\leq a$. For example, $\lfloor 5.4 \rfloor = 5$, $\lfloor -1.8 \rfloor = -2$, and $\lfloor 3 \rfloor = 3$. The fractional part is always ≥ 0. For simplicity, we let

$$f_{rk} = \bar{g}_{rk} - \lfloor \bar{g}_{rk} \rfloor$$
$$f_{r0} = \bar{b}_r - \lfloor \bar{b}_r \rfloor$$

be the fractional parts of tableau coefficients and the RHS of row r. Rearranging the terms, we have

$$y_{B_r} + \sum_k \lfloor \bar{g}_{rk} \rfloor y_k - \lfloor \bar{b}_r \rfloor = f_{r0} - \sum_k f_{rk} y_k \qquad (11.2)$$

Now in order for y_{B_r} and y_k $(k = 1, \ldots, p)$ to be integer, both the left-hand and right-hand sides of (11.2) must be integer. By the definition of congruence, we have

$$f_{r0} - \sum_k f_{rk} y_k = 0 \; (\text{mod } 1) \qquad (11.3)$$

But since $f_{r0} - \sum_k f_{rk} y_k \leq f_{r0} < 1$, the necessary condition for integrality becomes

$$f_{r0} - \sum_k f_{rk} y_k \leq 0$$

$$\text{or} \quad \sum_k f_{rk} y_k = f_{r0} \quad (\text{Gomory fractional cutt}) \qquad (11.4)$$

$$\text{or} \quad \sum_k -f_{rk} y_k + s = -f_{r0} \quad (\text{Gomory fractional cut}) \qquad (11.5)$$

where $s \geq 0$ is Gomory's slack variable associated with the fractional cut. Note that all f_{rk} and f_{r0} must be nonnegative fractions, that is, $0 \leq f_{rk}$ $(k = 1, \ldots, p)$ and $f_{r0} \leq 1$.

Example 11.3 Solve the pure IP problem in (11.1) using the cutting plane method.

TABLE 11.1 Tableau 1 for Example 11.3

Basic Variable	z	y_1	y_2	y_3	y_4	y_5	RHS
z	1	0	0	0	17/7	16/7	179/7
y_1	0	1	0	0	3/7	2/7	39/7
y_2	0	0	1	0	−1/7	−3/7	8/7
y_3	0	0	0	1	5/7	8/7	58/7

We first add a nonnegative slack variable to each inequality constraint:

$$\text{Maximize} \quad z = 5y_1 - 2y_2$$
$$\text{subject to} \quad -y_1 + 2y_2 + y_3 = 5$$
$$3y_1 + 2y_2 + y_4 = 19$$
$$-y_1 - 3y_2 + y_5 = -9$$
$$y_1, y_2, y_3, y_4, y_5 \geq 0 \text{ and integer}$$

We then solve the LP relaxation, yielding the optimal simplex tableau shown in Table 11.1.

The optimum solution is noninteger: $y_1 = 39/7$, $y_2 = 8/7$, $y_3 = 58/7$, $y_4 = y_5 = 0$, and $z = 179/7$. We arbitrarily select y_1 row as the source row and generate the following fractional cut.

$$\frac{3}{7}y_4 + \frac{2}{7}y_5 \geq \frac{4}{7}$$

or

$$-\frac{3}{7}y_4 - \frac{2}{7}y_5 + s_1 = -\frac{4}{7}$$

where s_1 is called *Gomory's slack variable* to differentiate from the ordinary slack variable. Appending the equation to tableau 1, we obtain tableau 2 (Table 11.2).

Applying the dual simplex iteration, s_1 is replaced by y_4 yielding tableau 3 (Table 11.3).

Repeating generation of fractional cuts and application of dual simplex iterations, the reader may verify the subsequent simplex tableaux 4 through 6 (Tables 11.4–11.6).

TABLE 11.2 Tableau 2 for Example 11.3

Basic Variable	z	y_1	y_2	y_3	y_4	y_5	s_1	RHS
z	1	0	0	0	17/7	16/7	0	179/7
y_1	0	1	0	0	3/7	2/7	0	39/7
y_2	0	0	1	0	−1/7	−3/7	0	8/7
y_3	0	0	0	1	5/7	8/7	0	58/7
s_1	0	0	0	0	−3/7	−2/7	1	−4/7

TABLE 11.3 Tableau 3 for Example 11.3

Basic Variable	z	y_1	y_2	y_3	y_4	y_5	s_1	RHS
z	1	0	0	0	0	2/3	17/3	67/3
y_1	0	1	0	0	0	0	1	5
y_2	0	0	1	0	0	$-1/3$	$-1/3$	4/3
y_3	0	0	0	1	0	2/3	5/3	22/3
y_4	0	0	0	0	1	2/3	$-7/3$	4/3

TABLE 11.4 Tableau 4 for Example 11.3

Basic Variable	z	y_1	y_2	y_3	y_4	y_5	s_1	s_2	RHS
z	1	0	0	0	0	2/3	17/3	0	67/3
y_1	0	1	0	0	0	0	1	0	5
y_2	0	0	1	0	0	$-1/3$	$-1/3$	0	4/3
y_3	0	0	0	1	0	2/3	5/3	0	22/3
y_4	0	0	0	0	1	2/3	$-7/3$	0	4/3
s_2	0	0	0	0	0	$-2/3$	$-2/3$	1	$-1/3$

TABLE 11.5 Tableau 5 for Example 11.3

Basic Variable	z	y_1	y_2	y_3	y_4	y_5	s_1	s_2	RHS
z	1	0	0	0	0	0	5	1	22
y_1	0	1	0	0	0	0	1	0	5
y_2	0	0	1	0	0	0	0	$-1/2$	3/2
y_3	0	0	0	1	0	0	1	1	3
y_4	0	0	0	0	1	0	-3	1	1
y_5	0	0	0	0	0	1	1	$-3/2$	1/2
s_3	0	0	0	0	0	0	0	$-1/2$	$-1/2$

TABLE 11.6 Tableau 6 for Example 11.3

Basic Variable	z	y_1	y_2	y_3	y_4	y_5	s_1	s_2	s_3	RHS
z	1	0	0	0	0	0	5	0	2	21
y_1	0	1	0	0	0	0	1	0	0	5
y_2	0	0	1	0	0	0	0	0	1	2
y_3	0	0	0	1	0	0	1	0	2	2
y_4	0	0	0	0	1	0	-3	0	2	0
y_5	0	0	0	0	0	1	1	0	-3	2
s_2	0	0	0	0	0	0	0	1	-2	1

Because all basic variables are integer, we have an integer optimum $y_1 = 5$, $y_2 = 2$, and $z = 21$. The solution is the same as that obtained by branch-and-bound approach.

The cutting plane approach often takes a large number of cuts to reach an integer solution even for a small or moderate sized IP problem, although it can be shown that the fractional cutting plane method is ensured to converge to an IP optimum after a finite number of cuts. Here, we arbitrarily select a source row, although alternative rules may be applied to select other source rows. For example, we may select a source row r with f_{r0} closest to 0.5, or select a row with the largest f_{r0}. However, no evidence shows that a certain selection rule is better than the others in all cases.

11.2.3 Mixed Integer Cutting Plane Method

The mixed integer cutting plane method, also due to Gomory, can be used to solve the following MIP problem:

$$\text{Maximize} \quad z = \sum_j c_j x_j + \sum_k d_k y_k$$

$$\text{subject to} \quad \sum_j a_{ij} x_j + \sum_k g_{ik} y_k \leq b_i \quad (i = 1, 2, \ldots, m)$$

$$x_j \geq 0 \qquad\qquad\qquad (j = 1, 2, \ldots, n)$$

$$y_k \geq 0 \text{ and integer} \quad (k = 1, 2, \ldots, p)$$

Essentially, the solution procedure is similar to that of the fractional cutting plane method. It generates a valid cut from the optimal simplex tableau of the LP relaxation of the MIP problem. Any row r associated with y_k, which is basic but has fractional right-hand side, may be chosen to generate the cut. Just like the fractional cuts, each of the generated mixed cuts will violate primal feasibility and will be restored to primal feasibility after applying the dual simplex method.

Let the nonzero coefficients $(\bar{a}_{rj})$ of the continuous variables x_j $(j \in J)$ be partitioned into two sets: positive coefficients $(\bar{a}_{rj} > 0)$ and negative coefficients $(\bar{a}_{rj} < 0)$. Also, let $f_{rk} = \bar{g}_{rk} - \lfloor \bar{g}_{rk} \rfloor$ and $f_{r0} = \bar{b}_r - \lfloor \bar{b}_r \rfloor$ as before. It can be shown that a mixed cut due to Gomory can be derived (see Section 11.5) as

$$\sum_{j: \bar{a}_{rj} > 0} \bar{a}_{rj} x_j + \sum_{j: \bar{a}_{rj} < 0} \left(\frac{f_{r0}}{f_{r0} - 1} \right) \bar{a}_{rj} x_j + \sum_{k: f_{rk} \leq f_{r0}} f_{rk} y_k + \sum_{k: f_{rk} > f_{r0}} \frac{f_{r0}(1 - f_{rk})}{1 - f_{r0}} y_k \geq f_{r0}$$

We use the following numerical problem to show how to generate a mixed integer cut. The remaining procedure will be similar to that in Example 11.3, except that no rows corresponding to continuous variables are used for source rows to generate cuts.

Example 11.4   Solve the given MIP problem using a cutting plane approach.

$$\text{Maximize} \quad z = 5x_1 + 3x_2 + 7y_1 + 2y_2$$
$$\text{subject to} \quad 7x_1 + 8x_2 + 9y_1 + 3y_2 \leq 43$$
$$11x_1 + 4x_2 + 4y_1 + 5y_2 \leq 51$$
$$\mathbf{x} \geq \mathbf{0}$$
$$\mathbf{y} \geq \mathbf{0} \text{ and integer}$$

Solving the LP relaxation, we obtain an LP optimum $y_1 = 43/9$, with the following source row:

$$\frac{7}{9}x_1 + \frac{8}{9}x_2 + y_1 + \frac{1}{3}y_2 + \frac{1}{9}s_1 = \frac{43}{9}$$

Here, we have all positive coefficients and no negative coefficients for the continuous variables. Compute

$$f_{r0} = \frac{7}{9}$$
$$f_{r1} = 0$$
$$f_{r2} = \frac{1}{3}$$

and we obtain a mixed integer cut

$$\frac{7}{9}x_1 + \frac{8}{9}x_2 + \frac{1}{3}y_2 + \frac{1}{9}s_1 \geq \frac{7}{9}$$

In Exercise 11.10, the reader is asked to continue on this example.

11.3   GROUP THEORETIC APPROACH

Gomory showed that the coefficient *row* vectors of the derived inequalities form a finite set that is closed under the operation of addition when the arithmetic operations are taken modulo 1 (i.e., integer parts are dropped). Such a set forms what is called a *group*. Furthermore, this group can have at most D elements, where D is the absolute value of the determinant of the current LP basis. If the starting basis is an identity matrix, then this group contains exactly D elements.

Gomory also showed that by relaxing nonnegative (but not integer) requirements of the current basic variables, an integer program can be transformed into one in which the *columns* of constraint coefficients and the right-hand side are elements of an abelian group. If this group problem (in terms of nonbasic variables only) is solved and a solution containing nonnegative values for all variables is obtained, then the original integer program is solved. Before describing the role of group theory in integer programming, we need some definitions from group theory.

11.3.1   Group Theory Terminology

Definition 11.1   A *group* is a set of *elements* with a single operation (e.g., ordinary addition is taken modulo 1) defined on pairs of elements such that the operation is *closed, associative,* and for each element there exist unique *identity* and *inverse* elements.

Specifically, elements here are column vectors. *Closure* means that the sum of any two elements is also an element in the group. *Associativity* means that the operation satisfies the law of association such that $\mathbf{a} + (\mathbf{b} + \mathbf{c}) = (\mathbf{a} + \mathbf{b}) + \mathbf{c}$. The identity element has the property that any element in the group added to the identity (on the left or right) will result in itself. The inverse of an element has the property that the sum of the inverse and this element will result in the identity element.

Definition 11.2   A group is *finite* if it contains a finite number of elements. The *order* of a finite group is the number of elements comprising the group.

Definition 11.3   An *additive group* is a group whose operation is an ordinary addition (with modulo 1). An *abelian group* is one in which the operation is *commutative* such that $\mathbf{a} + \mathbf{b} = \mathbf{b} + \mathbf{a}$ for all elements $\mathbf{a}$ and $\mathbf{b}$ in the group.

Definition 11.4   A cyclic *group* is one in which there exists an element such that successive additions of the element to itself (or a scalar multiple of itself) will generate the entire group. If a group does not have such an element, then the group is called *noncyclic* or *acyclic.*

Definition 11.5   The number a is *congruent* to b modulo c if there exists an integer n such that $a - b = nc$. The *congruence relationship* (or simply *congruence*) is written as $a = b \pmod{c}$. For example, $2 = 5 \pmod 3$, $-1 = 3 \pmod 2$, and $1 = 0 \pmod 1$.

To explain the above group definitions, we utilize tableau 1 of Example 11.3. The group contains three *elements*: $(1/3, 2/3, 2/3)^T$, $(2/3, 1/3, 1/3)^T$, and $(0, 0, 0)^T$. Hence, the group is *finite* with the *order* of 3. The group is *closed* because the sum (modulo 1) of any two elements will also result in an element in the group. For example, $(1/3, 1/3, 1/3)^T + (0, 0, 0)^T = (1/3, 1/3, 1/3)^T \pmod 1$; $(2/3, 2/3, 2/3)^T + (0, 0, 0)^T = (2/3, 2/3, 2/3)^T \pmod 1$; and $(1/3, 1/3, 1/3)^T + (2/3, 2/3, 2/3)^T = (0, 0, 0)^T \pmod 1$. It can be easily shown that the *associative law* also holds:

$$[(1/3, 1/3, 1/3)^T + (2/3, 2/3, 2/3)^T] + (0, 0, 0)^T$$
$$= (1/3, 1/3, 1/3)^T + [(2/3, 2/3, 2/3)^T + (0, 0, 0)^T]$$
$$= (0, 0, 0)^T \pmod 1$$

Element $(0, 0, 0)^T$ is the identity element for the entire group because any other element added to it will result in itself. Element $(1/3, 1/3, 1/3)^T$ is the *inverse element* of $(2/3, 2/3, 2/3)^T$ and vice versa because $(1/3, 1/3, 1/3)^T + (2/3, 2/3, 2/3)^T = (1, 1, 1)^T \pmod 1 = (0, 0, 0)^T$. The group is *additive* because its single operation on any coordinate is an ordinary addition. The group is abelian because the operation is communicative:

$$[(1/3, 1/3, 1/3)^T + (2/3, 2/3, 2/3)^T] \pmod 1$$
$$= [(2/3, 2/3, 2/3)^T + (1/3, 1/3, 1/3)^T] \pmod 1 = (0, 0, 0)^T$$

11.3.2 Deriving the Group (Minimization) Problem

Consider the integer program with *inequality* constraints

$$
\begin{aligned}
\text{Maximize} \quad & z = \mathbf{c}'^T \mathbf{y}' \\
\text{subject to} \quad & \mathbf{A}' \mathbf{y}' \leq \mathbf{b} \\
& \mathbf{y}' \geq \mathbf{0} \text{ and integer}
\end{aligned}
\tag{11.6}
$$

where $\mathbf{A}'$ is an $m \times n$ integer matrix, $\mathbf{y}'$ an n column vector, $\mathbf{b}$ an m integer column vector, and $\mathbf{c}'^T$ an n integer row vector. After adding m nonnegative slack variables, $y_{n+1}, y_{n+2}, \ldots, y_{n+m}$, one to each inequality, we have the equivalent IP problem with *equality* constraints

$$
\begin{aligned}
\text{Maximize} \quad & z = \mathbf{c}^T \mathbf{y} \\
\text{subject to} \quad & \mathbf{A} \mathbf{y} = \mathbf{b} \\
& \mathbf{y} \geq \mathbf{0} \text{ and integer}
\end{aligned}
\tag{11.7}
$$

where $\mathbf{A} = (\mathbf{A}', \mathbf{I})$ is an $m \times (n + m)$ integer matrix, $\mathbf{I}$ an $m \times m$ identity matrix, $\mathbf{y}$ an $(n + m)$ column vector, $\mathbf{b}$ an m integer column vector, and $\mathbf{c}^T = (\mathbf{c}'^T, \mathbf{0}^T)$ an $(n + m)$ integer row vector.

Partitioning variables into sets of basic and nonbasic such that $\mathbf{y} = (\mathbf{y_B}, \mathbf{y_N})^T$ with associated coefficients $\mathbf{c}^T = (\mathbf{c_B}^T, \mathbf{c_N}^T)$ and $\mathbf{A} = (\mathbf{B}, \mathbf{N})$, we have the partitioned problem

$$
\begin{aligned}
\text{Maximize} \quad & z = \mathbf{c_B}^T \mathbf{y_B} + \mathbf{c_N}^T \mathbf{y_N} \\
\text{subject to} \quad & \mathbf{B} \mathbf{y_B} + \mathbf{N} \mathbf{y_N} = \mathbf{b} \\
& \mathbf{y_B}, \mathbf{y_N} \geq \mathbf{0} \text{ and integer}
\end{aligned}
\tag{11.8}
$$

where $\mathbf{B}$ is an $m \times m$ basis matrix and $\mathbf{N}$ an $m \times n$ nonbasis matrix. To express $\mathbf{y_B}$ in terms of $\mathbf{y_N}$, we premultiply $\mathbf{B}^{-1}$ on the equalities in problem (11.8):

$$\mathbf{I} \mathbf{y_B} + \mathbf{B}^{-1} \mathbf{N} \mathbf{y_N} = \mathbf{B}^{-1} \mathbf{b}$$

or

$$\mathbf{y_B} = \mathbf{B}^{-1} \mathbf{b} - \mathbf{B}^{-1} \mathbf{N} \mathbf{y_N}$$

Substituting $\mathbf{y_B}$ into the objective function, we obtain the IP problem in terms of $\mathbf{y_N}$:

$$\text{Maximize} \quad z = \mathbf{c_B^T B^{-1} b} - (\mathbf{c_B^T B^{-1} N} - \mathbf{c_N^T})\mathbf{y_N}$$
$$\text{subject to} \quad \mathbf{Iy_B} + \mathbf{B^{-1} Ny_N} = \mathbf{B^{-1} b} \qquad (11.9)$$
$$\mathbf{y_B}, \mathbf{y_N} \geq \mathbf{0} \text{ and integer}$$

Relaxing integer requirements and letting $\mathbf{B}$ be the LP optimum basis, we have the optimum solution $\mathbf{y_B} = \mathbf{B^{-1} b}$, $\mathbf{y_N} = \mathbf{0}$ with $z = \mathbf{c_B^T B^{-1} b}$. The optimality conditions are $\mathbf{c_B^T B^{-1} N} - \mathbf{c_N^T} \geq \mathbf{0}$. If $\mathbf{B^{-1} b}$ happens to be an integer vector, the LP optimum solution is also an IP optimum. When the solution vector $\mathbf{B^{-1} b}$ contains noninteger components, we must increase some $y_j \in \mathbf{y_N}$ from value 0 to some positive amount while maintaining the following conditions:

$$\mathbf{y_B} = \mathbf{B^{-1} b} - \mathbf{B^{-1} Ny_N} \geq \mathbf{0} \text{ and integer}$$

This poses two questions concerning the vector of basic variables $\mathbf{y_B}$:

(1) When is $\mathbf{y_B}$ an integer vector?

(2) When is $\mathbf{B^{-1} b} - \mathbf{B^{-1} Ny_N} \geq \mathbf{0}$?

We first address question (1) concerning integer vector. Denote the columns of matrix $\mathbf{B^{-1} N}$ as $(\boldsymbol{\alpha}_1, \boldsymbol{\alpha}_2, \ldots, \boldsymbol{\alpha}_n)^T$ and the right-hand side column $\mathbf{B^{-1} b}$ as $\boldsymbol{\alpha}_0$. Then question (1) can be posed as a problem of finding some nonnegative components y_j $(j = 1, 2, \ldots, n)$ of $\mathbf{y_N}$ such that

$$\sum_j \boldsymbol{\alpha}_j y_j = \boldsymbol{\alpha}_0 (\text{mod } 1) \qquad (11.10)$$
$$y_j \geq 0 \text{ and integer } y_j \in \mathbf{y_N}$$

Note that an integer vector $\boldsymbol{\alpha}_j$ multiplied by any integer scalar y_j will yield a null vector $\mathbf{0}(\text{mod } 1)$ and that the addition or subtraction of multiples of $y_j = 0(\text{mod } 1)$ will not destroy the congruence relationships. Thus, (11.10) can be reduced to one that contains only nonnegative fractional parts of $\boldsymbol{\alpha}_j$ and $\boldsymbol{\alpha}_0$, denoted by $\hat{\boldsymbol{\alpha}}_j$ and $\hat{\boldsymbol{\alpha}}_0$. We have the equivalent form

$$\sum_j \hat{\boldsymbol{\alpha}}_j y_j = \hat{\boldsymbol{\alpha}}_0 (\text{mod } 1) \qquad (11.11)$$
$$y_j \geq 0 \text{ and integer } y_j \in \mathbf{y_N}$$

with the objective function

$$\text{Maximize } z = \mathbf{c_B^T B^{-1} b} - (\mathbf{c_B^T B^{-1} N} - \mathbf{c_N^T})\mathbf{y_N}$$

Dropping $\mathbf{y_B} \geq \mathbf{0}$, letting $\hat{z} = -z + \mathbf{c_B^T B^{-1} b}$ and $\hat{\mathbf{c}}_\mathbf{N} = \mathbf{c_B^T B^{-1} N} - \mathbf{c_N^T}$, and changing to a minimization problem, we obtain

$$\text{Minimize} \quad \hat{z} = \sum_j \hat{c}_j y_j$$

$$\text{subject to} \quad \sum_j \hat{\boldsymbol{\alpha}}_j y_j - \hat{\boldsymbol{\alpha}}_0 \,(\text{mod } 1) \tag{11.12}$$

$$y_j \geq 0 \text{ and integer } y_j \in \mathbf{y_N}$$

where $\hat{c}_j \in \hat{\mathbf{c}}_\mathbf{N}$. Note that all components of $\hat{\boldsymbol{\alpha}}_j$ and $\hat{\boldsymbol{\alpha}}_0$ are nonnegative fractions and $\hat{c}_j \geq 0$ for all j. Equation (11.12) is termed a *group minimization problem* (or *group problem*, for short).

11.3.3 Formulating a Group Problem

Example 11.5   Construct the group minimization problem for IP problem (11.1). For ease of reference, we restate it here.

$$\begin{aligned}
\text{Maximize} \quad & z = 5y_1 - 2y_2 \\
\text{subject to} \quad & -y_1 + 2y_2 \leq 5 \\
& 3y_1 + 2y_2 \leq 19 \\
& y_1 + 3y_2 \geq 9 \\
& y_1, y_2 \geq 0 \text{ and integer}
\end{aligned}$$

Adding a nonnegative slack variable to each inequality constraint, we have

$$\begin{aligned}
\text{Maximize} \quad & z = 5y_1 - 2y_2 \\
\text{subject to} \quad & -y_1 + 2y_2 + y_3 = 5 \\
& 3y_1 + 2y_2 + y_4 = 19 \\
& -y_1 - 3y_2 + y_5 = -9 \\
& y_1, y_2, y_3, y_4 \geq 0 \text{ and integer}
\end{aligned}$$

After solving the LP relaxation, we have the optimal basis **B**. basic variables $\mathbf{y_B}$, and nonbasic variables $\mathbf{y_N}$. Representing in partitioned form, we have

$$\mathbf{y_B} = \begin{pmatrix} y_1 \\ y_2 \\ y_3 \end{pmatrix} = \begin{pmatrix} 39/7 \\ 8/7 \\ 58/7 \end{pmatrix}, \mathbf{y_N} = \begin{pmatrix} y_4 \\ y_5 \end{pmatrix} = \begin{pmatrix} 0 \\ 0 \end{pmatrix}, \mathbf{B} = \begin{pmatrix} -1 & 2 & 1 \\ 3 & 2 & 0 \\ -1 & -3 & 0 \end{pmatrix},$$

$$\mathbf{N} = \begin{pmatrix} 0 & 0 \\ 1 & 0 \\ 0 & 1 \end{pmatrix}, \mathbf{b} = \begin{pmatrix} 5 \\ 19 \\ -9 \end{pmatrix}, \mathbf{c_B^T} = (5, -2, 0), \mathbf{c_N^T} = (0, 0)$$

Calculating the components of the optimum tableau, we have

$$\mathbf{B}^{-1} = \begin{pmatrix} 0 & 3/7 & 2/7 \\ 0 & -1/7 & -3/7 \\ 1 & 5/7 & 8/7 \end{pmatrix}, \mathbf{B}^{-1}\mathbf{N} = \begin{pmatrix} 3/7 & 2/7 \\ -1/7 & -3/7 \\ 5/7 & 8/7 \end{pmatrix} = (\boldsymbol{\alpha}_4, \boldsymbol{\alpha}_5),$$

$$\mathbf{B}^{-1}\mathbf{b} = \begin{pmatrix} 39/7 \\ 8/7 \\ 58/7 \end{pmatrix} = \boldsymbol{\alpha}_0, \mathbf{c}_B^T\mathbf{B}^{-1}\mathbf{N} - \mathbf{c}_N^T = (17/7, 16/7), \text{ and } \mathbf{c}_B^T\mathbf{B}^{-1}\mathbf{b} = 179/7$$

Converting to the notation for the group problem, we have

$$\hat{\boldsymbol{\alpha}}_4 = \begin{pmatrix} 3/7 \\ 6/7 \\ 5/7 \end{pmatrix}, \hat{\boldsymbol{\alpha}}_5 = \begin{pmatrix} 2/7 \\ 4/7 \\ 1/7 \end{pmatrix}, \hat{\boldsymbol{\alpha}}_0 = \begin{pmatrix} 4/7 \\ 1/7 \\ 2/7 \end{pmatrix}, \hat{\mathbf{c}}_N^T = \mathbf{c}_B^T\mathbf{B}^{-1}\mathbf{N} - \mathbf{c}_N^T = (17/7, 16/7)$$

$$\hat{z} = -z + 179/7$$

Thus, we have the following associated group problem:

$$\text{Minimize} \quad \hat{z} = \frac{17}{7}y_4 + \frac{16}{7}y_5$$

$$\text{subject to} \quad \begin{pmatrix} 3/7 \\ 6/7 \\ 5/7 \end{pmatrix} y_4 + \begin{pmatrix} 2/7 \\ 4/7 \\ 1/7 \end{pmatrix} y_5 = \begin{pmatrix} 4/7 \\ 1/7 \\ 2/7 \end{pmatrix} \text{(mod 1)} \qquad (11.13)$$

$$y_4, y_5 \geq 0 \text{ and integer}$$

Note that the system of congruence equations in (11.13) is equivalent to

$$\begin{pmatrix} 3 \\ 6 \\ 5 \end{pmatrix} y_4 + \begin{pmatrix} 2 \\ 4 \\ 1 \end{pmatrix} y_5 = \begin{pmatrix} 4 \\ 1 \\ 2 \end{pmatrix} \text{(mod 7)} \qquad (11.14)$$

11.3.4   Solving Group Problem as a Shortest Route Problem

Essentially there are two approaches available for solving the group minimization problem: (1) treating it as a special shortest route problem and solving it by a more efficient algorithm than the standard one, and (2) treating it as a variant of knapsack problem and solving it by a dynamic programming algorithm. Here, we shall describe how to construct the group problem as a special shortest route problem and leave the knapsack approach for the interested reader (see references in Section 11.5).

Now we want to construct a shortest route problem from the group minimization problem (11.13). First, we construct a directed network with nodes equal to the number of group elements D, where D is the absolute value of the determinant of

TABLE 11.7   Group Elements for Example 11.5

Group Element	g_1	g_2	g_3	g_4	g_5	g_6	g_0
	$\begin{pmatrix}3\\6\\5\end{pmatrix}$	$\begin{pmatrix}6\\5\\3\end{pmatrix}$	$\begin{pmatrix}2\\4\\1\end{pmatrix}$	$\begin{pmatrix}5\\3\\6\end{pmatrix}$	$\begin{pmatrix}1\\2\\4\end{pmatrix}$	$\begin{pmatrix}4\\1\\2\end{pmatrix}$	$\begin{pmatrix}0\\0\\0\end{pmatrix}$
$K_4\hat{\alpha}_4$	$1\hat{\alpha}_4$	$2\hat{\alpha}_4$	$3\hat{\alpha}_4$	$4\hat{\alpha}_4$	$5\hat{\alpha}_4$	$6\hat{\alpha}_4$	$7\hat{\alpha}_4$
$K_5\hat{\alpha}_5$	$5\hat{\alpha}_5$	$3\hat{\alpha}_5$	$1\hat{\alpha}_5$	$6\hat{\alpha}_5$	$4\hat{\alpha}_5$	$2\hat{\alpha}_5$	$7\hat{\alpha}_5$

the optimal basis **B**. For (11.13), or $D = |\det \mathbf{B}| = 7$. Each group element is a 3-tuple column vector. To generate the entire group elements, we calculate

$$K_4\hat{\alpha}_4 (\bmod\ 7) \text{ for } K_4 = 1, 2, \dots 7$$
$$K_5\hat{\alpha}_5 (\bmod\ 7) \text{ for } K_5 = 1, 2, \dots 7$$

where

$$\hat{\alpha}_4 = \begin{pmatrix}3\\6\\5\end{pmatrix} \text{ and } \hat{\alpha}_5 = \begin{pmatrix}2\\4\\1\end{pmatrix}$$

resulting in Table 11.7. Here, each $\hat{\alpha}_j$ can generate the entire group $(g_0, g_1, \dots, g_6)$. Thus, the group is cyclic and contains no cyclic subgroups.

Each directed arc $(\mathbf{g}_i, \mathbf{g}_j)$ is constructed by connecting the initial node $\mathbf{g}_i$ to the terminal node by setting $y_j = 1$ in the congruence relationship

$$\mathbf{g}_j = (\mathbf{g}_i + \hat{\alpha}_j y_j)(\bmod\ D)$$

which incurs a distance $\hat{c}_j$. Examples are

$$\mathbf{g}_1 = (\mathbf{g}_0 + \hat{\alpha}_4 y_4)(\bmod\ 7) \text{ with a cost } \hat{c}_4 = 17/7$$
$$\mathbf{g}_5 = (\mathbf{g}_0 + \hat{\alpha}_5 y_5)(\bmod\ 7) \text{ with a cost } \hat{c}_5 = 16/7$$
$$\mathbf{g}_2 = (\mathbf{g}_1 + \hat{\alpha}_4 y_4)(\bmod\ 7) \text{ with a cost } \hat{c}_4 = 17/7$$
$$\mathbf{g}_6 = (\mathbf{g}_3 + \hat{\alpha}_5 y_5)(\bmod\ 7) \text{ with a cost } \hat{c}_5 = 16/7$$

Similarly, we can construct the remaining arcs of a complete directed network as shown in Figure 11.7. Let node $\mathbf{g}_0$ be the origin (or source) node and $\mathbf{g}_6 (= \hat{\alpha}_0)$ be the destination (or sink) node of the shortest route problem. The objective of the problem is to find a route from the origin to the destination such that the total distance is minimal. To solve this problem, any standard shortest route algorithm will do. However, due to the special structure of the network, standard algorithms can be simplified in order to drastically increase the computational efficiency. For details about the algorithms, see Section 11.5.

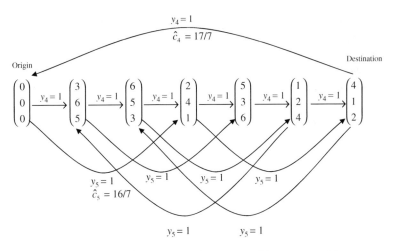

FIGURE 11.7   Directed network constructed by group problem.

By inspection of the above network, we can easily find the shortest route with $y_5 = 2$, $y_4 = 0$, and $\hat{z} = 32/7$.

11.3.5   Solving the Original Integer Program

Recall that when the group problem was derived from an integer program, the constraints $\mathbf{y_B} \geq \mathbf{0}$ were ignored. This implies that if the optimal solution $\mathbf{y_N}$ (which is a nonnegative integer) to the group problem also yields $\mathbf{y_B} \geq \mathbf{0}$, then we have already solved the integer program. If some basic variable turns out to be negative, then the optimal solution to the group problem is not an optimum to the original IP problem. To verify this for our example, we calculate

$$\mathbf{y_B} = \begin{pmatrix} y_1 \\ y_2 \\ y_3 \end{pmatrix} = \mathbf{B}^{-1}\mathbf{b} - \mathbf{B}^{-1}\mathbf{N}\mathbf{y_N} = \begin{pmatrix} 39/7 \\ 8/7 \\ 58/7 \end{pmatrix} - \begin{pmatrix} 3/7 & 2/7 \\ -1/7 & -3/7 \\ 5/7 & 8/7 \end{pmatrix} \begin{pmatrix} 0 \\ 2 \end{pmatrix} = \begin{pmatrix} 5 \\ 2 \\ 6 \end{pmatrix}$$

Since $\mathbf{y_B} \geq \mathbf{0}$, this optimal solution to the group problem is also an optimal solution to the original IP problem with $\mathbf{y} = (\mathbf{y_B}, \mathbf{y_N})^{\mathrm{T}} = (5, 2, 6, 0, 2)^{\mathrm{T}}$. Should any component of $\mathbf{y}$ turns out to be negative, we continue to find the next shortest route and test for $\mathbf{y_B} \geq \mathbf{0}$ until such condition is obtained or no more route can be found, in which case the original IP problem has no feasible solution. However, this process is quite complicated and difficult to implement.

An alternative method is apply a branch-and-bound enumeration scheme that implicitly examines all possible integer solutions to the group problem by successively adding constraints of the form $y_j \geq C$ for each j, where C begins with value 0 and is incremented by 1. Morito (1976) utilized this basic bookkeeping scheme with a particular branching rule and information from the optimal solution to the group problem to create an efficient branch-and-bound algorithm. Again, the scheme is not implemented in practice.

In conclusion, there are still several technical hurdles to overcome before the group theoretic approach can be implemented as a general, stand-alone IP solver. However, the faces of corner polyhedron generated by the group problem are strong cuts (or valid inequalities) that can be incorporated into the novel branch-and-cut approach, in the same manner as other cutting planes are utilized.

11.4 GEOMETRIC CONCEPTS

Now we discuss the solution space to the pure integer program defined by the structural variables $\mathbf{y'}$ in (11.6), and the solution space to the group problem defined by nonbasic variables $\mathbf{y_N}$ in (11.12). Many important geometric concepts will be covered, including polyhedron, convex hull, corner polyhedron, and faces. For illustration, we reduce Example 11.5 to a two-constraint problem by dropping the first constraint.

Example 11.6

$$\begin{aligned} \text{Maximize} \quad & z = 5y_1 - 2y_2 \\ \text{subject to} \quad & 3y_1 + 2y_2 \le 19 \\ & -y_1 - 3y_2 \le -9 \\ & y_1, y_2 \ge 0 \text{ and integer} \end{aligned} \tag{11.15}$$

Adding nonnegative slack variables y_3 and y_4 to constraints 1 and 2, respectively, we find the optimum LP solution containing basic variables y_1 and y_2, and so

$$\mathbf{y_B} = \begin{pmatrix} y_1 \\ y_2 \end{pmatrix} = \begin{pmatrix} 39/7 \\ 8/7 \end{pmatrix}, \mathbf{y_N} = \begin{pmatrix} y_3 \\ y_4 \end{pmatrix} = \begin{pmatrix} 0 \\ 0 \end{pmatrix}, \mathbf{B} = \begin{pmatrix} 3 & 2 \\ -1 & -3 \end{pmatrix},$$

$$\mathbf{N} = \begin{pmatrix} 1 & 0 \\ 0 & 1 \end{pmatrix}, \mathbf{b} = \begin{pmatrix} 19 \\ -9 \end{pmatrix}, \mathbf{B^{-1}} = \begin{pmatrix} 3/7 & 2/7 \\ -1/7 & -3/7 \end{pmatrix},$$

$$\mathbf{B^{-1}N} = \begin{pmatrix} 3/7 & 2/7 \\ -1/7 & -3/7 \end{pmatrix}, \mathbf{B^{-1}b} = \begin{pmatrix} 39/7 \\ 8/7 \end{pmatrix}, D = |\det \mathbf{B}| = 7$$

$$\mathbf{c_B^T B^{-1}N - c_N^T} = (5, -2) \begin{pmatrix} 3/7 & 2/7 \\ -1/7 & -3/7 \end{pmatrix} - (0, 0) = (17/7, 16/7),$$

$$z = \mathbf{c_B^T B^{-1}b} = 179/7$$

From (11.12), we obtain the associated group problem

$$\begin{aligned} \text{Minimize} \quad & \hat{z} = \frac{17}{7}y_3 + \frac{16}{7}y_4 \\ \text{subject to} \quad & \begin{pmatrix} 3/7 \\ 6/7 \end{pmatrix}y_3 + \begin{pmatrix} 2/7 \\ 4/7 \end{pmatrix}y_4 = \begin{pmatrix} 4/7 \\ 1/7 \end{pmatrix}(\bmod 1) \\ & y_3, y_4 \ge 0 \text{ and integer} \end{aligned} \tag{11.16}$$

Solving the problem as a shortest route problem, we obtain the solution $y_3 = 0$, $y_4 = 2$, which gives

$$\mathbf{y_B} = \mathbf{B}^{-1}\mathbf{b} - \mathbf{B}^{-1}\mathbf{N}\mathbf{y_N} = \begin{pmatrix} 39/7 \\ 8/7 \end{pmatrix} - \begin{pmatrix} 3/7 & 2/7 \\ -1/7 & -3/7 \end{pmatrix}\begin{pmatrix} 0 \\ 2 \end{pmatrix} = \begin{pmatrix} 5 \\ 2 \end{pmatrix} \geq \begin{pmatrix} 0 \\ 0 \end{pmatrix}$$

Since all components of vector $\mathbf{y_B}$ are nonnegative, the optimal solution to the original integer program is $\mathbf{y} = (5, 2, 0, 2)^T$ with objective value

$$z = -\hat{z} + \mathbf{c_B^T}\mathbf{B}^{-1}\mathbf{b} = -32/7 + 179/7 = 21$$

11.4.1 Various Polyhedrons in Original Space

From Chapter 8, we know that the solution region of the LP relaxation defined by the constraints and the nonnegative restrictions in (11.15) is a polyhedron, say P. In our example, P is the triangle area enclosed by points $\mathbf{a}'$, $\mathbf{b}'$, and $\mathbf{c}'$, shown in Figure 11.8. A basis $\mathbf{B}$ selected from the coefficient matrix yields the basic feasible solution $\mathbf{y} = (\mathbf{y_B}, \mathbf{y_N})^T = (\mathbf{B}^{-1}\mathbf{b}, \mathbf{0})^T$, which is an extreme point to P. In our example, $\mathbf{a}'$, $\mathbf{b}'$, and $\mathbf{c}'$ are extreme points of P. Taking the objective function into consideration, we obtain an optimal extreme point $\mathbf{c}' = (39/7, 8/7)^T$ with objective value $z = 25.57$. The objective function, indicated by a dotted line and passing through vertex $\mathbf{c}'$, is plotted in Figure 11.8.

Recall that the congruence constraints of the group problem (11.16) are obtained from those of the IP program by ignoring the nonnegative requirements on the basic

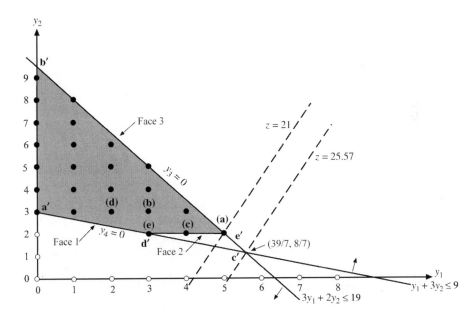

FIGURE 11.8 Corner polyhedron in original space.

variables $\mathbf{y_B}$. In the original solution space, this is equivalent to dropping those inequalities from $\mathbf{A'y'} \leq \mathbf{b}$ and $\mathbf{y'} \geq \mathbf{0}$ in (11.6), which are *nonbinding* at vertex $\mathbf{e'}$. In this example, there are two *binding* constraints at $\mathbf{c'}$, namely,

$$3y_1 + 2y_2 < 19$$
$$-y_1 - 3y_2 < -9$$

and two nonbinding constraints $y_3 > 0$, $y_4 > 0$.

An inequality constraint that is nonbinding at $\mathbf{v'}$ implies that it takes on a strict inequality ($<$) when substituting $\mathbf{v'}$ into it. As a result, the associated slack variable is strictly positive because only a basic variable can be positive. In general, when constraint i is binding at $\mathbf{v'}$, the associated slack variable y_{n+i} must take on value 0 and the corresponding hyperplane $y_{n+i} = 0$ must pass through $\mathbf{v'}$. In our example, the hyperplanes $y_3 = 0$ and $y_4 = 0$ must pass through $\mathbf{c'}$. This implies that these hyperplanes intersect at $\mathbf{c'}$. As shown in Figure 11.8, the hyperplanes $3y_1 + 2y_2 = 19$ and $y_1 + 3y_3 = 9$ in the original solution space correspond to the hyperplanes $y_3 = 0$ and $y_4 = 0$ in the space of nonbasic variables $\mathbf{x_N}$.

On the other hand, if $y_i > 0$ or y_i is a basic variable (where i: $y_i \in \mathbf{y_B}$), then the hyperplane $y_i = 0$ does not intersect the vertex $\mathbf{v'}$. Thus, ignoring the nonnegativity requirement on a basic variable y_i is equivalent to allowing y_i to take on any value when the corresponding hyperplane does not pass through $\mathbf{v'}$. In our example, we have strict inequalities $y_1 > 0$ and $y_2 > 0$, so hyperplanes $y_1 = 0$ and $y_2 = 0$ do not intersect at $\mathbf{c'}$. Therefore, these hyperplanes can be dropped with affecting the solution.

After dropping these hyperplanes, we obtain a polyhedron P with vertex $\mathbf{c'} = (39/7, 8/7)^T$ as shown in Figure 11.8. Adding the integer requirements yields the lattice points inside P, indicated by dots. The convex hull of these lattice points may be viewed as the feasible region to the group problem. The *corner polyhedron* is obtained by dropping the hyperplanes $y_1 = 0$ and $y_2 = 0$, and taking the convex hull of the integer solutions. In Figure 11.8, the polyhedron bounded by integer points $\mathbf{a'}$, $\mathbf{d'}$, $\mathbf{e'}$, and $\mathbf{b'}$ forms a corner polyhedron, which is denoted by $P^{y'}$ and indicated by the shaded area. The hyperplanes that bound $P^{y'}$ are the *faces* of the corner polyhedron. Recall that in n-dimensional space, a face is an $(n-1)$-dimensional hyperplane. Since $n = 2$ in our example, so a face must be a line such as face 1, face 2, and face 3, indicated in Figure 11.8.

Note that any integer point $\mathbf{y'}$ in $P^{y'}$ will produce nonnegative integer values for the nonbasic variables $\mathbf{y_N}$ that satisfy the congruence relationship in (11.16). Furthermore, a vertex of the corner polyhedron is an optimal solution for the integer program. The vertex $\mathbf{e'}$ is at the intersection of faces 2 and 3 in Figure 11.8. These inequalities are the strongest that can be obtained from the integer program without using the requirements of $\mathbf{y_B} \geq \mathbf{0}$. Therefore, it is worthwhile finding the face inequalities and using them in a cutting plane algorithm or in a branch-and-bound algorithm. To enable generating these valid inequalities, we next investigate the corner polyhedron in $\mathbf{y_N}$ space.

11.4.2 Corner Polyhedron in Solution Space of Nonbasic Variables

The solution region defined by the constraints in (11.12), or (11.16) in particular for Example 11.6, can also be plotted in $\mathbf{y_N}$ space. To do this, recall that the congruence relationship

$$\sum_j \hat{\alpha}_j y_j = \hat{\alpha}_0 (\text{mod } 1)$$

is equivalent to

$$\sum_j f_{ij} y_j = f_{i0} \ (\text{mod } 1) \quad (i = 1, 2, \dots, m) \tag{11.17}$$

where $y_j \in \mathbf{y_N}$, and f_{ij} is the ith component $(i = 1, 2, \dots, m)$ of $\hat{\alpha}_j (j = 1, 2, \dots, n)$. Because the left-hand side in (11.12) must differ from the right-hand side by an integer amount and because $y_j \geq 0$ and integer, the congruence can hold only when the left-hand side is f_{i0}, $1 + f_{i0}$, $2 + f_{i0}$, and so on. That is, equation (11.12) implies

$$\sum_{j=1}^{n} f_{ij} y_j \geq f_{i0} \quad (i = 1, 2, \dots, m) \tag{11.18}$$

Note that the inequalities (11.18) are the Gomory fractional cuts obtained from the m equalities $\mathbf{y_B} = \mathbf{B}^{-1}\mathbf{b} - \mathbf{B}^{-1}\mathbf{N}\mathbf{y_N}$. For Example 11.6, we have

$$3/7 y_3 + 2/7 y_4 \geq 4/7$$
$$6/7 y_3 + 4/7 y_4 \geq 1/7 \tag{11.19}$$
$$y_3, y_4 \geq 0 \text{ and integer}$$

Note that the second constraint is redundant and may be dropped without affecting the solution.

Every y_j that satisfies (11.17) will also satisfy (11.18). However, the converse is not true because $y_1, y_2, \dots, y_n$ satisfying (11.18) may yield a value for $\sum_j \hat{\alpha}_j y_j$ that is not an integer plus $\hat{\alpha}_0$. Therefore, the constraints

$$\sum_j \hat{\alpha}_j y_j = \hat{\alpha}_0 (\text{mod } 1)$$

$$\mathbf{y_N} \geq 0 \text{ and integer}$$

for the group problem may be viewed in $\mathbf{y_N}$ space by plotting the inequalities (11.18) along with $\mathbf{y_N} \geq \mathbf{0}$ and integer, and taking the convex hull of the points satisfying (11.18). This region is termed the *corner polyhedron* P_N.

We plot (11.19), or (11.16) to be exact, in the solution space of $\mathbf{y_N} = (y_3, y_4)$ shown in Figure 11.9. The solution set is an unbounded polyhedron with two extreme points at (4/3, 0) and (0, 2). The circled lattice points are those that satisfy

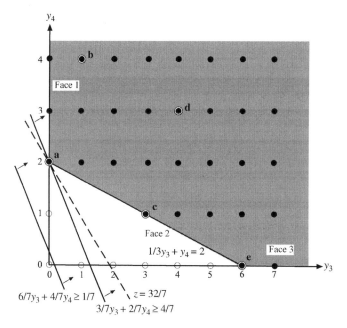

FIGURE 11.9 Corner polyhedron in $\mathbf{y_N}$ space.

the congruence relationship. Some of them are marked by **a**, **b**, **c**, and **d**, and **e** in Figure 11.9:

$$\mathbf{a} = \begin{pmatrix} 0 \\ 2 \end{pmatrix}, \mathbf{b} = \begin{pmatrix} 1 \\ 4 \end{pmatrix}, \mathbf{c} = \begin{pmatrix} 3 \\ 1 \end{pmatrix}, \mathbf{d} = \begin{pmatrix} 4 \\ 3 \end{pmatrix}, \mathbf{e} = \begin{pmatrix} 6 \\ 0 \end{pmatrix}$$

Using

$$\mathbf{y_B} = \mathbf{B}^{-1}\mathbf{b} - \mathbf{B}^{-1}\mathbf{N}\mathbf{y_N}$$

we obtain $\mathbf{y_B}$ components, as shown in Table 11.8, for the corresponding points in the original solution space.

TABLE 11.8 Integer Solution Points

Point	a	b	c	d	e
$\mathbf{y_N} = \begin{pmatrix} y_3 \\ y_4 \end{pmatrix}$	$\begin{pmatrix} 0 \\ 2 \end{pmatrix}$	$\begin{pmatrix} 1 \\ 4 \end{pmatrix}$	$\begin{pmatrix} 3 \\ 1 \end{pmatrix}$	$\begin{pmatrix} 4 \\ 3 \end{pmatrix}$	$\begin{pmatrix} 6 \\ 0 \end{pmatrix}$
$\mathbf{y_B} = \begin{pmatrix} y_1 \\ y_2 \end{pmatrix}$	$\begin{pmatrix} 5 \\ 2 \end{pmatrix}$	$\begin{pmatrix} 4 \\ 3 \end{pmatrix}$	$\begin{pmatrix} 4 \\ 2 \end{pmatrix}$	$\begin{pmatrix} 3 \\ 3 \end{pmatrix}$	$\begin{pmatrix} 3 \\ 2 \end{pmatrix}$

To see the correspondence of these points in the original solution space in Figure 11.8, we label such points with a bracket **(a)**, **(b)**, **(c)**, **(d)**, **(e)**.

A convex set may be formed by the convex combinations of these circled lattice points (integer solutions for both $\mathbf{y_B}$ and $\mathbf{y_N}$) in the unbounded polyhedral cone. That set is a corner polyhedron, denoted by P_N. The corner polyhedron for this example problem is the shaded area in Figure 11.9. Unlike the corner polyhedron in the original space, the corner polyhedron in $\mathbf{y_N}$ space always satisfies $\mathbf{y_N} \geq \mathbf{0}$. Also, note that this corner polyhedron is an unbounded convex set. This is always true because if $\mathbf{y_N}$ satisfies (11.17), then for any integer K, the value KD (where D is the absolute value of the determinant of $\mathbf{B}$) added to each of its components will also satisfy (11.18). To show their correspondences, we use the same labels for the faces in Figure 11.8. The faces (faces 1 through 3) of this corner polyhedron are marked in Figure 11.9.

It can be shown (Salkin and Marthur, 1989) that $\mathbf{y_B}$ is an extreme point of corner polyhedron $P^{y'}$ if and only if the corresponding $\mathbf{y_N}$ is an extreme point of P_N using the relationship between faces. In our example, $\mathbf{y_N} = (0, 2)^T$ is an extreme point in P_N denoted by $\mathbf{a}$, which corresponds to $\mathbf{y_B} = (5, 2)^T$, an extreme point $\mathbf{e}'$ in $P^{y'}$. Similarly, extreme point $\mathbf{e}$ in P_N corresponds to extreme point $\mathbf{d}'$ in $P^{y'}$.

11.5   NOTES

Sections 11.1

The branch-and-bound method was originated by Land and Doig (1960) and refined by Dakin (1965), which is regarded as the basis of the current algorithm.

Sections 11.2

Three fundamental cutting plane methods, dual fractional, dual mixed integer, and primal all-integer are developed by Gomory (1958, 1960, 1963). For details on these and other cutting plane methods, see Garfinkel and Nemhauser (1972), and Salkin and Mathur (1989).

Sections 11.3

Gomory (1960, 1963, 1965) developed group properties of IP program. For exposition of group theory approach to solving integer programs, see Hu (1969), Chen (1970), Chen and Zionts (1972), and Salkin and Mathur (1989).

Section 11.4

Special shortest route algorithms for solving the group problem are given in Hu (1968) and Chen and Zionts (1976). Shapiro (1968) treated the group problem as a variant of knapsack problem that is solved by a dynamic programming algorithm.

An equivalent group problem can be obtained by transforming the LP optimal basis into a Smith's normal form (Hu, 1969), which can be solved more efficiently. The form is a diagonal matrix (all elements not on the diagonal are zero) denoted by diag $(d_1, d_2, \ldots, d_n)$, such that d_i ($i = 1, 2, \ldots, m$) is a positive integer, d_i dives d_{i+1} ($i = 1, 2, \ldots, m-1$), and $D = \det \mathbf{B}$ can be determined by the product of diagonal elements d_i. The diagonalization process is performed by the elementary row and column operations.

There are three types of group problems, depending on the number of $\hat{\alpha}_j$ elements that can be used to generate the entire group: (1) each element (like Example 11.5), (2) some but not each element, and (3) no single element. The groups determined by types 1 and 2 are cyclic. The group type 3 is acyclic, whose elements are determined by the *direct sum* of cyclic subgroups. For details see Salkin and Mathur (1989), for example.

A sufficient condition for $\mathbf{y_B} \geq \mathbf{0}$ is used to ensure that an optimum solution $\mathbf{y_N}$ to the group problem is also an optimum solution $\mathbf{y} = (\mathbf{y_B}, \mathbf{y_N})$ to the original integer program. Several sufficient conditions have been developed. See Gomory (1965), Hu (1968), and Zionts (1974), for example. Sufficiency conditions have little use in application although are of theoretical value.

11.6   EXERCISES

11.1   Solve the problem in Example 11.1 again using branch-and-bound. This time, start your branching with y_2. Graphically show the changes in the feasible region at each node. Apply the depth-first rule.

11.2   Use the branch-and-bound method to solve the following IP problem. Show your solution procedure graphically as in Figures 11.1–11.3.

$$\begin{array}{ll} \text{Maximize} & y_1 + 3y_2 \\ \text{subject to} & y_1 + 5y_2 \leq 12 \\ & y_1 + 2y_2 \leq 8 \\ & y_1, y_2 \geq 0 \text{ and integer} \end{array}$$

11.3   Solve the following IP problem using the branch-and-bound method. Apply the best-bound-first rule. At each node, branch on the variable with fraction value closest to 0.5 first. Label the nodes in the order they are generated.

$$\begin{array}{ll} \text{Maximize} & 3y_1 + 2y_2 + y_3 + 2y_4 \\ \text{subject to} & y_1 - y_2 + 2y_3 + y_4 \leq 11 \\ & y_2 + y_3 + y_4 \leq 7 \\ & 3y_1 - y_3 - 3y_4 \leq 5 \\ & \mathbf{y} \geq \mathbf{0} \text{ and integer} \end{array}$$

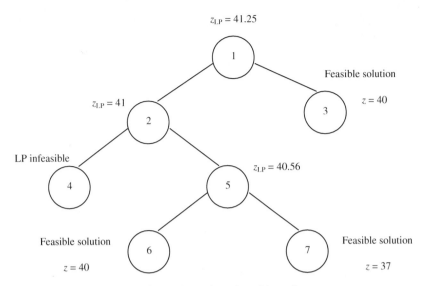

$z_{LP} = 41.25$

Feasible solution

$z = 40$

$z_{LP} = 41$

LP infeasible

$z_{LP} = 40.56$

Feasible solution

$z = 40$

Feasible solution

$z = 37$

FIGURE 11.10   A branch-and-bound tree.

11.4 Given a branch-and-bound tree for a maximization problem (as shown in Figure 11.10), where the number of the nodes shows the order they are generated, write the reason for each pruning.

11.5 Mary is the cashier of a fast food restaurant. Coins of four values are used to get customers the change: 1¢, 5¢, 10¢, and 25¢. Suppose now she is to give a customer 91¢ of change. What is the minimum number of coins needed? Formulate as an IP model and solve using branch-and-bound method. (*Hint:* convert to maximization problem first). Verify your solution with LINGO®.

11.6 Solve the following IP problem using branch-and-bound. Apply the rule of best-bound-first. At each node, branch on the variable with least index first.

$$\text{Maximize} \quad 10y_1 - y_2 + 5y_3 + 3y_4 - 7y_5$$
$$\text{subject to} \quad 3y_1 + 2y_2 - y_3 - y_4 - 3y_5 \leq 3$$
$$3y_1 - 2y_2 + 2y_3 - 2y_4 + 5y_5 \leq 7$$
$$y_1 - y_2 + y_4 - y_5 \leq 3$$
$$y_1, y_2, y_3, y_5 \geq 0 \text{ and integer}, y_4 = \{0, 1\}$$

11.7 Assume in Exercise 11.6 that y_3, y_5 are continuous. Solve the problem again using branch-and-bound method. Apply the rule of depth-first.

11.8 Consider the IP problem in Exercise 11.3. Which of the two cutting planes (fractional or mixed) could be applied to it? Why? No need to solve the problem to optimum.

11.9 Let $S = \{y \geq 0$ and integer $y_1 - y_2 \leq 2,\ 3y_1 + y_2 \leq 21,\ y_1 + 5y_2 \leq 34\}$.

   (a) Find an inequality description of Conv(S).

   (b) Find the extreme points of Conv(S).

11.10 Consider the problem in Example 11.4. Generate another Gomory fractional cut using the row corresponding to the basic variable of y_2.

11.11 Consider the integer problem

$$\begin{aligned}
\text{Maximize} \quad & 5y_1 - 2y_2 \\
\text{subject to} \quad & 11y_1 - y_2 \leq 21 \\
& 3y_1 - 3y_2 \leq 5 \\
& y_1 \leq 4 \\
& y \geq 0 \text{ and integer}
\end{aligned}$$

   (a) Solve the LP relaxation and show the optimal simplex tableau.

   (b) Is there any basic variable fractional in the optimal tableau? If yes, find a Gomory fractional cut based on it.

11.12 Consider the following IP problem.

$$\begin{aligned}
\text{Maximize} \quad & 3y_1 + y_2 + 2y_3 + 3y_4 \\
\text{subject to} \quad & -y_1 + 3y_2 + y_3 - 2y_4 \leq 17 \\
& 7y_1 + 3y_3 + y_4 \leq 23 \\
& y_1 + 2y_2 \leq 11 \\
& y_2 + 3y_4 \leq 13 \\
& y \geq 0 \text{ and integer}
\end{aligned}$$

Solve it using Gomory's fractional cutting plane. Show the cutting planes you generated at each step.

11.13 Generate a group of fractional cuts based on variable y_2 in Exercise 11.2.

11.14 Assuming in Exercise 11.12 that y_2 is continuous, solve the problem using Gomory mixed integer cut.

11.15 Complete solving Example 11.4 using Gomory mixed integer cutting plane algorithm. Compare your results with the solution obtained in Example 11.2 using a branch-and- bound method.

For each of the problems in Exercises 11.16–11.18, (a) formulate the group problem, (b) construct the group problem as a directed network, and (c) solve the shortest route problem by inspection.

TABLE 11.9 An Optimal Tableau

Basic Variable	z	y_1	y_2	y_3	y_4	y_5	y_6	y_7	RHS
z	1	7/10	0	69/10	0	4/10	0	19/10	139/10
y_2	0	5/10	1	5/10	0	0	0	5/10	25/10
y_4	0	16/10	0	2/10	1	2/10	0	2/10	32/10
y_6	0	16/10	0	−8/10	0	2/10	1	2/10	2/10

11.16

$$\text{Maximize} \quad z = 4y_1 + 3y_2 - 5y_3 + 2y_4$$
$$\text{subject to} \quad 7y_1 - 2y_2 + 5y_4 \leq 11$$
$$-y_3 - y_4 \leq -3$$
$$y_1 + 2y_2 + y_3 \leq 5$$
$$y_1, y_2, y_3, y_4 \geq 0 \text{ and integer}$$

Add slack variables y_5, y_6, and y_7 to the inequality constraints, respectively. Then solve the LP relaxation and obtain the following optimum tableau (Table 11.9).

11.17 (Balinski, 1965)[1]

$$\text{Maximize} \quad z = 4y_1 + 5y_2$$
$$\text{subject to} \quad -y_1 - y_2 \leq -5$$
$$-3y_1 - 2y_2 \leq -7$$
$$y_1, y_2 \geq 0 \text{ and integer}$$

Adding nonnegative slack variables y_3 and y_4 to constraints 1 and 2, respectively, we find the optimum LP solution containing basic variables y_1 and y_2.

11.18 (Zionts, 1974) (used with permission)

$$\text{Maximize} \quad z = 5y_1 + 2y_2$$
$$\text{subject to} \quad 2y_1 + 2y_2 \leq 9$$
$$3y_1 + y_2 \leq 11$$
$$y_1, y_2 \geq 0 \text{ and integer}$$

---

[1] Reprinted with permission from author (see Bibliography). Copyright 1965 The Institute for Operations Research and Management Sciences, 7240 Parkway Drive, Suite 300, Hanover, MD 21076.

Adding slack variables y_3 and y_4 to constraints 1 and 2, respectively, and ignoring the integer requirements, we solve the LP relaxation and obtain a noninteger optimum solution, written in equation form, as follows:

$$\text{Maximize} \quad z - 1/4y_3 - 6/4y_4 = 75/4$$
$$\text{subject to} \quad y_2 + 3/4y_3 - 2/4y_4 = 5/4$$
$$y_1 - 1/4y_3 + 2/4y_4 = 13/4$$
$$y_1, y_2, y_3, y_4 \geq 0$$

where $\mathbf{y_B} = (y_2, y_1)$.

11.19 For the problem in Exercise 11.17, draw the solution space of the IP program and identify the LP polyhedron, convex hull, corner polyhedron, and faces.

11.20 For the problem in Exercise 11.17, draw the solution space of nonbasic variables and identify faces and corner polyhedron.

11.21 For the problem in Exercise 11.17, relate all faces between the solution space of the structural variables and the solution space of the nonbasic variables.

11.22 For the problem in Exercise 11.18, draw the solution space of the IP program and identify the LP polyhedron, convex hull, corner polyhedron, and faces.

11.23 For the problem in Exercise 11.18, draw the solution space of nonbasic variables and identify faces and corner polyhedron.

11.24 For the problem in Exercise 11.18, relate all faces between the solution space of the structural variables and the solution space of the nonbasic variables.

12

BRANCH-AND-CUT APPROACH

Since the development of the branch-and-bound (B&B), cutting plane, and group theoretic approaches in the 1960s, progress on methods for solving large-scale IP or MIP problems was very limited for two decades. Then in the mid-1980s, a novel solution approach known as *branch-and-cut* (B&C) was introduced, which marked a breakthrough milestone in the power of MIP solution algorithms. This approach and its variations, coupled with the advances in modeling techniques, preprocessing techniques, LP software, and computer hardware, make the solution of large-scale MIP problems possible. As of today, the solution power has leapt from solving problems with up to one hundred integer variables in the early 1980s to solving problems with thousands of integer variables, and even in many instances with millions of 0–1 variables.

This textbook aims at addressing four of the five major factors that contributed to the advances in MIP: modeling techniques, preprocessing techniques, solution algorithms, and commercial software. Only computer hardware is outside our scope; modeling, transformation, and preprocessing techniques were addressed in Chapter 2 through Chapter 4. This chapter will address the solution algorithm known as *branch-and-cut*. A substantial portion of the discussion will focus on the generation of valid cuts capable of solving general and special MIP programs efficiently. Branch-and-cut as a feature of MIP commercial software will be addressed in Chapter 15.

Applied Integer Programming: Modeling and Solution, By Der-San Chen, Robert G. Batson, and Yu Dang
Copyright © 2010 John Wiley & Sons, Inc.

12.1 INTRODUCTION

12.1.1 Basic Concept

Conceptually, the branch-and-cut method can be viewed as a generalization of the branch-and-bound method. Basically, it builds upon the same branch-and-bound framework with additional cuts generated and imposed on each node of the branch-and-bound tree, prior to pruning and branching processes.

Although both methods solve a series of LP relaxation problems at various nodes, their solution philosophies are different. B&B applies two simple bound cuts at each node and takes advantage of fast reoptimization of the LP at each node. The B&C philosophy is to do as much work as necessary to get a "tight bound" at the node before pruning and branching. The work at each node may include generating strong cuts, improving formulations, problem preprocessing, and applying a primal heuristic. In practice, *many* cuts may be added at each node, which may slow down the reoptimization. For a given large-scale problem, an empirical investigation is usually used to determine the proper number of cuts to be imposed on the root and other nodes.

12.1.2 Branch-and-Cut Algorithm

We now describe the branch-and-cut algorithm below.
  Let:

$S =$ the given IP problem
$S_{LP} =$ the LP relaxation of S
$y_{LP} =$ the solution to the LP relaxation of the given IP
$\bar{z} =$ lowest (best) upper bound on z of the given IP problem
$\underline{z} =$ highest (best) lower bound on z of the given IP problem
$S^k =$ subproblem k of problem S
$S_{LP}^k =$ LP relaxation of subproblem k
$S^k(t) =$ subproblem k at iteration t
$S_{LP}^k(t) =$ LP relaxation of subproblem k at iteration t as an LP problem
$y_{LP}^k(t) =$ the optimum solution of the LP subproblem $S_{LP}^k(t)$
$y^* =$ the current incumbent solution
$z_{LP}^k(t) =$ the optimum objective value of $S_{LP}^k(t)$
$\bar{z}^k =$ lowest upper bound of subproblem S^k
$\underline{z}^k =$ highest lower bound of subproblem S^k

Step 0 (Initialization). Preprocess the given IP formulation. Solve its LP relaxation (S_{LP}). If S_{LP} is infeasible, so is the IP problem. Terminate. If the LP optimum solution satisfies the integer requirement, the IP problem S is also optimized. Terminate. Otherwise, set the best lower bound $\underline{z} = -\infty$, and $\bar{z} = z^*$ for S_{LP}.

Set $k = 1$. Let $S^k = S$. Place S^k on the active list of nodes (subproblems). Initially, there is no incumbent solution.

Step 1 (Choosing a Node). If the active list is empty, terminate. The current incumbent solution y^* is optimal. Otherwise, choose a node (subproblem) k with S^k. Remove S^k from the active list. Set iteration number $t = 1$. Denote the current subproblem by $S^k(t)$, which has LP relaxation $S^k_{LP}(t)$. Go to step 2.

Step 2. (Solving LP Relaxation of Subproblem). Solve $S^k_{LP}(t)$. If it is infeasible, prune node k and go to step 1. Otherwise, keep the optimal LP solution to $S^k_{LP}(t)$, which is $y^k_{LP}(t)$, and the optimal objective value $z^k_{LP}(t)$. Go to step 3.

Step 3 (Generating Cuts). Try to generate cuts to the optimized problem $S^k_{LP}(t)$ to cut off the point $y^k_{LP}(t)$. If no cut can be added, go to step 4. Otherwise, add a cut to $S^k_{LP}(t)$, resulting in a new LP problem $S^k_{LP}(t+1)$. Increase iteration number t by 1. Go to step 2.

Step 4 (Pruning). If $z^k_{LP}(t) \leq \underline{z}$, prune node k and go to step 1. Otherwise, if $y^k_{LP}(t)$ satisfies all the integer requirements of the given IP problem, go to step 5. If $y^k_{LP}(t)$ violates some integer requirements, go to step 6.

Step 5 (Updating Lower Bound). Since the optimal LP solution $y^k_{LP}(t)$ satisfies all integer requirements, a feasible solution to S is found and $y^k_{LP}(t)$ becomes a candidate solution. Set $\underline{z}^k$ to the optimal objective value of $S^k_{LP}(t)$, that is, $\underline{z}^k = z^k_{LP}(t)$, and compare $\underline{z}^k$ with $\underline{z}$. If $\underline{z}^k > \underline{z}$, set $\underline{z} = \underline{z}^k$, and $y^* = y^k_{LP}(t)$ becomes the incumbent; otherwise, $\underline{z}$ does not change. Node k is pruned because no better solution can be branched down from this node. Go to step 1.

Step 6 (Branching). Branch on the current node k to create more subproblems S^{k+1}, S^{k+2}, and so on. Place these new subproblems in the active list and go to step 1.

12.1.3   Generating Valid Cuts and Preprocessing

In Chapter 4, we learned about how to preprocess a given MIP model to obtain a "better" formulation. That exposition and the experience of practitioners support the belief that any original MIP formulation can almost always be improved. By improvement, we mean the new formulation has a smaller difference between the space of feasible continuous solutions and the space of feasible integer solutions. It may also mean that a new formulation has fewer variables (especially integer), less or no redundant constraints, and smaller differences between upper and lower bounds for variables. In short, a new formulation with such properties is called a "tighter" formulation. For the similarities and differences between preprocessing and cut generation, readers are suggested to make a quick review of Chapter 4.

Both preprocessing and cut generation share the similar steps for "tightening" formulation by adding and/or replacing constraints in the model so that the integer solution space remains unchanged but has a smaller continuous solution space. For this reason, they sometimes look alike. In fact, some software programs (such as LINGO®) include both options of preprocessing and cut generating in one place.

However, preprocessing and cut generation do have fundamental differences. First, preprocessing is applied to the original model to create a new model (independent of the relaxation problem), while cuts are generated and added to cut off a relaxation solution. Second, preprocessing introduces tighter constraints that *dominate* existing constraints, while cut generating introduces tighter constraints that cut off part of a particular relaxation solution.

12.2   VALID INEQUALITIES

12.2.1   Valid Inequalities for Linear Programs

A linear inequality is called a *valid inequality* for an LP problem if it is satisfied by *all* feasible solutions to the problem. Here, we are interested in when an inequality is valid. This is answered by the following theorem.

Theorem 12.1   A linear inequality $\pi^T \mathbf{x} \leq \pi_0$ is valid for a nonempty polyhedron $P = \{\mathbf{x}: \mathbf{A}\mathbf{x} \leq \mathbf{b}, \mathbf{x} \geq \mathbf{0}\}$ if and only if there exists $\mathbf{u} \geq \mathbf{0}$ such that $\mathbf{u}^T \mathbf{A} \geq \pi$ and $\mathbf{u}^T \mathbf{b} \leq \pi_0$.

To show this, we treat $\mathbf{u}$ as the dual variables to a maximization problem. Then by the LP duality theorem, we have $\max\{\pi^T \mathbf{x}: \mathbf{A}\mathbf{x} \leq \mathbf{b}, \mathbf{x} \geq \mathbf{0}\} \leq \pi_0$ if and only if $\min\{\mathbf{u}^T \mathbf{b}: \mathbf{u}^T \mathbf{A} \geq \pi, \mathbf{u} \geq \mathbf{0}\} \leq \pi_0$. That is, $\pi^T \mathbf{x} \leq \pi_0$ for all $\mathbf{x}$ in P and there exists at least one $\mathbf{u}$ such that $\mathbf{u}^T \mathbf{b} \leq \pi_0$.

12.2.2   Valid Inequalities for Integer Programs

A linear inequality is *valid for an MIP problem* if it is satisfied by the set of *all feasible* solutions of the MIP, in particular with the integer restrictions in place. Given a polyhedron P and an optimal LP solution $\mathbf{x}^* \in P$, we are interested in the particular set of valid inequalities that cuts off $\mathbf{x}^*$. Such particular valid inequalities are sometimes called *violated cuts*. The problem consisting of the determination of whether $\mathbf{x}^*$ is in the new polyhedron, and if not to find an inequality cutting off $\mathbf{x}^*$, is called a *separation problem*.

A violated cut $\mathbf{p}^T \mathbf{x} \leq b_1$ is said to be *stronger* than a violated cut $\mathbf{q}^T \mathbf{x} \leq b_2$ if the resultant polyhedron of $\mathbf{p}^T \mathbf{x} \leq b_1$ is a proper subset of the resultant polyhedron of $\mathbf{q}^T \mathbf{x} \leq b_2$. By Chapter 4 terminology, $\mathbf{p}^T \mathbf{x} \leq b_1$ results in a "better" formulation.

Consider Figure 12.1. The solid dots in the graph represent the feasible integer points. Inequalities *a, b,* and *c* are valid, while *d* is not because it excludes an integer point. Inequalities *b* and *c* are violated cuts, but *a* is not. Inequality *c* is stronger cut than *b* because the new polyhedron formed by *c* is a subset of that by *b*.

12.2.3   Types of Valid Inequalities

Based on the types of MIP problems, we will present valid inequalities of three types. Type 1 valid inequalities are generated from pure and mixed IP problems with no

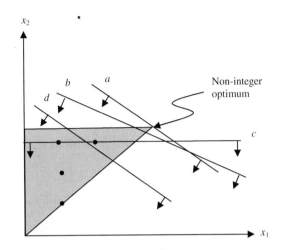

FIGURE 12.1   Valid inequalities for a pure integer program.

special structure. The advantage of this type is that they can always be used to *separate* a fractional point and can be applied to all IP or MIP problems, but the disadvantage is that the derived cuts are usually *very weak*. We will address inequalities for pure and mixed IP problems in Sections 12.4 and 12.5, respectively.

Type 2 inequalities are basically derived from problems with some local structure by considering a single constraint (such as knapsack sets) or a subset of the problem constraints (such as set packing). The inequalities thus derived can at best only separate fractional points that are infeasible to the convex hull of the relaxation. Frequently, type 2 inequalities are *facets* of the convex hull of the relaxation and should be stronger than type 1 inequalities. We will discuss the 0–1 knapsack sets and sets with 0–1 coefficients in Sections 12.6 and 12.7, respectively.

Type 3 inequalities are typically derived from a large part or full set of a *specific* problem structure such as the flow-conservation constraints in the network flow problem. Usually these inequalities are very strong because they may come from certain known classes of *facets of the convex hull* of feasible regions. However, their applications are limited to the particular type of problem. These cuts are very useful and widely implemented in most MIP software, because many hard combinatorial optimization problems possess some or all constraints of special structure. We shall discuss this type of problem in Section 12.8.

12.3   CUT GENERATING TECHNIQUES

Rounding, disjunction, and *lifting* are three powerful, widely used techniques to generate cuts from constraints. In this section, we introduce the basic concept of each technique, while the specific applications of various classes will be discussed in the subsequent sections.

12.3.1 Rounding Technique

The rounding technique has been applied to model preprocessing in Chapter 4. For example, a fractional upper bound on an integer variable can be rounded down and a fractional lower bound can be rounded up as seen below:

$$y \leq 3.8 \text{ implies } y \leq 3$$

and

$$y \geq 1.1 \text{ implies } y \geq 2$$

Another rounding example in model preprocessing is GCD (greatest common divisor) reduction, in which a constraint involves all integer variables with integer coefficients such as $6y_1 + 3y_2 + 12y_3 \leq 17$. The GCD of all coefficients is 3. Divide the constraint by 3 and round down the right-hand side resulting in $2y_1 + y_2 + 4y_3 \leq 5$.

The rounding technique for cut generation is in a more relaxed manner (hence, weaker) and applied more locally. For example, a rounding cut may be applied to a constraint involving nonnegative integer variables

$$\sum a_j y_j \leq b$$

where a_j may be any number. We may divide the constraint by some (arbitrarily) positive constant c and round down the right-hand side to obtain a rounding cut

$$\sum \left\lfloor \frac{a_j}{c} \right\rfloor y_j \leq \left\lfloor \frac{b}{c} \right\rfloor$$

This rounding cut is relevant for the general pure integer programs. But unfortunately, the cut thus generated may be very weak. We need additional information about problem structure to generate stronger cuts. There are several problems, such as node packing, for which strong cuts have been developed.

A strong rounding cut, called *mixed integer rounding(MIR) cut*, can be derived from a constraint involving multiple integer variables and a single continuous variable. This will be covered in Section 12.5.

12.3.2 Disjunction Technique

The disjunction technique is one of the most widely used techniques for constructing a cut for problems involving both continuous and integer variables. For ease of

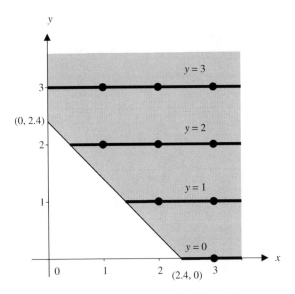

FIGURE 12.2   Mixed IP feasible region.

exposition, we consider a constraint set involving a single continuous variable and a single integer variable:

$$S = \{(x, y) : x + y \geq 2.4, x \geq 0; y \geq 0 \text{ and integer}\}$$

View this MIP feasible region in Figure 12.2. The shaded area represents the feasible region for the LP relaxation of S. The solid lines represent the feasible points to S. Assume point $(0, 2.4)$ is the point upon which a disjunctive cut is to be derived. First, we partition S into two separate sets by adding the following two either–or constraints:

$$y \leq \lfloor 2.4 \rfloor$$

and

$$y \geq \lceil 2.4 \rceil$$

The shaded areas in Figure 12.3 show the two feasible regions represented by the two disconnected sets. Either region is feasible to the original MIP problem. To combine both feasible regions, we apply the disjunction technique to obtain a "union" of the two disjoint sets. To achieve this, consider point $A = (0, 3)$ that intersects $y = \lceil 2.4 \rceil = 3$ and $x = 0$, and point $B = (0.4, 2)$ that intersects $y = \lfloor 2.4 \rfloor = 2$ and $x + y \geq 2.4$. Joining points A and B by a line and determining the inequality sign for

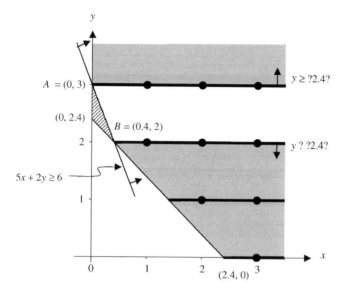

FIGURE 12.3   A disjunctive cut.

feasibility, we obtain a dotted line in Figure 12.3, which represents a disjunctive cut. The crosshatched region represents the infeasible region being cut off. Note that this inequality is valid for the original MIP because all feasible solutions to the MIP are unchanged, and this is a violated cut because it violates the points in the crosshatched region.

To represent this disjunctive cut algebraically, we first write a line equation passing through two points $A = (0, 3)$ and $B = (0.4, 2)$,

$$\frac{y-3}{x-0} = \frac{2-3}{0.4-0}$$

resulting in $5x + 2y = 6$. After checking with the origin for feasibility, we obtain the following inequality:

$$5x + 2y \geq 6$$

12.3.3   Lifting Technique

Lifting is a technique for strengthening valid inequalities and obtaining facet-defining inequalities especially for binary IP programs. There are two types of lifting: sequential versus sequence independent. In sequential lifting, the cut coefficients $\{\pi_j\}$ are evaluated one by one, while in sequence-independent lifting, the π_j's are evaluated simultaneously. The details of lifting procedure will be discussed later in Section 12.6.1.

12.4   CUTS GENERATED FROM SETS INVOLVING PURE INTEGER VARIABLES

This section introduces several violated cuts that are applicable only to pure integer programs. These cuts cannot be applied to problems that contain both integer and continuous variables.

12.4.1   Gomory Fractional Cut

Recall the fractional cutting plane method for pure integer programs described in Chapter 11. Suppose an LP optimum contains a fractional value of some basic variable (say y_{B_r}) in row r

$$y_{B_r} + \sum_{k \in J} \bar{g}_{rk} y_k = \bar{b}_r \qquad (12.1)$$

where J is the set of indices of all nonbasic variables. Row (12.1) can be used to derive a Gomory fractional cut

$$\sum_{k \in J} f_{rk} y_k \geq f_{r0} \qquad (12.2)$$

where $f_{rk} = \bar{g}_{rk} - \lfloor \bar{g}_{rk} \rfloor \geq 0$ and $f_{r0} = \bar{b}_r - \lfloor \bar{b}_r \rfloor \geq 0$.

12.4.2   Chvátal–Gomory Cut

The above Gomory fractional cut (12.2) is derived from the updated coefficients $\bar{g}_j$ of the optimum simplex tableau. To find the same fractional cut but in terms of the original variables and coefficients g_j, we use the relation

$$\bar{g}_j = B^{-1} g_j$$

where B^{-1} is the inverse of the optimum basis B. Matrix B^{-1} may be obtained by direct calculations from B or by extracting it from the optimum simplex tableau. Recall that if the initial simplex tableau utilizes slack variables as basic variables, then the associated coefficients of the slack variables form an identity matrix and hence we have the updated matrix, $B^{-1} I = B^{-1}$. The result implies that we can immediately find B^{-1} in the columns located under the starting basic variables in the corresponding simplex tableau, without additional calculations. One can show that a Gomory fractional cut is the *Chvátal-Gomory cut*.

Theorem 12.2   Let β denote row r of B^{-1}, β_i denote the ith element of β, and $f_i' = \beta_i - \lfloor \beta_i \rfloor$ for $i = 1, 2, \ldots, m$. The Gomory fractional cut $\sum_{k \in J} f_{rk} y_k \geq f_{r0}$, when written in terms of the original variables and coefficients, is the Chvátal–Gomory cut

$$\sum_{j=1}^{n} \left\lfloor \mathbf{f}'^{\mathrm{T}} \mathbf{g}_j \right\rfloor y_j \leq \left\lfloor \mathbf{f}'^{\mathrm{T}} \mathbf{b} \right\rfloor \tag{12.3}$$

where n is the number of original variables, $\mathbf{f}'^{\mathrm{T}} = (f_1', f_2', \ldots, f_n')$, $\mathbf{g}_j$ is the original jth coefficient column, and $\mathbf{b}$ is the original right-hand side column.

A family of Chvátal–Gomory cuts may be rewritten in terms of the original data of the given pure IP problem defined by $\max\{\mathbf{d}^{\mathrm{T}}\mathbf{y}: \sum_j \mathbf{g}_j y_j \leq \mathbf{b}, \mathbf{y} \geq \mathbf{0}$ and integer$\}$, where $\mathbf{g}_j$ ($j = 1, 2, \ldots, n$) are column vectors of matrix $\mathbf{G}$ of dimension $m \times n$.

Let multipliers $\mathbf{u} \geq \mathbf{0}$ and polyhedron $P = \{\mathbf{y}: \sum_j \mathbf{g}_j y_j \leq \mathbf{b}, \mathbf{y} \geq \mathbf{0}\}$ be defined for the IP. Then the inequality

$$\sum_j \mathbf{u}^{\mathrm{T}} \mathbf{g}_j y_j \leq \mathbf{u}^{\mathrm{T}} \mathbf{b} \tag{12.4}$$

is a valid inequality for P because $\mathbf{u} \geq \mathbf{0}$, $\sum_j \mathbf{g}_j y_j \leq \mathbf{b}$ and $\mathbf{y} \geq \mathbf{0}$. Rounding down the noninteger coefficients on the left-hand side of (12.4), we obtain a valid inequality,

$$\sum_j \left\lfloor \mathbf{u}^{\mathrm{T}} \mathbf{g}_j \right\rfloor y_j \leq \mathbf{u}^{\mathrm{T}} \mathbf{b} \tag{12.5}$$

The right-hand side of (12.5) can be further rounded down because y_j are nonnegative integer resulting in the integrality of the left-hand side,

$$\sum_j \left\lfloor \mathbf{u}^{\mathrm{T}} \mathbf{g}_j \right\rfloor y_j \leq \left\lfloor \mathbf{u}^{\mathrm{T}} \mathbf{b} \right\rfloor \tag{12.6}$$

The inequality (12.6), defined for any $\mathbf{u}$, generates a valid inequality for a given pure IP problem. The three-step Chvátal–Gomory procedure, (12.4)–(12.6), can be used to construct a valid inequality for a pure integer program. The optimum dual solution $\mathbf{u}$ for an LP relaxation can be used for the values in (12.4)–(12.6).

12.4.3 Pure Integer Rounding Cut

"Rounding" is a widely used technique for generating valid cuts for pure general integer and pure binary programs. We saw a similar rounding technique in Chapter 4, when a preprocessing method called GCD reduction is used. Here, we apply the rounding technique to a $\leq$ constraint involving nonnegative integer variables,

$$\sum a_j y_j \leq b$$

The rounding procedure is rather simple:

1. Divide the constraint by some positive constant d: $\sum (a_j/d) y_j \leq b/d$.
2. Round down the coefficients on the left-hand side: $\sum \lfloor a_j/d \rfloor y_j \leq \sum (a_j/d) y_j \leq b/d$.
3. Round down the right-hand side: $\sum \lfloor a_j/d \rfloor y_j \leq \lfloor b/d \rfloor$.

If the given constraint is of $\geq$ form, the procedure is similar except that the rounding-down operations are changed to rounding-up operations.

Example 12.1   Derive an integer rounding cut for the integer feasible region of a pure IP problem defined by the following inequality and $\mathbf{y} \geq \mathbf{0}$ and integer:

$$7y_1 + 3y_2 + 4y_3 + 10y_4 + 9y_5 \leq 20$$

Dividing both sides by a coefficient of arbitrary selection (say, 9), we obtain an equivalent inequality:

$$\frac{7}{9}y_1 + \frac{1}{3}y_2 + \frac{4}{9}y_3 + \frac{10}{9}y_4 + y_5 \geq 2\frac{2}{9}$$

Since $\mathbf{y} \geq \mathbf{0}$ and integer, rounding up the coefficients of all terms on the left-hand side will give an upper bound on the left-hand side. That is,

$$y_1 + y_2 + y_3 + 2y_4 + y_5 \geq \frac{7}{9}y_1 + \frac{1}{3}y_2 + \frac{4}{9}y_3 + \frac{10}{9}y_4 + y_5 \geq 2\frac{2}{9}$$

Then the inequality

$$y_1 + y_2 + y_3 + 2y_4 + y_5 \geq 2\frac{2}{9} \qquad (12.7)$$

is a weaker formulation of the feasible region. Since the left-hand side is integer, hence we can round up the right-hand side to obtain an integer rounding cut for the original IP problem:

$$y_1 + y_2 + y_3 + 2y_4 + y_5 \geq 3$$

12.4.4   Objective Integrality Cut

During the solution process for a pure IP problem, upper bounds on the optimum objective value are obtained and updated. These upper bounds can also be used to generate cuts. If the objective coefficients $\mathbf{d}$ are integer, then the entire left-hand side must be integer because variables $\mathbf{y}$ are required to be integer. Let $\bar{z}$ be the best upper bound found so far, we have an objective integrality cut

$$\mathbf{d}^\mathsf{T}\mathbf{y} \geq \lceil \bar{z} \rceil$$

12.5   CUTS GENERATED FROM SETS INVOLVING MIXED INTEGER VARIABLES

12.5.1   Gomory Mixed Integer Cut

Recall the Gomory mixed integer cutting plane method for mixed integer programs in Chapter 11. If the LP optimum contains a basic integer variable with a fractional value, the corresponding row (say r) is selected for generating a mixed integer cut. Let the generating row r be

$$\sum_j \bar{a}_{rj} x_j + \sum_k \bar{g}_{rk} y_k \leq \bar{b}_r$$

We partition the continuous variables x_j into two cases: those with positive coefficients and those with negative coefficients. We also partition integer variables y_k into two cases: those with $f_{rk} \leq f_{r0}$ and those with $f_{rk} > f_{r0}$. Then a Gomory mixed integer cut can be generated by

$$\sum_{j:\bar{a}_{rj}>0} \bar{a}_{rj} x_j + \sum_{j:\bar{a}_{rj}<0} \left(\frac{f_{r0}}{f_{r0}-1}\right) \bar{a}_{rj} x_j + \sum_{k:f_{rk}\leq f_{r0}} f_{rk} y_k + \sum_{k:f_{rk}>f_{r0}} \frac{f_{r0}(1-f_{rk})}{1-f_{r0}} y_k \geq f_{r0}$$

where $f_{rk} = \bar{g}_{rk} - \lfloor \bar{g}_{rk} \rfloor$ and $f_{r0} = \bar{b}_r - \lfloor \bar{b}_r \rfloor$. For simplicity, we drop subscript r to obtain

$$\sum_{j:\bar{a}_j>0} \bar{a}_j x_j + \sum_{j:\bar{a}_j<0} \left(\frac{f_0}{f_0-1}\right) \bar{a}_j x_j + \sum_{k:f_k\leq f_0} f_k y_k + \sum_{k:f_k>f_0} \frac{f_0(1-f_k)}{1-f_0} y_k \geq f_0 \qquad (12.8)$$

Example 12.2   Find a Gomory mixed integer cut for the following mixed integer program:

$$\begin{aligned}
\text{Maximize} \quad & 2x_1 + 5x_2 + 3x_3 + 4x_4 + y_1 + 7y_2 + 2y_3 \\
\text{subject to} \quad & x_1 + 2x_2 + 11x_3 + x_4 + 3y_1 + 2y_2 + y_3 \leq 23 \\
& -x_1 + x_2 + x_3 + 2x_4 - 5y_1 + y_2 + 3y_3 \leq 23 \\
& x_1, x_2, x_3, x_4, s_1, s_2 \geq 0; y_1, y_2, y_3 \geq 0 \text{ and integer}
\end{aligned}$$

The optimal tableau of its LP relaxation contains the following row, in which the variable y_2 is basic and fractional and s_1 and s_2 are slack variables:

$$x_1 + x_2 + 7x_3 + \frac{11}{3}y_1 + y_2 - \frac{1}{3}y_3 + \frac{2}{3}s_1 - \frac{1}{3}s_2 = \frac{23}{3}$$

To construct a mixed integer cut, we must first classify the variables. Among noninteger variables, x_1, x_2, x_3, and s_1 have positive coefficients, while s_2 has a negative coefficient. Integer variables are y_1, y_2, and y_3, whose nonnegative fractions are f_1, f_2, and f_3, respectively. After computing $f_1 = 2/3, f_2 = 0, f_3 = 2/3$, and $f_0 = \frac{23}{3} - \lfloor \frac{23}{3} \rfloor = \frac{2}{3}$, we conclude $f_k \leq f_0$ for all $k = 1$, 2, and 3. That is, there are no integer variables y_k whose $f_k > f_0$. Applying (12.8), we obtain the following mixed integer cut:

$$\left(x_1 + x_2 + 7x_3 + \frac{2}{3}s_1\right) + (-2)\left(\frac{-1}{3}\right)s_2 + \frac{2}{3}y_1 + \frac{2}{3}y_3 \geq \frac{2}{3}$$

In the previous section, we gave (12.8) without elaborating how it is obtained. Here we will show how. Consider the following LP optimum simplex tableau,

$$x_{B_r} + \sum_{j \in J} \bar{a}_{rj} x_j = \bar{b}_r \tag{12.9}$$

For simplicity, we omit row subscripts in (12.9) to get (12.10) and assume that x_j may be either an integer or a continuous variable.

$$x + \sum_{j \in J} \bar{a}_j x_j = \bar{b} \tag{12.10}$$

Since basic variable $x \geq 0$ is required to be an integer, it follows $x = 0 \pmod{1}$. Since $\bar{b}$ by assumption is not an integer, it follows $\bar{b} = f_0 \pmod{1}$. Hence any integer solution to (12.9) must satisfy

$$\sum_{j \in J} \bar{a}_j x_j = f_0 \pmod{1} \tag{12.11}$$

Let the coefficients on the left-hand side of (12.11) be partitioned into two sets, $J^+ = \{j | \bar{a}_j \geq 0\}$ and $J^- = \{j | \bar{a}_j < 0\}$. Then,

$$\sum_{j \in J^+} \bar{a}_j x_j + \sum_{j \in J^-} \bar{a}_j x_j = f_0 \pmod{1} \tag{12.12}$$

where $0 < f_0 < 1$.

The left-hand side of (12.12) is either positive or negative. If it is positive, then it must be one of $f_0, f_0 + 1, f_0 + 2, \ldots$, and we have

$$\sum_{j \in J^+} \bar{a}_j x_j \geq \sum_{j \in J^+} \bar{a}_j x_j + \sum_{j \in J^-} \bar{a}_j x_j \geq f_0 \tag{12.13}$$

If the left-hand side is negative, then it must be one of $-1 + f_0, -2 + f_0, \ldots$, and we have

$$\sum_{j \in J^-} \bar{a}_j x_j \leq \sum_{j \in J^+} \bar{a}_j x_j + \sum_{j \in J^-} \bar{a}_j x_j \leq -1 + f_0 \tag{12.14}$$

Multiplying both sides of (12.14) by $f_0/(-1 + f_0)$, we have

$$\sum_{j \in J^-} \frac{f_0}{f_0 - 1} \bar{a}_j x_j \geq f_0 \tag{12.15}$$

Note that either (12.13) or (12.15) must hold. Because the left-hand sides of (12.13) and (12.15) are both nonnegative, and one of them is $\geq f_0$, then by union (disjunction) of two sets we conclude

$$\sum_{j \in J^+} \bar{a}_j x_j + \sum_{j \in J^-} \frac{f_0}{f_0 - 1} \bar{a}_j x_j \geq f_0 \qquad (12.16)$$

The above inequality must be satisfied by every integer solution, but will violate the current LP solution, because substituting all nonbasic variables x_j to 0 makes the left-hand side zero, which cannot be $\geq f_0$ (a positive fraction).

Note that in the process of deriving (12.16), we use only the fact that the basic variable x on the left-hand side of (12.10) must be an *integer* and that the nonbasic variables x_j on the right-hand side must be *nonnegative*. Therefore, if some nonbasic variables are not required to be integers, (12.16) still represents a valid inequality.

However, we can further utilize the integer requirement of some nonbasic variables x_j to improve (12.16) to a *stronger inequality*. To achieve this goal, we make the coefficients $\bar{a}_j$ for $j \in J^+$ and $((\bar{a}_j f_0)/(f_0-1))$ for $j \in J^-$ to be as small as possible.

Consider a certain term $\bar{a}_q x_q$ in (12.11) for which x_q is required to be integer. Because (12.16) is derived from (12.11), any increasing or decreasing by an integer multiple clearly will still satisfy the congruence relation (12.11). Among all $\bar{a}_q \geq 0$, the smallest coefficient that can be obtained is f_q. Among all $\bar{a}_q < 0$, setting $\bar{a}_q$ to $f_q - 1$ will give the smallest value to $f_0/(f_0 - 1)\, \bar{a}_q$. Therefore, the smallest coefficient in (12.16) must be

$$\min\{f_q, f_0(1-f_q)/(1-f_0)\}$$

Clearly, if $f_q \leq f_0$, then

$$f_q(1-f_0) \leq f_0(1-f_q)$$

or

$$f_q \leq f_0(1-f_q)/(1-f_0)$$

If $f_q > f_0$, then

$$f_q > f_0(1-f_q)/(1-f_0)$$

Combining all four cases, we obtain Gomory mixed integer cut in the form

$$\sum_j f_j^* x_j \geq f_0$$

where

$$
\begin{aligned}
f_j^* &= \bar{a}_j & &\text{if } \bar{a}_j \geq 0 \text{ and } x_j \text{ noninteger}\\[4pt]
f_j^* &= \frac{f_0}{f_0-1}\bar{a}_j & &\text{if } \bar{a}_j < 0 \text{ and } x_j \text{ noninteger}\\[4pt]
f_j^* &= f_q & &\text{if } f_q \leq f_0 \text{ and } x_j \text{ integer}\\[4pt]
f_j^* &= \frac{f_0}{1-f_0}(1-f_q) & &\text{if } f_q > f_0 \text{ and } x_j \text{ integer}
\end{aligned}
$$

12.5.2 Mixed Integer Rounding Cut

Here we intend to derive a rounding cut for the mixed integer set in the following form:

$$
S = \{(\mathbf{y}, x) : \mathbf{a}^T\mathbf{y} - x \leq b, \mathbf{y} \geq \mathbf{0} \text{ and integer}, x \geq 0\} \tag{12.17}
$$

We begin with the simplest case where the set contains a single integer variable, a single continuous variable, and a single inequality, symbolically

$$
S' = \{(y, x) : y - x \leq b, y \geq 0 \text{ and integer}, x \geq 0\}
$$

Then the following inequality, due to Nemhauser and Wolsey (1988), is valid for Conv(S').

$$
y \leq \lfloor b \rfloor + \frac{x}{1-f} \tag{12.18}
$$

where $f = b - \lfloor b \rfloor$. To prove (12.18), we consider the disjunction (union) of the following two sets: $S^1 = S' \cap \{(y, x): y \leq \lfloor b \rfloor\}$ and $S^2 = S' \cap \{(y, x): y \geq \lfloor b \rfloor + 1\}$. For S^1, we multiply $y \leq \lfloor b \rfloor$ by $(1-f)$, multiply $0 \leq x$ by 1, and sum the two resultant inequalities, yielding

$$
\left(y - \lfloor b \rfloor\right)(1-f) \leq x \tag{12.19}
$$

equivalent to (12.18). For S^2, we multiply $-(y - \lfloor b \rfloor) \leq -1$ by f, multiply $y - b \leq x$ by 1, and sum the two resultant inequalities, yielding (12.19) and then (12.18). Therefore, (12.18) is valid for Conv($S^1 \cup S^2$) = Conv(S').

The above single-integer variable case can be extended to derive a *mixed integer rounding cut* for the two-integer variable case

$$
S'' = \{(y_1, y_2, x) : g_1 y_1 + g_2 y_2 - x \leq b; x \geq 0; y_1, y_2 \geq 0 \text{ and integer}\}
$$

where g_1, g_2, and b are scalars with fractional b.

Let $f = b - \lfloor b \rfloor > 0$ and $f_k = g_k - \lfloor g_k \rfloor \geq 0$ for $k = 1, 2$. Then it can be shown that the inequality

$$\left(\lfloor g_1 \rfloor + \frac{(f_1 - f)^+}{1 - f}\right) y_1 + \left(\lfloor g_2 \rfloor + \frac{(f_2 - f)^+}{1 - f}\right) y_2 \le \lfloor b \rfloor + \frac{x}{1 - f}$$

where $(f_k - f)^+ = \max(0, f_k - f)$ is valid for $\mathrm{Conv}(S'')$. This can be generalized to the set containing p integer variables and a single continuous variable.

Theorem 12.3 Let the set $S = \{(\mathbf{y}, x): \mathbf{g}^{\mathrm{T}}\mathbf{y} - x \le b, \mathbf{y} \ge \mathbf{0} \text{ and integer}, x \ge 0\}$, the inequality

$$\sum_k \left(\lfloor g_k \rfloor + \frac{(f_k - f)^+}{1 - f}\right) y_k \le \lfloor b \rfloor + \frac{x}{1 - f} \tag{12.20}$$

is valid for $\mathrm{Conv}(S)$, where $f = b - \lfloor b \rfloor > 0$, $f_k = g_k - \lfloor g_k \rfloor \ge 0$, and $(f_k - f)^+ = \max(0, f_k - f)$.

12.6 CUTS GENERATED FROM 0–1 KNAPSACK SETS

12.6.1 Knapsack Cover

Consider a knapsack constraint

$$K = \{\mathbf{y} \in (0, 1)^n : \sum_j a_j y_j \le b, a_j > 0; b > 0\} \tag{12.21}$$

Any negative coefficient can be converted to a positive coefficient by substituting y_j for a new variable $y'_j = 1 - y_j$.

A set C is a *cover* if $\sum_{j \in C} a_j > b$, or $\Delta = \sum_{j \in C} a_j - b > 0$. The cover C is said to be *minimal* if $a_j \ge \Delta$ for all $j \in C$.

Theorem 12.4 Let $C \subseteq N$ be a *cover* of K and $|C|$ be the number of elements in C, then the *cover inequality*

$$\sum_{j \in C} y_j \le |C| - 1 \tag{12.22}$$

is valid for K. Moreover, if C is a *minimal cover*, then the inequality (12.22) defines a *facet* of $\mathrm{Conv}(K_c)$

$$K_c = K \cap \{\mathbf{y} : y_j = 0, j \in N \backslash C\} \tag{12.23}$$

where $N \backslash C$ is the difference of sets N and C, and $\mathrm{Conv}(K_c)$ is the convex hull of K_c.

Example 12.3 Construct a cut for $K = \{\mathbf{y} \in (0, 1)^5: 2y_1 + y_2 + 5y_3 + 2y_4 + 3y_5 \le 9\}$.

$C = \{3, 4, 5\}$ is a cover because $\Delta = 5 + 2 + 3 - 9 = 1 > 0$, and is a minimal cover for K because $5 > 1$, $2 > 1$ and $3 > 1$. We obtain the corresponding cover inequality $y_3 + y_4 + y_5 \leq 2$, which defines a facet of $\mathrm{Conv}(\{\mathbf{y} \in (0, 1)^3 : 5y_3 + 2y_4 + 3y_5 \leq 9\})$.

If a cover C is not minimal, then it is clearly seen that the corresponding cover inequality is redundant because it is the sum of a minimal cover inequality and some upper bound constraints.

12.6.2 Lifted Knapsack Cover

Lifting can be used to strengthen knapsack cover inequalities and to obtain a large class of facet-defining inequalities for $\mathrm{Conv}(K)$ called *lifted cover inequalities*.

Consider the knapsack set K defined in (12.21) and let M be a subset of N. Suppose we have an inequality,

$$\sum_{j \in M} \pi_j y_j \leq \pi_0 \tag{12.24}$$

which is valid for $K_M = K \cap \{\mathbf{y}: y_j = 0, j \in N \backslash M\}$. The lifting problem is to find the lifting coefficients $\{\pi_j\}, j \in N \backslash M$, so that

$$\sum_{j \in N} \pi_j y_j \leq \pi_0 \tag{12.25}$$

is valid for K. Ideally, we would like inequality (12.25) to be "strong." That is, if inequality (12.24) defines a face of high dimension of $\mathrm{Conv}(K_M)$, we would like the inequality (12.25) to define a face of high dimension of $\mathrm{Conv}(K)$.

There are two types of lifting: (1) *sequential lifting* and (2) *sequence independent lifting*. We first describe sequential lifting. The sequential lifting obtains coefficients $\{\pi_j\}, j \in N \backslash M$, one at a time. Specifically, the coefficient π_k is computed for a given $k \in N \backslash M$ so that

$$\sum_{j \in M} \pi_j y_j + \pi_k y_k \leq \pi_0 \tag{12.26}$$

is valid for $K_{M \cup \{k\}}$. This can be done by computing the *lifting function*

$$F_M(a_k) = \min\{\pi_0 - \sum_{j \in M} \pi_j y_j : \sum_{j \in M} a_j y_j \leq b - a_k, \mathbf{y} \in (0, 1)^M\} \tag{12.27}$$

For a given $k \in N \backslash M$, suppose $K_{M \cup \{k\}} \cap \{\mathbf{y}: y_k = 1\} \neq \emptyset$. Then inequality

$$\sum_{j \in M} \pi_j y_j + \pi_k y_k \leq \pi_0 \tag{12.28}$$

is valid for $K_{M \cup \{k\}}$ if $\pi_k \le F_M(a_k)$. Once variable y_k is lifted, M is updated by including k, and lifting can be done to a second variable by repeating the lifting function. The lifting procedure for a cover inequality is given below.

Assume the given cover $C = \{r + 1, \ldots, n\}$ and $N \backslash C = \{1, \ldots, r\}$ after reordering the variables.

1. Let $k = 1$.
2. Compute
$$F_C(a_k) = \max\{\textstyle\sum_{j=1}^{k-1} \pi_j y_j + \sum_{j \in C} y_j : \sum_{j=1}^{k-1} \pi_j y_j + \sum_{j \in C} a_j y_j \le b - a_k,$$
$$\mathbf{y} \in (0, 1)\}$$
3. Compute $\pi_k = |C| - 1 - F_C(a_k)$.
4. $k \leftarrow k + 1$. Stop if $k = r + 1$, otherwise go to step 2.

Example 12.4   Find a lifted knapsack cover for the knapsack set

$$K = \{\mathbf{y} \in (0, 1)^7 : 4y_1 + 7y_2 + 5y_3 + y_4 + 2y_5 + 3y_6 + 4y_7 \le 15\}$$

Let $M = \{2, 3, 7\}$, then

$$K_M = K \cap \{\mathbf{y} : y_1 = y_4 = y_5 = y_6 = 0\} = \{\mathbf{y} \in (0, 1)^3 : 7y_2 + 5y_3 + 4y_7 \le 15\}$$

The inequality $y_2 + y_3 + y_7 \le 2$ is valid for K_M. So is $y_2 + y_3 + y_7 \le 3$ used below.

The lifting procedure is as follows:

1. Let $k = 1$. $F_M(a_k) = F_M(4) = \min\{3 - (y_2 + y_3 + y_7): \quad 7y_2 + 5y_3 + 4y_7 \le 11\} = 1$.
2. $K_{M \cup \{1\}} = \{\mathbf{y} \in (0, 1): 4y_1 + 7y_2 + 5y_3 + 4y_7 \le 15\}$.
3. $K_{M \cup \{1\}} \cap \{\mathbf{y}: y_1 = 1\} = \{7y_2 + 5y_3 + 4y_7 \le 11\} \ne \emptyset$.
4. Let $\pi_1 = 1$. Then the inequality, $y_1 + y_2 + y_3 + y_7 \le 3$, is valid for $K_{M \cup \{1\}}$.
5. $M \leftarrow M \cup \{1\}$. Select another k. Repeat the procedure until all variables are lifted.

Now we apply the method of sequence independent lifting to the knapsack cover. Before doing this, we need to define the term *superadditive*.

Definition 12.1   A function $F: \mathbb{R} \to \mathbb{R}$ is *superadditive* on $\mathbb{R}$ if $F(c_1) + F(c_2) \le F(c_1 + c_2)$ for all real c_1, c_2.

Let $F: \mathbb{R} \to \mathbb{R}$ be a function. The inequality

$$\sum_{j \in M} \pi_j y_j + \sum_{j \in N \backslash M} F(a_j) y_j \le \pi_0$$

is valid for K if the following two conditions are satisfied:

(i) $F(c) \leq F_M(c)$ for all real c

(ii) $F(c)$ is superadditive

Example 12.5 Consider Example 12.4. Let

$$F_M(a_k) = \min\{\pi_0 - \sum_{j \in M} \pi_j y_j : \sum_{j \in M} a_j y_j \leq b - a_k, \mathbf{y} \in (0, 1)\}$$

$$= \min\{3 - (y_2 + y_3 + y_7) : 7y_2 + 5y_3 + 4y_7 \leq 17 - a_k, \mathbf{y} \in (0, 1)\}$$

We have

$$F_M(a_k) = \begin{cases} 0 & 0 \leq a_k \leq 1 \\ 1 & 1 < a_k \leq 8 \\ 2 & 8 < a_k \leq 13 \\ 3 & a_k > 13 \end{cases}$$

Let $F(a_k) = a_k - 1$, then $F(a_k) \leq F_M(a_k)$ for any a_k. (i) is satisfied.

Since $F(a_k) + F(a_j) = a_k + a_j - 2 \leq F(a_k + a_j) = a_k + a_j - 1$, (ii) is satisfied.

In this problem, $\mathbf{NM} = \{1, 4, 5, 6\}$, $\{a_k\} = \{a_1, a_4, a_5, a_6\} = \{4, 1, 2, 3\}$, we have $F(a_1) = 3$, $F(a_4) = 0$, $F(a_5) = 1$, $F(a_6) = 2$, and the inequality

$$3y_1 + y_2 + y_3 + y_5 + 2y_6 + y_7 \leq 3$$

is valid for $\mathbf{K}$.

12.6.3 GUB Cover

A GUB (generalized upper bound) cover inequality is derived from the following GUB set:

$$S = \{\mathbf{y} \in (0, 1)^N : \sum_j a_j y_j \leq b, \sum_{j \in Q_i} y_j \leq 1, Q_i \cap Q_j = \varnothing \text{ for all } i \neq j, \cup_i Q_i = N\}$$

A strong cut can be derived by

$$\sum_{j \in C} y_j \leq J$$

where C is a GUB cover defined by no two elements of C belonging to the same Q_i.

Example 12.6 Construct a GUB cut for $S = \{\mathbf{y} \in (0, 1): 2y_1 + y_2 + 5y_3 + 2y_4$ $3y_5 + 6y_6 + 4y_7 + y_8 \leq 9, y_1 + y_3 + y_4 \leq 1, y_2 + y_7 \leq 1, y_5 + y_8 \leq 1, y_6 \leq 1\}$.

Select a variable from each of the constraints 2–5 so that no two variables are the same. Then, we obtain an inequality $y_3 + y_6 + y_7 + y_8 \leq 4$ as a GUB cover. There are many other combinations that can be used to form GUB covers.

12.7   CUTS GENERATED FROM SETS CONTAINING
0–1 COEFFICIENTS AND 0–1 VARIABLES

The constraint sets comprising 0–1 coefficients and binary variables frequently arise in graph- or network-related problems and in combinatorial problems, for example, node packing and traveling salesman. Here, we will discuss the polyhedron of the node packing problem and the construction of strong valid inequalities.

Definition 12.2   Given is a graph $G = (V, E)$, where V and E are sets of vertices (or nodes) and edges (or arcs), respectively. A *node packing* (set) S is a set of nodes such that no two nodes have a common edge, that is,

$$S = \{y \in \{0, 1\}^{|V|} : y_i + y_j \leq 1 \text{ for all } (i, j) \in E\}$$

where $i, j \in V$ and $|V|$ is the number of vertices. The node packing problem is to find the maximum-cardinality node packing (independent node set) in G

$$\max\left\{\sum_{j=1}^{|V|} y_j\right\}$$

Consider the graph for a node packing problem with five nodes and seven arcs in Figure 12.4.

To represent in matrix notation, let $\mathbf{A}$ be an arc–node incidence matrix and $\mathbf{y}$ be a column of 0–1 variables. Then, a node packing set must satisfy $\mathbf{Ay} \leq \mathbf{1}$, where $\mathbf{A}$ is a 0–1 coefficient matrix and $\mathbf{y}$ is a 0–1 vector. For example, in Figure 12.4, $|V| = 5$, $|E| = 7$, and the arc–node matrix is

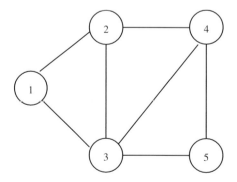

FIGURE 12.4   Graph for node packing problem.

$$A = \begin{bmatrix} 1 & 1 & 0 & 0 & 0 \\ 1 & 0 & 1 & 0 & 0 \\ 0 & 1 & 1 & 0 & 0 \\ 0 & 1 & 0 & 1 & 0 \\ 0 & 0 & 1 & 1 & 0 \\ 0 & 0 & 1 & 0 & 1 \\ 0 & 0 & 0 & 1 & 1 \end{bmatrix} \quad y = \begin{bmatrix} y_1 \\ y_2 \\ y_3 \\ y_4 \\ y_5 \end{bmatrix} \quad 1 = \begin{bmatrix} 1 \\ 1 \\ 1 \\ 1 \\ 1 \\ 1 \end{bmatrix}$$

To find a node packing set, we can "directly" solve $\max\{\sum_{j=1}^{|V|} y_j : Ay \leq 1$, where $y_j = 1$ if node j is in the set, $y_j = 0$ otherwise$\}$. However, we can solve it easier by constructing strong valid inequalities to the problem. To achieve this, we introduce the concept of *maximal clique*.

Definition 12.3 Given is the graph $G = (V, E)$, where V and E are sets of vertices and edges, respectively. A set of nodes, $C_q \subseteq V$, is called a *clique* if every pair of nodes in C_q is joined by an edge. A clique is *maximal* if it contains a maximal number of nodes.

Consider Figure 12.4. The two-node cliques are $\{1, 2\}, \{1, 3\}, \{2, 3\}, \{2, 4\}, \{3, 4\}$, $\{3, 5\}$, and $\{4, 5\}$. The three-node cliques are $\{1, 2, 3\}, \{2, 3, 4\}$, and $\{3, 4, 5\}$. The set $\{1, 2, 3, 4\}$ is not a clique because there is no edge between nodes 1 and 4. The set $\{2, 3, 4, 5\}$ is not a clique because there is no edge between nodes 2 and 5. In fact, there is no clique containing four or more nodes. Therefore, the maximal clique is three. The maximal clique can be used to construct a strong cut for a node pack set called *clique cut (inequality)*.

Theorem 12.5 Let C_q be a maximal clique, the clique inequality or cut $\sum_{j \in C_q} y_j \leq 1$ defines a facet of $Conv(\{y \in \{0, 1\}^{|V|} : Ay \leq 1\})$.

Consider Figure 12.4. For the maximal clique $C_q = \{1, 2, 3\}$, the clique cut is $y_1 + y_2 + y_3 \leq 1$. For the maximal clique $C_q = \{2, 3, 4\}$, the clique cut is $y_2 + y_3 + y_4 \leq 1$. For the maximal clique $C_q = \{3, 4, 5\}$, the clique cut is $y_3 + y_4 + y_5 \leq 1$. These three clique cuts provide strong valid inequalities for solving the node packing problem in Figure 12.4. In this particular problem, the reader can list all the seven constraints in $Ay \leq 1$ and verify that they all are *dominated* by these three clique cuts.

$$y_1 + y_2 + y_3 \leq 1 \tag{12.29}$$

$$y_2 + y_3 + y_4 \leq 1 \tag{12.30}$$

$$y_3 + y_4 + y_5 \leq 1 \tag{12.31}$$

Now we utilize these cuts to solve the node packing problem. From (12.29), $y_1 = 1$ implies $y_2 = y_3 = 0$. From (12.30), $y_4 = 1$ because we want to maximize

$\sum_{j=1}^{|V|} y_j$. From (12.31), $y_4 = 1$ implies $y_5 = 0$. Now we have a solution $y_1 = y_4 = 1$, $y_2 = y_3 = y_5 = 0$. Similarly, we can generate two alternative solutions.

$$y_2 = y_5 = 1, \quad y_1 = y_3 = y_4 = 0$$

$$y_1 = y_5 = 1, \quad y_2 = y_3 = y_4 = 0$$

These three solutions correspond to the respective node packing sets $\{1,4\}$, $\{2,5\}$, and $\{1,5\}$, and there are no packing sets including three or more nodes. This example shows how a maximal clique can be used to generate strong cuts to help solve a problem containing 0–1 coefficients and binary variables. For this example, it is coincidental that the generated cuts can immediately find a solution. Nevertheless, the generated strong cuts are useful for making the branch-and-cut method more efficient.

12.8   CUTS GENERATED FROM SETS WITH SPECIAL STRUCTURES

This section discusses several well-known special-structure constraint sets from which strong cuts are generated, including

- Flow cover from a simple fixed-charge flow network
- Plant location

12.8.1   Flow Cover from Fixed-Charge Flow Network

Consider the following simple fixed-charge network flow problem in mixed integer variables:

$$S = \left\{ (\mathbf{x}, \mathbf{y}) : \sum_{j \in N_1} x_j - \sum_{j \in N_2} x_j \leq b, \quad x_j \leq a_j y_j \text{ for all } j, \ \mathbf{x} \geq \mathbf{0}, \ \mathbf{y} \text{ binary} \right\}$$

The constraints contain two sets: the flow-conservation constraints and fixed-charge constraints imposed on each node j. A flow cover cut is derived below:

$$\sum_{j \in C_1} x_j + \sum_{j \in C_2} \left(a_j - \sum_{j \in C_1} a_j + \sum_{j \in C_2} a_j + b \right)^+ (1 - y_j)$$

$$- \sum_{j \in C_1} a_j - \left(\sum_{j \in C_1} a_j - \sum_{j \in C_2} a_j - b \right) \sum_{j \in L_2} y_j - \sum_{j \in N_2 \setminus (C_2 \cup L_2)} x_j \leq b$$

where $C_1 \subseteq N_1, C_2 \subseteq N_2$ are generalized covers defined by

$$\sum_{j \in C_1} a_j - \sum_{j \in C_2} a_j > b, \quad \text{where } L_2 = N_2 \backslash C_2.$$

Example 12.7 Assume the demand on a certain node is $b = 5$, and the capacities of the supply and demand nodes are $(13, 15, 8)$ and $(9, 14, 10, 12)$, respectively. Then the constraint set is as follows:

$$x_1 + x_2 + x_3 - x_4 - x_5 - x_6 - 4x_7 \le 5$$
$$x_1 \le 13y_1, x_2 \le 15y_2, x_3 \le 8y_3, x_4 \le 9y_4, x_5 \le 14y_5, x_6 \le 10y_6, x_7 \le 12y_7$$

Let $C_1 \subseteq N_1 = \{1, 2, 3\}$, $C_2 \subseteq N_2 = \{4, 7\}$, $L_2 \subseteq (N_2 \backslash C_2) = \{5\}$, with $\sum_{j \in C_1} a_j = 36$ and $\sum_{j \in C_2} a_j = 21$. Then the inequality

$$x_1 + x_2 + x_3 + 3(1 - y_1) + 5(1 - y_2) - 21 - 10y_5 - x_6 \le 5$$

or equivalently,

$$x_1 + x_2 + x_3 - x_6 - 3y_1 - 5y_2 - 10y_5 \le 18$$

is a valid cut for this problem.

12.8.2 Plant/Facility Location (Fixed-Charge Transportation)

Consider the following constraint set of the fixed-charge transportation problem

$$S = \left\{ (\mathbf{x}, \mathbf{y}) : \sum_j x_{ij} = a_i, \sum_i x_{ij} \le b_j y_j, x_{ij} \right.$$

$$\left. \le \min(a_i, b_j) y_j, \mathbf{x} \ge \mathbf{0}, \mathbf{y} \text{ binary}, i \in M, j \in N \right\}$$

A strong cut can be derived as follows:

$$\sum_{j \in C} \sum_{i \in K_j} x_{ij} + \sum_{j \in C} \left(\tilde{b}_j - \sum_{j \in C} \tilde{b}_j + \sum_{i \in K} a_i \right)^{+} (1 - y_j) \le \sum_{i \in K} a_i$$

where $C \subseteq N$ is a subset of locations, and a cover such that $\sum_{j \in C} \tilde{b}_j > \sum_{i \in K} a_i, K \subseteq M$ is a subset of clients, $K_j \subseteq K$ is possibly a smaller subset of K, $\tilde{b}_j = \min(b_j, \sum_{i \in K_j} a_i)$ is the "effective" capacity of location j.

Example 12.8 Products are manufactured in four plants to supply customers from five different cities, as shown in Figure 12.5. The capacities b_j for the four plants are 30, 20, 40, and 30, respectively. The demands from the five cities a_i are 15, 12, 18, 10,

Plant $j \in N$ Customer $i \in M$

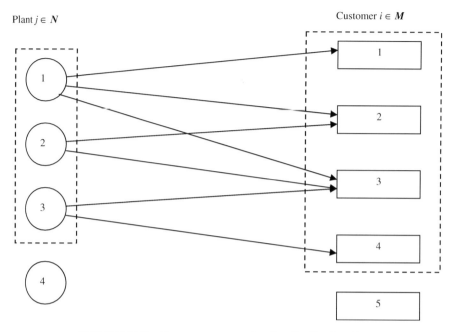

FIGURE 12.5   Constructing a cut for plant/facility location.

and 16, respectively. The set of all constraints are

$$x_{11} + x_{21} + x_{31} + x_{41} = 15$$
$$x_{12} + x_{22} + x_{32} + x_{42} = 12$$
$$x_{13} + x_{23} + x_{33} + x_{43} = 18$$
$$x_{14} + x_{24} + x_{34} + x_{44} = 10$$
$$x_{15} + x_{25} + x_{35} + x_{45} = 16$$
$$x_{11} + x_{12} + x_{13} + x_{14} + x_{15} \leq 30y_1$$
$$x_{21} + x_{22} + x_{23} + x_{24} + x_{25} \leq 20y_2$$
$$x_{31} + x_{32} + x_{33} + x_{34} + x_{35} \leq 40y_3$$
$$x_{41} + x_{42} + x_{43} + x_{44} + x_{45} \leq 30y_4$$
$$x_{ij} \leq \min(a_i, b_j)y_j$$

Let $K \subseteq M = \{1, 2, 3, 4\}$, $K_1 = \{1, 2, 3\}$, $K_2 = \{2, 3\}$, and $K_3 = \{3, 4\}$. Compute

$$\sum_{i \in K} a_i = 15 + 12 + 18 + 10 = 55$$

$$\tilde{b}_1 = \min\left(b_1, \sum_{i \in K_1} a_i\right) = \min(30, 45) = 30,$$

$$\tilde{b}_2 = \min\left(b_2, \sum_{i \in K_2} a_i\right) = \min(20, 30) = 20,$$

$$\tilde{b}_3 = \min\left(b_3, \sum_{i \in K_3} a_i\right) = \min(40, 28) = 28$$

Let $C \subseteq N = \{1, 2, 3\}$, ($\sum_{j \in C} \tilde{b}_j = 78 > \sum_{i \in K} a_i = 55$, so C is a cover), then inequality

$$x_{11} + x_{21} + x_{31} + x_{22} + x_{32} + x_{33} + x_{43} + 7(1-y_1) + 5(1-y_3) \leq 55$$

or equivalently

$$x_{11} + x_{21} + x_{31} + x_{22} + x_{32} + x_{33} + x_{43} - 7y_1 - 5y_3 \leq 43$$

is a valid cut for the problem.

12.9 NOTES

The branch-and-cut approach, a generalization of branch-and-bound (Land and Doig, 1960), follows a series of key contributing papers by Crowder et al. (1983), Johnson et al. (1985), Van Roy and Wolsey (1987), and Hoffman and Padberg (1991). In fact, the name was given by the authors in the last article, which claimed to solve a 0–1 IP instance containing as many as 6000 binary variables. Johnson et al. (2000) present an excellent survey paper about modeling and solving mixed integer programs using LP-based algorithms. Marchand et al. (2002) provide a complete treatment of various types of cutting planes that are useful or potentially useful in solving pure integer and mixed integer programs. Cordier et al. (1999) describe a branch-and-cut MIP software system called *bc-opt*.

Section 12.4

Gomory fractional (Gomory, 1960) and Chvátal–Gomory cuts for pure integer programs are rarely implemented and are usually replaced by the Gomory mixed integer cut (Gomory, 1960), even for pure integer programs.

Section 12.5

The Gomory mixed integer cut (Gomory, 1960) in (12.8) is the most implemented cut for general IPs and MIPs. An alternative form of this cut is obtained by dropping the first two terms corresponding to the continuous variables x_j in (12.8). Mixed integer rounding cut is due to Nemhauser and Wolsey (1988).

Section 12.6

Knapsack covers (Crowder et al., 1983, Weisemantel, 1997) are the first cuts to find extensive use in general purpose solvers and have been successfully used in commercial codes for many years. GUB covers are due to Gu et al. (1998).

Section 12.7

Cliques are due to Johnson and Padberg (1983) and Atamturk et al. (1988).

Section 12.8

Flow covers are due to Padberg et al. (1985) and Gu et al. (1999).

12.10   EXERCISES

12.1 Consider the integer set $S = \{y \geq 0$ and integer: $2y_1 + 3y_2 \leq 23, -5y_1 + 5y_2 \geq 8\}$ and two given valid inequalities for S: $y_1 \leq 4$ and $-y_1 + y_2 \geq 2$. Which one is stronger?

12.2 Graphically represent the feasible region of $S = \{x \geq 0: 3x_1 + x_2 \leq 6, x_1 - x_2 \geq 1\}$. Check if $7x_1 + x_2 \leq 10$ is valid for S with no calculation involved.

12.3 Consider the set $S = \{y \in (0, 1): 5y_1 + 2y_2 - 3y_3 - y_4 + 4y_5 \leq 6\}$. Check if the following inequalities are valid for S:

   (a) $y_1 = 1$

   (b) $y_3 = 0$

   (c) $y_1 + y_2 + y_5 \leq 2$

   (d) $y_3 + y_4 \geq 1$

12.4 In each of the following problems, a set S and a point are given. Find a valid inequality for S that cuts off the point.

   (a) $S = \{y \geq 0$ and integer: $7y_1 + 10y_2 + 5y_3 + 13y_4 \geq 30\}$, $y = (0, 0, 0, 13/30)$

   (b) $S = \{x \geq 0, y \geq 0$ and integer: $x_1 + x_2 \leq 41, x_1 + x_2 \leq 15y\}$, (x, y) $(10, 16, 15/26)$

   (c) $S = \{x \geq 0, y \in (0, 1): x_1 + x_2 \leq 3y, x_1 \leq 2, x_2 \leq 1\}$, $(x, y) = (1, 1, 2/3)$

   (d) $S = \{y \geq 0$ and integer: $2y_1 + y_2 + 7y_3 + 5y_4 + 3y_5 \leq 25\}$, $y = (25/2, 0, 0, 0, 0)$

12.5 Generate a valid inequality using Chvátal–Gomory cut procedure for the following IP:

$$\text{Maximize} \quad y_1 + y_2 + y_3$$
$$3y_1 + 5y_2 - y_3 \leq 12$$
$$y_1 + y_3 \leq 7$$
$$y_1 - y_2 + 2y_3 \leq 9$$
$$y \geq 0 \text{ and integer}$$

12.6 Let $S = \{x \geq 0, y \geq 0$ and integer: $-\alpha x_0 + ax + gy \leq b, bx_0 + ax + gy \leq b + \beta, \alpha, \beta > 0\}$. Show that $ax + gy \leq b$ is an MIR inequality for S. (*Hint*: Scale each inequality in S by $1/(\alpha + \beta)$.)

12.7 Consider the set $S = \{x \geq 0, y \geq 0$ and integer: $x \leq My, 0 \leq x \leq b\}$. Show that $x \leq b - \alpha(\beta - y)$, where $\alpha = b - (\lceil \frac{b}{M} \rceil - 1)M$, $\beta = \lceil \frac{b}{M} \rceil$ is valid for S.

12.8 Consider the mixed integer problem

$$\text{Maximize} \quad y_1 + y_2 + y_3 - 2x$$

$$\text{subject to} \quad 3.1y_1 + 1.3y_2 + 1.4y_3 - x \leq 19.7$$

$$y \geq 0 \text{ and integer}, \ x \geq 0$$

(a) Solve this problem using MIR cuts.

(b) Solve this problem using Gomory mixed integer cuts.

12.9 Consider Example 12.5. Suppose the objective function for S_1 and S_2 is max $11x_1 + 6x_2$. Let $\mathbf{u}_1 = (1/2, 1/3)$, $\mathbf{u}_2 = (1, 1/4)$. Generate a disjunctive cut for $S_1 \cup S_2$ using these parameters.

12.10 Show that the valid inequality in Example 12.1 is a Chvátal–Gomory cut.

12.11 Find two valid cuts for $S = \{y \in (0, 1): 4y_1 + 3y_2 + 2y_3 + 7y_4 + 5y_5 \leq 11\}$ using two different algorithms learned in Chapters 11 and 12.

12.12 (Mixed Integer Knapsack) Let $S = \{x \geq 0, y \in (0, 1): \sum_{j \in N} a_j x_j - y \leq b, b \geq 0, a_j \geq 0\}$, $C \subseteq N$ is a cover. Then, the *mixed integer knapsack inequality*

$$\sum_{j \in C} \min\left(a_j, \sum_{j \in C} a_j - b\right) x_j \leq \sum_{j \in C} \min\left(a_j, \sum_{j \in C} a_j - b\right) - \sum_{j \in C} a_j + b + y$$

is valid for S. Consider the instance where $S = \{x \geq 0, y \in (0, 1): 5x_1 + 3x_2 + 5x_3 + 4x_4 + 7x_5 - y \leq 15\}$. Find at least three mixed integer knapsack covers for S.

12.13 Consider the following generalized assignment problem:

$$\text{Maximize} \quad \sum_i \sum_j c_{ij} y_{ij}$$

$$\text{subject to} \quad \sum_j y_{ij} = 1 \qquad i = 1, \ldots, 4$$

$$\sum_i a_{ij} y_{ij} \leq b_j \quad j = 1, \ldots, 3$$

$$y \in (0, 1)$$

where $\mathbf{b} = (6, 12, 10)$ and

$$A = \begin{pmatrix} 3 & 1 & 2 & 4 \\ 1 & 5 & 4 & 7 \\ 2 & 9 & 3 & 5 \end{pmatrix}, \quad C = \begin{pmatrix} 2 & 4 & 2 & 1 \\ 3 & 2 & 1 & 2 \\ 4 & 1 & 3 & 2 \end{pmatrix}$$

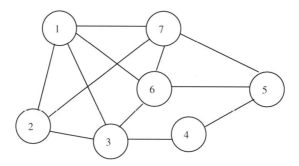

FIGURE 12.6   A simple graph.

(a) Generate two knapsack covers.

(b) Find a Gomory fractional cut.

(c) Find two disjunctive cuts.

12.14   Consider the flow conservation set $S = \{\mathbf{x} \geq \mathbf{0}, \mathbf{y} \in (0, 1): x_1 + x_2 + x_3 + x_4 - (x_5 + x_6 + x_7) \leq 6,\ x_1 \leq 5y_1,\ x_3 \leq 4y_2,\ x_4 \leq 7y_4,\ x_5 \leq 3y_5,\ x_6 \leq 2y_6,\ x_7 \leq 5y_7\}$. Derive a flow cover inequality with $C_1 = \{1, 4\}$.

12.15   Given the following graph $G(V, E)$ and the set $S = \{y_i + y_j \leq 1, \text{for } (i, j) \in E\}$,

(a) Find two cliques of G.

(b) Generate two clique cuts associated with the two cliques in (1). Figure 12.6

12.16   (The Stable Set) Given a graph $G(V, E)$, a stable set is a subgraph induced by a subset of vertices G', so that no pair of vertices in G' defines an edge of E. For each stable set G' in G we can define a point y_i in the following way:

$$y_i = \begin{cases} 1 & \text{if } i \in G' \\ 0 & \text{otherwise} \end{cases}$$

The convex hull of all these points, called the *stable set polytope*, associated with G will be denoted by STAB(G). Mathematically, it can be expressed as

$$\text{STAB}(G) = \text{Conv}\{\mathbf{y} \in (0, 1) : y_i + y_j \leq 1, (i, j) \in E\}$$

Let $C \subseteq E$ be a cycle of odd cardinality in G. The odd cycle inequality $\sum_{i \in V(C)} x_i \leq (|V(C)| - 1)/2$ is valid for STAB(G). Consider the graph $G(V, E)$ in Exercise 12.15.

(a) Find two stable sets in G.

(b) For each stable set found in part (a), find an odd cycle inequality for STAB (G).

12.17 Consider the instance of a capacitated plant location problem where there are three plants and four customers. The problem data are as follows: $c = (100, 120, 110)$ and $d = (90, 70, 80, 60)$.

(a) Formulate the problem with input data.

(b) Find a valid plant location inequality.

13

BRANCH-AND-PRICE APPROACH

In the previous chapter, branch-and-bound is generalized to include generation of cuts or rows, hence the name *branch-and-cut*. In this chapter, branch-and-bound is first generalized to include generation of *columns* by solving pricing problems, hence the name *branch-and-price*, and yet another generalization includes generation of columns and rows, hence the name *branch-and-price-and-cut*. Basically, all these generalizations solve a sequence of LP relaxations of a given IP. Branch-and-cut tightens the LP relaxations (or polyhedra) by adding cuts or constraints (rows). Branch-and-price tightens the LP relaxations by generating a subset of profitable columns associated with variables to join the current basis. These columns are generated iteratively by solving subproblems or *pricing problems.*

13.1 CONCEPTS OF BRANCH-AND-PRICE

Branch-and-price builds upon the branch-and-bound framework. It applies column generation throughout the branch-and-bound tree prior to branching. Branching occurs when no profitable columns can be found and the LP solution does not satisfy the integrality conditions. The concept of column generation is outlined below.

- The column generation approach is used when the LP relaxation of a given IP formulation contains too many columns (associated with variables) to handle explicitly and simultaneously.

Applied Integer Programming: Modeling and Solution, By Der-San Chen, Robert G. Batson, and Yu Dang
Copyright © 2010 John Wiley & Sons, Inc.

- Instead of handling all columns of a given master problem explicitly, a restricted version of the master problem that contains only a subset of columns (usually associated with the basic variables) is maintained and updated, while the remaining huge number of columns (usually associated with nonbasic variables) are left out of the LP relaxation.

- Because most of these columns will likely have their associated variables equal to zero in an optimal solution, only profitable columns (associated with nonbasic variables) are generated and added to the current restricted master problem to improve its current LP solution. Such columns can be generated iteratively by solving subproblems (or pricing problems).

- The column generation approach to integer programming is closely related to Dantzig–Wolfe decomposition in linear programming. Initially, the restricted master problem is represented by the revised simplex tableau that contains the current basis inverse, primal solution, and dual solution. Then the dual solution is passed to update the objective function of a subproblem, which in turn is solved to determine if the LP solution of the master problem is optimal—and if not, to identify a pivot column to enter the basis to improve the current LP solution.

- An LP optimum is found when there is no column that can be generated with a profitable reduced cost. The LP optimum may or may not satisfy the integrality conditions.

- If the LP optimum satisfies the integrality conditions, a lower bound for the IP is found. Otherwise, the noninteger LP optimum (or approximation) can be used as an upper bound and then branching occurs.

- A special (problem-specific) branching scheme is usually needed because column generation may destroy the original problem structure.

In the next section, Dantzig–Wolfe decomposition for linear programs will be introduced via the revised simplex method to provide the necessary background for the development of column generation for integer programs.

13.2 DANTZIG–WOLFE DECOMPOSITION

Consider the following linear program containing two sets of constraints: $\mathbf{Ax} \le \mathbf{b}$ and $\mathbf{Gx} \le \mathbf{d}$. Usually, the first set of constraints is of general structure and the second set is of special structure.

$$
\begin{array}{rl}
\text{(LP)} \quad \text{Maximize} & z = \mathbf{c}^T\mathbf{x} \\
\text{subject to} & \mathbf{Ax} \le \mathbf{b} \\
& \mathbf{Gx} \le \mathbf{d} \\
& \mathbf{x} \ge \mathbf{0}
\end{array}
$$

where $\mathbf{c}$ is a "profit" vector in the objective function to be maximized. Let $S = \{\mathbf{x}: \mathbf{Gx} \le \mathbf{d},\ \mathbf{x} \ge \mathbf{0}\}$. For ease of exposition, we assume that S is bounded (this assumption can be relaxed). Since S is a bounded polyhedron, any point $\mathbf{x} \in S$ can be represented as a convex combination of all (say t) extreme points of S. Denoting these extreme points by $\mathbf{x}^1, \mathbf{x}^2, \ldots, \mathbf{x}^t$, any $\mathbf{x} \in S$ can be represented as

$$\mathbf{x} = \sum_{j=1}^{t} \lambda_j \mathbf{x}^j$$

$$\sum_{j=1}^{t} \lambda_j = 1$$

$$\lambda_j \geq 0 \qquad j = 1, 2, \ldots, t$$

Substituting for $\mathbf{x}$, LP can be transformed into the following so-called *master problem* ($\mathbf{P_M}$) in the variables $\lambda_1 \ldots \lambda_t$.

$$(\mathbf{P}_M) \quad \text{maximize} \quad z = \sum_{j=1}^{t} (\mathbf{c}^T \mathbf{x}^j) \lambda_j$$

$$\text{subject to} \quad \sum_{j=1}^{t} (\mathbf{A} \mathbf{x}^j) \lambda_j \leq \mathbf{b}$$

$$\sum_{j=1}^{t} \lambda_j = 1$$

$$\lambda_j \geq 0 \qquad j = 1, 2, \ldots, t$$

Let $\hat{c}_j = \mathbf{c}^T \mathbf{x}^j$ associated with basic variable λ_j, $\hat{\mathbf{c}}_B = (\hat{c}_1 \ldots \hat{c}_j \ldots \hat{c}_{m+1})$, and $\lambda_B = (\lambda_1, \ldots, \lambda_j, \ldots, \lambda_{m+1})$. Let $\mathbf{u}$ denote the vector of dual variables corresponding to the constraint set $\sum_{j=1}^{t} (\mathbf{A} \mathbf{x}^j) \lambda_j \leq b$, and α denote the dual variable corresponding to the convexity constraint $\sum_{j=1}^{t} \lambda_j = 1$. The right-hand side column of the master problem is $\mathbf{b}_{m+1} = \begin{bmatrix} \mathbf{b} \\ 1 \end{bmatrix}$, and $\mathbf{B}_{m+1}$ is a basis for $\mathbf{P_M}$.

In Figure 13.1, the left box depicts the original LP and the right box depicts the transformed LP. There are three columns to the right of the transformed LP. In the coefficient column j, the first entry is the negative of the jth coefficient of the objective function, the next m entries are coefficients associated with the jth variable of the general constraints, and the last entry is the jth coefficient of the convexity constraint. The RHS column contains one entry from the objective function, m entries from the general constraints, and one entry from the convexity constraint.

In the dual variable column, the first m entries are dual variables (denoted by $\mathbf{u}$) corresponding to m general constraints and the last entry is the dual variable α corresponding to the convexity constraint.

Because of t, the number of extreme points of set S, is usually very large and intractable to explicitly enumerate all possible extreme points and explicitly solve this problem. Instead, we solve the transformed problem (and hence the original problem) by simply maintaining a revised simplex tableau of size $(m + 2) \times (m + 2)$, usually a small subset of all possible extreme points. The revised simplex tableau of the master

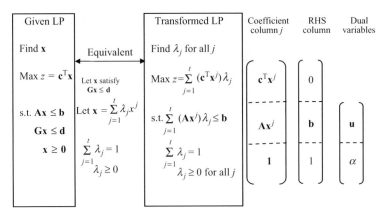

FIGURE 13.1 Dantzig–Wolfe decomposition.

problem is updated iteratively by generating pivot columns from the solutions of subproblems.

From the transformed problem, a revised simplex tableau for the master problem of size $(m + 1) \times (m + 1)$ can be constructed as shown in the left box of Figure 13.2. Suppose that we have a basic feasible solution $\lambda = (\lambda_B, \lambda_N)$ and that $\mathbf{B}^{-1}$ of size $(m + 1) \times (m + 1)$ is known. Then the primal solution can be obtained by calculating $\mathbf{B}^{-1}\mathbf{b}$, the dual solution by $\hat{\mathbf{c}}_B^T \mathbf{B}^{-1} = (\mathbf{u}^T, \alpha)$, and the objective value by $\hat{\mathbf{c}}_B^T \mathbf{B}^{-1}\mathbf{b}$, where $\hat{\mathbf{c}}_B^T$ is the profit vector of the basic variables with a profit of $\hat{c}_j = \hat{\mathbf{c}}_B^T \mathbf{x}^j$ for each basic variable λ_j.

Consider Figure 13.2. The left box contains the subproblem subject to the constraints of special structure and the right box contains the master problem. The master problem passes the values of the current dual solution, $(\mathbf{u}^T, \alpha) = \hat{\mathbf{c}}_B^T \mathbf{B}^{-1}$, to the subproblem for constructing its objective function. After the subproblem has been solved, a pivot column is formed and passed to the master program. The interaction between the master problem and the subproblem are repeated until the dual solution is nonnegative.

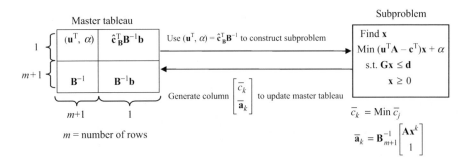

FIGURE 13.2 Interaction between master problem and subproblem(s).

To find an entering variable, x_k, we choose a variable with a most negative reduced cost defined by

$$\bar{c}_k = \min_{j=1,2,\dots,t} \{\bar{c}_j < 0\}$$

$$= \min_{j=1,2,\dots,t} \left\{ (\mathbf{u}^T, \alpha) \begin{pmatrix} \mathbf{A}\mathbf{x}^j \\ 1 \end{pmatrix} - \mathbf{c}^T\mathbf{x}^j \right\} \tag{13.1}$$

$$= \min_{j=1,2,\dots,t} \{(\mathbf{u}^T\mathbf{A} - \mathbf{c}^T)\mathbf{x}^j\} + \alpha$$

Minimizing over all the extreme points in (13.1) is equivalent to solving over the entire polyhedron S,

$$\min_{x \in S}\{(\mathbf{u}^T\mathbf{A} - \mathbf{c}^T)\mathbf{x}\} + \alpha \tag{13.2}$$

Note that α is a constant and can be dropped from consideration for finding an optimum solution. That is, solving (13.2) is equivalent to solving the following subproblem ($\mathbf{P}_S$):

$$(\mathbf{P}_S) \quad \text{Minimize} \quad (\mathbf{u}^T\mathbf{A} - \mathbf{c}^T)\mathbf{x}$$
$$\text{subject to} \quad \mathbf{G}\mathbf{x} \leq \mathbf{d}$$
$$\mathbf{x} \geq \mathbf{0}$$

The following are some important remarks regarding the decomposition algorithm:

1. The constraint set of the subproblem remains unchanged from iteration to iteration, while the objective functions of the subproblem are different between iterations.
2. At each iteration, a different dual vector is passed from the master problem to the subproblem. Rather than solving the subproblem from scratch at each iteration, the optimal basis of the last iteration could be used by modifying the objective row.
3. At each iteration, the subproblem need not be completely optimized. It is only necessary that the current basic feasible solution $\mathbf{x}^k$ satisfies $(\mathbf{u}^T\mathbf{A} - \mathbf{c}) \mathbf{x}^k + \alpha > 0$. In this case, the corresponding λ_k is a candidate to enter the basis of the master problem.

Example 13.1 Solve the following problem by decomposition.

$$\text{Maximize} \quad 2x_1 + 2x_2 + 3x_3 - x_4$$
$$\text{subject to} \quad x_1 + x_2 + x_3 + 2x_4 \leq 17$$
$$-2x_1 + 2x_2 + x_3 + x_4 \leq 11$$
$$-x_1 + x_4 \leq 2$$
$$2x_1 + x_3 \leq 9$$
$$x_2 + x_4 \leq 5$$
$$\mathbf{x} \geq \mathbf{0}$$

Although there is no constraint set of special structure, we still can decompose the problem into two sets: one set contains the first two constraints and the other contains the remaining three. Here $m = 2$.

Let the starting basis, $\mathbf{B}_{m+1}$, consist of $\mathbf{s}$ (slack variables) and λ_1 (can be any one element of λ). Since $\mathbf{B}_{m+1} = \mathbf{I}$ (the identity matrix), we have $\mathbf{B}_{m+1}^{-1} = \mathbf{I}$. Then the entire vector $\mathbf{x}$ is nonbasic, hence $\mathbf{x}^1 = (0,\, 0,\, 0,\, 0)^T$, and $\hat{c}_1 = \mathbf{c}\mathbf{x}^1 = 0$ for all j. The revised simplex tableau for master problem $\mathbf{P}_M$ is as follows:

		z	0	0	0	0
$(\mathbf{u}^T, \alpha)$	$\hat{\mathbf{c}}_{\mathbf{B}}^T \mathbf{B}_{m+1}^{-1} \mathbf{b}_{m+1}$	s_1	1	0	0	17
$\mathbf{B}_{m+1}^{-1}$	$\mathbf{B}_{m+1}^{-1} \mathbf{b}_{m+1}$	s_2	0	1	0	11
		λ_2	0	0	1	1

Given the vector $(\mathbf{u}^T, \alpha)$ from $\mathbf{P}_M$, the corresponding subproblem $(\mathbf{P}_S)$ is

(PS)   Minimize $(\mathbf{u}^T\mathbf{A} - \mathbf{c}^T)\mathbf{x} + \alpha$ Minimize $-2x_1 - 2x_2 - 3x_3 + x_4 + 0$

Subject to $\mathbf{G}\mathbf{x} \le \mathbf{d}$ Subject to $(12.3)-(12.5)$

$\mathbf{x} \ge \mathbf{0}$.

The subproblem is optimized at $\mathbf{x}^2 = (0,\, 5,\, 9,\, 0)^T$, with objective function -37. Since $-37 < 0$, the coefficient of $\mathbf{x}^2$, λ_2, enters the basis. The solution $\mathbf{x}^2$ is passed back to the master problem. Calculate a new coefficient vector of λ_2,

$$\mathbf{B}_{m+1}^{-1} \begin{bmatrix} \mathbf{A}\mathbf{x}^2 \\ 1 \end{bmatrix} = \begin{bmatrix} 14 \\ 19 \\ 1 \end{bmatrix}$$

and the associated reduced cost -37. Adjoin this as a pivot column to the revised simplex tableau, and pivot

z	0	0	0	0		λ_2
						$\begin{bmatrix} -37 \end{bmatrix}$
s_1	1	0	0	17		14
s_2	0	1	0	11		19
λ_2	0	0	1	1		$\begin{bmatrix} 1 \end{bmatrix}$

The tableau after pivoting is as follows:

z	0	1.95	0	21.42
s_1	1	0.74	0	8.89
λ_2	0	0.05	0	0.58
λ_1	0	-0.05	1	0.42

The best feasible solution found so far is given by

$$\lambda_1 \mathbf{x}_1 + \lambda_2 \mathbf{x}_2 = 0.42(0,0,0,0)^T + 0.58(0,5,9,0)^T = (0,2.89,5.21,0)^T$$

with objective value 21.42. Note that the new $(\mathbf{u}^T, \alpha) = (0, 1.95, 0)$. This information is used to generate the new subproblem

$$\text{Minimize} \quad -5.89x_1 + 1.89x_2 - 1.05x_3 + 2.95x_4 + 0$$
$$\text{subject to} \quad (12.3) - (12.5)$$

The subproblem is optimized at $\mathbf{x}^3 = (4.5, 0, 0, 0)^T$ with objective -26.53. Since objective is still less than zero, the coefficient of $\mathbf{x}^3$, λ_3 is introduced into the basis, and solution $\mathbf{x}^3$ is passed back to the master problem.

$$\mathbf{B}_{m+1}^{-1}\begin{bmatrix} \mathbf{Ax}^3 \\ 1 \end{bmatrix} = \begin{bmatrix} 1 & -0.74 & 0 \\ 0 & 0.05 & 0 \\ 0 & -0.05 & 1 \end{bmatrix}\begin{bmatrix} 4.5 \\ -9 \\ 1 \end{bmatrix} = \begin{bmatrix} 11.13 \\ -0.47 \\ 1.47 \end{bmatrix}$$

Add the pivot column to the revised simplex tableau and pivot

z	0	19.5	0	21.42
s_1	1	-0.74	0	8.89
λ_2	0	0.05	0	0.58
λ_1	0	-0.05	1	0.42

$$\lambda_3 \begin{bmatrix} -26.53 \\ 11.13 \\ -0.47 \\ 1.47 \end{bmatrix}$$

After pivoting, the tableau becomes

z	0	1	18	29
s_1	1	-0.34	-7.55	40/7
λ_2	0	0.04	0.32	5/7
λ_3	0	-0.04	0.68	2/7

Because $\mathbf{u} \geq \mathbf{0}$, we can claim that the optimal solution has been found, with objective value 29. The optimal solution is given by $\lambda_2 \mathbf{x}^2 + \lambda_3 \mathbf{x}^3 = 5/7(0,5,9,0)^T + 2/7(4.5, 0, 0, 0)^T = (9/7, 25/7, 45/7, 0)^T$. If we solve this problem using another method, we can find that this is not the unique optimal solution. Now let us work on an example with two subproblems.

Example 13.2 Solve the following LP problem by decomposition.

$$\text{Maximize} \quad -3x_1 + 7x_2 + 5x_3 + 4x_4$$
$$\text{subject to} \quad 2x_1 - x_2 + 2x_3 + 2x_4 \leq 19 \tag{13.3}$$

$$-2x_1 + 2x_2 + x_3 - 3x_4 \leq 21 \qquad (13.4)$$

$$x_1 + x_2 \leq 12 \qquad (13.5)$$

$$3x_1 - x_2 \leq 15 \qquad (13.6)$$

$$x_3 + x_4 \leq 5 \qquad (13.7)$$

$$-x_3 + x_4 \leq 2 \qquad (13.8)$$

$$\mathbf{x} \geq \mathbf{0}$$

We can easily see that this problem has some constraints of special structures. That is, constraints (13.5) and (13.6) involve only variables x_1 and x_2, while constraints (13.7) and (13.8) involve only variables x_3 and x_4. Taking advantage of these specialties, we can decompose this problem into a master problem with constraints (13.3) and (13.4), and two subproblems, one subject to (13.5) and (13.6), and the other subject to (13.7) and (13.8).

For the convenience of notation, we partition matrix $\mathbf{A}$ into $(\mathbf{A}_1 \mathbf{A}_2)$, where

$$\mathbf{A}_1 = \begin{bmatrix} 2 & -1 \\ -2 & 2 \end{bmatrix}, \qquad \mathbf{A}_2 = \begin{bmatrix} 2 & 2 \\ 1 & -3 \end{bmatrix}$$

We also partition $\mathbf{x}$ into $\mathbf{x} = [\mathbf{x}^{(1)}\ \mathbf{x}^{(2)}]$, where $\mathbf{x}^{(1)} = (x_1 x_2)$ and $\mathbf{x}^{(2)} = (x_3 x_4)$. Then the problem, in terms of extreme points, can be expressed as follows:

$$\text{Maximize} \quad \sum_{j=1}^{t_1} \mathbf{c}_1 \mathbf{x}_j^{(1)} \lambda_j + \sum_{j=1}^{t_2} \mathbf{c}_2 \mathbf{x}_j^{(2)} \beta_j$$

$$\text{subject to} \quad \sum_{j=1}^{t_1} \mathbf{A}_1 \mathbf{x}_j^{(1)} \lambda_j + \sum_{j=1}^{t_2} \mathbf{A}_2 \mathbf{x}_j^{(2)} \beta_j \leq \mathbf{b}$$

$$\sum_{j=1}^{t_1} \lambda_j = 1$$

$$\sum_{j=1}^{t_2} \beta_j = 1$$

$$\lambda \geq \mathbf{0}$$

$$\beta \geq \mathbf{0}$$

Let α_1 and α_2, respectively, denote the dual variables corresponding to the constraints

$$\sum_{j=1}^{t_1} \lambda_j = 1 \quad \text{and} \quad \sum_{j=1}^{t_2} \beta_j = 1$$

Initialization step. Let the starting basis consists of s_1, s_2, λ_1, and β_1, which implies that $\mathbf{x}_1^{(1)} = (0, 0)$, $\mathbf{x}_1^{(2)} = (0, 0)$, and obtain the revised simplex tableau

z	0	0	0	0	0
s_1	1	0	0	0	19
s_2	0	1	0	0	21
λ_1	0	0	1	0	1
β_1	0	0	0	1	1

Iteration 1. Generate two subproblems using $(\mathbf{u}^T, \alpha_1)$ and $(\mathbf{u}^T, \alpha_2)$.

Subproblem 1

minimize $\quad (\mathbf{u}^T\mathbf{A}_1 - \mathbf{c}_1)\mathbf{x}^{(1)} + \alpha_1$

$\qquad\qquad = 3x_1 - 7x_2$

subject to $\quad x_1 + x_2 \leq 12$

$\qquad\qquad 3x_1 - x_2 \leq 15$

$\qquad\qquad x_1, x_2 \geq 0$

Subproblem 2

minimize $\quad (\mathbf{u}^T\mathbf{A}_2 - \mathbf{c}_2)\mathbf{x}^{(2)} + \alpha_2$

$\qquad\qquad = -5x_3 - 4x_4$

subject to $\quad x_3 + x_4 \leq 5$

$\qquad\qquad -x_3 + x_4 \leq 15$

$\qquad\qquad x_3, x_4 \geq 0$

The two respective subproblems are optimized at $(0, 12)$ and $(5, 0)$ with objectives -84 and -25. Since both objective values are negative, λ_2 and β_2 are both eligible to enter the basis. Here, we choose λ_2 as the entering variable.

$$
\mathbf{B}_{m+2}^{-1}
\begin{bmatrix} \mathbf{A}_1\mathbf{x}_2^{(1)} \\ 1 \\ 0 \end{bmatrix}
=
\begin{bmatrix} 1 & 0 & 0 & 0 \\ 0 & 1 & 0 & 0 \\ 0 & 0 & 1 & 0 \\ 0 & 0 & 0 & 1 \end{bmatrix}
\begin{bmatrix} -12 \\ 24 \\ 1 \\ 0 \end{bmatrix}
=
\begin{bmatrix} -12 \\ 24 \\ 1 \\ 0 \end{bmatrix}
$$

Master step. Adjoin the objective value of -84 at the top of the vector and append this column λ_2 to the simplex tableau, and then pivot.

z	0	0	0	0	0
s_1	1	0	0	0	19
s_2	0	1	0	0	21
λ_1	0	0	1	0	1
β_1	0	0	0	1	1

$$
\lambda_2
\begin{bmatrix} -84 \\ -12 \\ 24 \\ 0 \\ 1 \end{bmatrix}
$$

The updated tableau is as follows:

z	0	3.5	0	0	73.5
s_1	1	0.5	0	0	29.5
λ_2	0	0.04	0	0	0.875
λ_1	0	-0.04	1	0	0.125
β_1	0	0	0	1	1

Iteration 2. Generate two subproblems with the new objective functions:

$$\text{Subproblem 1}: \text{minimize} -4x_1 + 0$$

$$\text{and subproblem 2}: \text{minimize} -1.5x_3 - 14.5x_4 + 0$$

The two subproblems are optimized at $(5, 0)$ and $(0, 5)$, respectively, with objectives -20 and -72.5. Since λ_2 just entered the basis, it is not qualified to be the entering variable, so this time β_3 enters the basis.

$$\mathbf{B}_{m+2}^{-1}\begin{bmatrix} \mathbf{A}_2\mathbf{x}_3^{(2)} \\ 0 \\ 1 \end{bmatrix} = \begin{bmatrix} 1 & 0.5 & 0 & 0 \\ 0 & 0.04 & 0 & 0 \\ 0 & -0.04 & 1 & 0 \\ 0 & 0 & 0 & 1 \end{bmatrix}\begin{bmatrix} 10 \\ -15 \\ 0 \\ 1 \end{bmatrix} = \begin{bmatrix} 2.5 \\ -0.625 \\ 0.625 \\ 1 \end{bmatrix}$$

Master step. Adjoin the objective value of -72.5 and append this column β_3 to the simplex tableau, and then pivot.

						β_3
z	0	3.5	0	0	73.5	-72.5
s_1	1	0.5	0	0	29.5	2.5
λ_2	0	0.04	0	0	0.875	-0.625
λ_1	0	-0.04	1	0	0.125	0.625
β_1	0	0	0	1	1	1

The updated tableau is as follows:

z	0	-1.33	116	0	88
s_1	1	0.67	-4	0	29
λ_2	0	0	1	0	1
β_3	0	-0.07	1.6	0	0.2
β_1	0	0.07	-1.6	1	0.8

Iteration 3. Generate two subproblems with the new objective functions:

$$\text{Subproblem 1}: \text{minimize } 5.67x_1 - 9.67x_2 + 116$$

$$\text{and subproblem 2}: \text{minimize } -6.33x_3 + 0$$

Subproblem 1 is optimized at $(0, 12)$, with objective value 0, and subproblem 2 is optimized at $(5, 0)$, with objective -31.67. Since the objective value of subproblem 1 is 0, λ is not qualified to enter basis, so β_4 becomes the entering variable.

$$\mathbf{B}_{m+2}^{-1}\begin{bmatrix}\mathbf{A}_2\mathbf{x}_4^{(2)}\\0\\1\end{bmatrix}=\begin{bmatrix}1&0.67&-4&0\\0&0&1&0\\0&-0.07&1.6&0\\0&0.07&-1.6&1\end{bmatrix}\begin{bmatrix}10\\5\\0\\1\end{bmatrix}=\begin{bmatrix}13.33\\0\\-0.33\\1.33\end{bmatrix}$$

Master step. Add the objective value of -31.67 and append this column β_4 to the simplex tableau, and then pivot.

z	0	-1.33	116	0	88
s_1	1	0.67	-4	0	29
λ_2	0	0	1	0	1
β_3	0	-0.07	1.6	0	0.2
β_1	0	0.07	-1.6	1	0.8

$$\beta_4\begin{bmatrix}-31.67\\13.33\\0\\-0.33\\1.33\end{bmatrix}$$

The updated tableau is as follows:

z	0	0.25	78	23.75	107
s_1	1	0	12	-10	21
λ_2	0	0	1	0	1
β_3	0	-0.05	1.2	0.25	0.4
β_4	0	0.05	-1.2	0.75	0.6

Iteration 4. Generate two subproblems with the new objective functions:

$$\text{Subproblem 1}: \text{minimize } 2.5x_1-6.5x_2+78, \text{ and}$$
$$\text{Subproblem 2}: \text{minimize } -4.75x_3-4.75x_4+23.75$$

Both subproblems have optimal objective value of 0, so we can claim that the entire model is optimized with objective value of 107. The optimal solution $(x_1, x_2, x_3, x_4)^{\mathrm{T}}$ is given by

$$(\lambda_2\mathbf{x}_2^{(1)}, \beta_3\mathbf{x}_3^{(2)}+\beta_4\mathbf{x}_4^{(2)})=[0, 12, 0.4(0, 5)+0.6(5, 0)]^{\mathrm{T}}=(0, 12, 2, 3)^{\mathrm{T}}$$

Now we apply the column generation scheme to a specific problem called the generalized assignment problem (GAP).

13.3   GENERALIZED ASSIGNMENT PROBLEM

The assignment problem is to find a maximum profit assignment of n tasks to n machines such that each task $(i=1, 2, \ldots, n)$ is assigned to exactly one machine

($j = 1, 2, \ldots, n$) and each machine is assigned to exactly one task. The GAP is a generalization of the assignment problem that finds a maximum profit assignment of m tasks to n ($m \geq n$) machines such that each task is assigned to exactly one machine and that each machine is allowed to be assigned to more than one task, subject to its capacity limitation.

13.3.1 Conventional Formulation

Let binary variable $y_{ij} = 1$ if task i is assigned to machine j and $y_{ij} = 0$ otherwise. Conventionally, GAP is formulated as

$$\text{Maximize} \quad z = \sum_{i=1}^{m} \sum_{j=1}^{n} p_{ij} y_{ij}$$

$$\text{subject to} \quad \sum_{j=1}^{n} y_{ij} = 1 \quad i = 1, 2, \ldots, m \quad \text{(assignment)}$$

(13.9)

$$\sum_{i=1}^{m} w_{ij} y_{ij} \leq d_j \quad j = 1, 2, \ldots, n \quad \text{(machine capacity)} \quad (13.10)$$

$$y_{ij} = 0 \text{ or } 1 \quad \text{for all} \quad i, j \quad (13.11)$$

where p_{ij} is the profit incurred if task i is assigned to machine j, d_j is the capacity of machine j, and w_{ij} is the amount of capacity of machine j used by task i. The assignment constraint (13.9) ensures that each task is assigned to exactly one machine. The capacity constraint (13.10) ensures that each machine capacity is not exceeded. To get an idea about the structure of this formulation, consider a GAP example of $m = 3$ tasks and $n = 2$ machines given below. *Note*: the size of the constraint matrix is $(m + n) \times mn$.

Machine 1			Machine 2			
y_{11}	y_{21}	y_{31}	y_{12}	y_{22}	y_{32}	RHS
1			1			$= 1$
	1			1		$= 1$
		1			1	$= 1$
w_{11}	w_{21}	w_{31}				$\leq d_1$
			w_{12}	w_{22}	w_{32}	$\leq d_2$

13.3.2 Column Generation Formulation

Consider the set of points satisfying $S = \{ \mathbf{y} : \sum_{i=1}^{m} w_{ij} y_{ij} \leq d_j, y_{ij} = 0 \text{ or } 1, j = 1, \ldots, n \}$. Clearly, S is just a finite set of points, say $S = \{ \mathbf{z}_1^1, \ldots, \mathbf{z}_1^{K_1}, \ldots, \mathbf{z}_n^1, \ldots, \mathbf{z}_n^{K_n} \}$, where $\mathbf{z}_j^k = (z_{1j}^k, z_{2j}^k, \ldots, z_{mj}^k)^{\mathrm{T}}$ and K_j is the number of feasible solutions for j.

Every point $\mathbf{y} \in \{0, 1\}^m$ can be represented by

$$y_{ij} = \sum_{k=1}^{K_j} z_{ij}^k \lambda_j^k \tag{13.12}$$

$$\sum_{k=1}^{K_j} \lambda_j^k = 1 \qquad j = 1, 2, \ldots, n$$

$$\lambda_j^k = 0 \text{ or } 1 \qquad k = 1, 2, \ldots, K_j, j = 1, 2, \ldots, n$$

where $y_{ij} = 1$ if task i is assigned to machine j and $y_{ij} = 0$, $\lambda_j^k = 1$ $(k = 1, \ldots, K_j)$ if the kth assignment of machine j is used and $\lambda_j^k = 0$ otherwise.

Substituting (13.12) and the convexity constraint into the objective function of the conventional GAP formulation and into the assignment constraints (13.9) that define the master problem, we obtain the master problem in the column generation formulation

$$\text{Maximize } z = \sum_{j=1}^{n} \sum_{k=1}^{K_j} \left(\sum_{i=1}^{m} p_{ij} z_{ij}^k \right) \lambda_j^k \tag{13.13}$$

$$\text{subject to } \sum_{j=1}^{n} \sum_{k=1}^{K_j} z_{ij}^k \lambda_j^k \qquad i = 1, 2, \ldots, m \quad \text{(assignment)} \tag{13.14}$$

$$\sum_{k=1}^{K_j} \lambda_j^k = 1 \quad j = 1, 2, \ldots, n \quad \text{(convexity)} \tag{13.15}$$

$$\lambda_j^k = 0 \text{ or } 1 \quad k = 1, 2, \ldots, K_j; j = 1, 2, \ldots, n \tag{13.16}$$

This alternative GAP is formulated in terms of columns representing *feasible assignments of tasks to machines*. The assignment constraints (13.14) ensure that each task is assigned to a machine. The convexity constraints (13.15)–(13.16) ensure that exactly one feasible assignment of tasks to machines is selected for each machine.

To get an idea about the structure of this formulation, consider a GAP example of $m = 3$ tasks and $n = 2$ machines given below. Symbol 0/1 in the entries represents 0 or 1 value of z_{ij}^k.

Variable		Machine j								RHS
		1				2				
		λ_1^1	λ_1^2	$\ldots$	$\lambda_1^{K_1}$	λ_2^1	λ_2^2	$\ldots$	$\lambda_2^{K_2}$	
Task i	1	0/1	0/1	$\ldots$	0/1	0/1	0/1	$\ldots$	0/1	1
	2	0/1	0/1	$\ldots$	0/1	0/1	0/1	$\ldots$	0/1	1
	3	0/1	0/1	$\ldots$	0/1	0/1	0/1	$\ldots$	0/1	1
Machine j	1	1	1	$\ldots$	1	0	0	$\ldots$	0	1
	2	0	0	$\ldots$	0	1	1	$\ldots$	1	1

The fundamental difference between the conventional formulation and the column generation formulation is that the IP feasible solution S is replaced by a finite set of points. Any fractional point to the LP relaxation of the conventional formulation is a feasible solution to the LP relaxation of the column generation formulation if and only if it can be represented by a convex combination of extreme points of Conv(S). Geoffrion (1974) has shown that if polyhedron Conv(S) does not have all integral extreme points, then the LP relaxation of the column generation formulation will be tighter than that of the conventional formulation for some objective functions.

However, the column generation formulation often contains a huge number of columns due to a huge number of extreme points of a bounded polyhedron. It may be necessary to work with restricted versions that contain only a subset of all columns and to generate additional columns only as needed. The column generation formulation is called *the master problem* and a restricted version of the master problem is called a *restricted master problem*. Column generation is carried out by solving subproblems or pricing problems of the form

$$\text{Max}\left\{\sum_i (p_{ij}-u_i)y_{ij}-v_j : \sum_i w_{ij}y_{ij} \le d_j, y_{ij} = 0 \text{ or } 1\right\}$$

where u_i and v_j are determined from optimum dual solutions to the LP relaxation of a restricted master problem.

Note that the column generation formulation contains an exponential number of variables, while the conventional formulation contains much less variables. Why bother developing a formulation with a huge number of variables? The reasons are as follows. First, a compact IP formulation such as the conventional GAP has a *weak* LP relaxation. The LP relaxation of the master problem in the column generation formulation is *tighter* than that of the conventional formulation because fractional solutions that are not convex combinations of 0–1 solutions to the knapsack constraints (13.12) are not feasible to the column generation formulation. The advantage of applying decomposition to the column generation formulation is not to speed up the solution of its LP relaxation but to improve the LP bound.

Second, instead of considering all possible feasible assignments, only a subset of feasible assignments are considered, whose columns are generated by solving a series of subproblems or pricing problems. In particular, the subproblem for GAP can be decomposed into n knapsack problems, for which efficient algorithms are available.

Unfortunately, it is intractable to directly solve the LP relaxation of the master problem in the column generation formulation due to exponential number of variables (columns). Instead, we solve the LP relaxation of a restricted version of the master problem that considers only $m + n$ columns, usually a small subset of all columns, which can be directly solved by the revised simplex method. Moreover, the pivot column associated with the entering variable can be generated by solving the subproblem, which in turn can be decomposed into a set of n knapsack problems.

Let u_i be the dual variable (price) associated with the assignment constraint (13.14) of task i and let v_j be the dual variable (price) associated with the convexity constraint (13.15) of machine j. The subproblem for machine j is a knapsack problem defined by

$$(\text{KP}_j) \quad \text{maximize} \quad z(\text{KP}_j) = \sum_i (p_{ij} - u_i) y_{ij} - v_j$$

$$\text{subject to} \quad \sum_i w_{ij} y_{ij} = d_j$$

$$y_{ij} = 0 \text{ or } 1 \qquad\qquad i = 1, \dots, m$$

If the optimum value of any pricing problem is positive, then we have identified a column with positive reduced cost that can be added to the restricted master problem to improve the solution. If the maximal reduced costs of all the knapsack problems are nonpositive, then the LP solution obtained is also maximal for the relaxation of the unrestricted master problem.

13.3.3   Initial Solution

Column generation begins with an initial restricted master problem, which must possess a feasible LP relaxation solution. This will assure that proper dual problem information is available for passage to the pricing problem. An initial LP relaxation solution, if exists, can be found by the two-phase simplex method.

13.4   GAP EXAMPLE

Here, we illustrate the column generation procedure to a GAP example with parameters $m = 3$, $n = 2$, $(d_1, d_2) = (11, 18)$,

$$\{p_{ij}\} = \begin{pmatrix} 10 & 6 \\ 7 & 8 \\ 5 & 11 \end{pmatrix}, \{w_{ij}\} = \begin{pmatrix} 9 & 5 \\ 6 & 7 \\ 3 & 9 \end{pmatrix}$$

From $9z_{11} + 6z_{21} + 3z_{31} \leq 11$, we have four ($=K_1$) possible feasible solutions:

$$\mathbf{z}_1^k = \{(1,0,0), (0,1,0), (0,0,1), (0,1,1)\}$$

From $5z_{12} + 7z_{22} + 9z_{32} \leq 18$, we have six ($=K_2$) possible feasible solutions:

$$\mathbf{z}_2^k = \{(1,0,0), (0,1,0), (0,0,1), (1,1,0), (0,1,1), (1,0,1)\}$$

Assuming λ_j^k be the variable associated with $\mathbf{z}_j^k$, we obtain the master problem in a column generation formulation:

$$\text{Maximize} \quad 10\lambda_1^1 + 7\lambda_1^2 + 5\lambda_1^3 + 12\lambda_1^4 + 6\lambda_2^1 + 8\lambda_2^2 + 11\lambda_2^3 + 14\lambda_2^4 + 19\lambda_2^5 + 17\lambda_2^6 \tag{13.17}$$

$$\text{subject to} \quad \lambda_1^1 + 0 + 0 + 0 + \lambda_2^1 + 0 + 0 + \lambda_2^4 + 0 + \lambda_2^6 = 1 \tag{13.18}$$

$$0 + \lambda_1^2 + 0 + \lambda_1^4 + 0 + \lambda_2^2 + 0 + \lambda_2^4 + \lambda_2^5 + 0 = 1 \tag{13.19}$$

$$0 + 0 + \lambda_1^3 + \lambda_1^4 + 0 + 0 + \lambda_2^3 + 0 + \lambda_2^5 + \lambda_2^6 = 1 \tag{13.20}$$

$$\lambda_1^1 + \lambda_1^2 + \lambda_1^3 + \lambda_1^4 + 0 + 0 + 0 + 0 + 0 + 0 = 1 \tag{13.21}$$

$$0 + 0 + 0 + 0 + \lambda_2^1 + \lambda_2^2 + \lambda_2^3 + \lambda_2^4 + \lambda_2^5 + \lambda_2^6 = 1 \tag{13.22}$$

$$\lambda_j^k \geq 0 \tag{13.23}$$

To get an identity matrix (i.e., $\mathbf{B} = \mathbf{B}^{-1} = \mathbf{I}$) as an initial basis for the "restricted" master problem, we add an artificial variable to each of equations (13.18)–(13.22) and apply the two-phase method. In phase I, the objective is to minimize the sum of all artificial variables (min $\sum_{i=1}^{m+n} \lambda_i^a$). At the end of phase I, a feasible solution is obtained if all artificial variables reduce to 0 (either become nonbasic or basic variables). In this case, phase 2 begins and its objective is to maximize the original objective function.

In phase I, we begin with constructing a starting revised simplex tableau using all artificial variables as basic variables $\boldsymbol{\lambda_B} = (\lambda_1^a, \lambda_2^a, \lambda_3^a, \lambda_4^a, \lambda_5^a)^T$ and minimizing $\sum_{i=1}^5 \lambda_i^a$. At iteration 1, variable λ_1^4 enters the basis to replace λ_2^a. The pivot column is $(0, 1, 0, 0, 1)^T$ with a reduced cost 3. At iteration 2, variable λ_1^1 enters the basis to replace λ_1^a. The pivot column is $(1, 1, 0, 0, 1)^T$ with a reduced cost 2. At iteration 3, variable λ_1^3 enters the basis to replace λ_3^a. The pivot column is $(0, 1, 1, 1, 0)^T$ with a reduced cost 2. At iteration 4, variable λ_1^2 enters the basis to replace λ_4^a. The pivot column is $(0, 1, -1, 1, 0)^T$ with a reduced cost 1. At iteration 5, variable λ_2^3 enters the basis to replace λ_5^a. The pivot column is $(0, 0, 0, -1, 1)^T$ with a reduced cost 1. Since there are no more positive reduced cost and all the artificial variables reduce to 0, we have obtained an initial feasible solution, $\boldsymbol{\lambda_B} = (\lambda_1^2, \lambda_1^3, \lambda_1^4, \lambda_2^1, \lambda_2^3)^T$, after rearrangement, with $\mathbf{c}_\mathbf{B}^T = (7, 5, 12, 6, 11)$,

$$\mathbf{B} = \begin{bmatrix} 0 & 0 & 0 & 1 & 0 \\ 1 & 0 & 1 & 0 & 0 \\ 0 & 1 & 1 & 0 & 1 \\ 1 & 1 & 1 & 0 & 0 \\ 0 & 0 & 0 & 1 & 1 \end{bmatrix}, \quad \mathbf{B}^{-1} = \begin{bmatrix} -1 & 0 & -1 & 1 & 1 \\ 0 & -1 & 0 & 1 & 0 \\ 1 & 1 & 1 & -1 & -1 \\ 1 & 0 & 0 & 0 & 0 \\ -1 & 0 & 0 & 0 & 1 \end{bmatrix}$$

$\hat{\mathbf{c}}_\mathbf{B}^T \mathbf{B}^{-1} = (0, 7, 5, 0, 6)$, and $\hat{\mathbf{c}}_\mathbf{B}^T B^{-1} \mathbf{b} = 18$. We have the following initial restricted master tableau.

z	0	7	5	0	6	18
λ_1^2	-1	0	-1	1	1	0
λ_1^3	0	-1	0	1	0	0
λ_1^4	1	1	1	-1	-1	1
λ_2^1	1	0	0	0	0	1
λ_2^3	-1	0	0	0	1	0

Passing the dual values u_i and v_j to the subproblems, we have

Subproblem 1:

$$\text{Max} \quad z(KP_1) = (10-0)z_{11} + (7-7)z_{21} + (5-5)z_{31} - 0$$
$$\text{s.t.} \quad 9z_{11} + 6z_{21} + 3z_{31} \leq 11$$
$$z_{11}, z_{21}, z_{31} \in \{0, 1\}$$

Subproblem 1 is optimized at $(1, 0, 0)^T = \mathbf{z}_1^1$ with $z(KP_1) = 10$.

Subproblem 2:

$$\text{Max} \quad z(KP_2) = (6-0)z_{12} + (8-7)z_{22} + (11-5)z_{32} - 6$$
$$\text{s.t.} \quad 5z_{12} + 7z_{22} + 9z_{32} \leq 18$$
$$z_{12}, z_{22}, z_{32} \in \{0, 1\}$$

Subproblem 2 is optimized at $(1, 0, 1)^T = \mathbf{z}_2^6$ with $z(KP_2) = 6$.

$z(KP_1) > z(KP_2) > 0$, so we choose λ_1^1 as the new column. Compute $\mathbf{B}^{-1}\mathbf{a}_1^1$ and append the column to the revised simplex tableau and pivot:

z	0	7	5	0	6	18	-10
λ_1^2	-1	0	-1	1	1	0	0
λ_1^3	0	-1	0	1	0	0	1
λ_1^4	1	1	1	-1	-1	1	0
λ_2^1	1	0	0	0	0	1	1
λ_2^3	-1	0	0	0	1	0	-1

The updated tableau is

z	0	3	5	10	6	18	0
λ_1^2	-1	0	-1	1	1	0	0
λ_1^1	0	-1	0	1	0	0	1
λ_1^4	1	1	1	-1	-1	1	0
λ_2^1	1	1	0	-1	0	1	0
λ_2^3	-1	-1	0	1	1	0	0

Subproblem 1:

$$\text{Max} \quad z(\text{KP}_1) = (10-0)z_{11} + (7+3)z_{21} + (5-5)z_{31} - 10$$
$$\text{s.t.} \quad 9z_{11} + 6z_{21} + 3z_{31} \le 11$$
$$z_{11}, z_{21}, z_{31} \in \{0, 1\}$$

Subproblem 1 is optimized at $(0, 1, 0)^\text{T} = \mathbf{z}_1^2$ with $z(\text{KP}_1) = 0$.

Subproblem 2:

$$\text{Max} \quad z(\text{KP}_2) = (6-0)z_{12} + (8+3)z_{22} + (11-5)z_{32} - 6$$
$$\text{s.t.} \quad 5z_{12} + 7z_{22} + 9z_{32} \le 18$$
$$z_{12}, z_{22}, z_{32} \in \{0, 1\}$$

Subproblem 2 is optimized at $(0, 1, 1)^\text{T} = \mathbf{z}_2^5$ with $z(\text{KP}_2) = 11$.
$z(\text{KP}_2) > 0$, so we choose λ_2^5 as the new column. Compute $\mathbf{B}^{-1}\mathbf{a}_2^5$ and append the column to the revised simplex tableau and pivot:

z	0	-3	5	10	6	18	-11
λ_1^2	-1	0	-1	1	1	0	0
λ_1^1	0	-1	0	1	0	0	-1
λ_1^4	1	1	1	-1	-1	1	1
λ_2^1	1	1	0	-1	0	1	1
λ_2^3	-1	-1	0	1	1	0	0

The updated tableau is

z	11	8	16	-1	-5	29	0
λ_1^2	-1	0	-1	1	1	1	0
λ_1^1	1	0	1	0	-1	1	0
λ_2^5	1	1	1	-1	-1	1	1
λ_2^1	0	0	-1	0	1	0	0
λ_2^3	-1	-1	0	1	1	0	0

Subproblem 1:

$$\text{Max} \quad z(\text{KP}_1) = (10-11)z_{11} + (7-8)z_{21} + (5-16)z_{31} + 1$$
$$\text{s.t.} \quad 9z_{11} + 6z_{21} + 3z_{31} \le 11$$
$$z_{11}, z_{21}, z_{31} \in \{0, 1\}$$

Subproblem 1 is optimized at $(0, 0, 0)^\text{T}$ with $z(\text{KP}_1) = 1$.

Subproblem 2:

$$\text{Max} \quad z(KP_2) = (6-11)z_{12} + (8-8)z_{22} + (11-16)z_{32} + 5$$
$$\text{s.t.} \quad 5z_{12} + 7z_{22} + 9z_{32} \leq 18$$
$$z_{12}, z_{22}, z_{32} \in \{0, 1\}$$

Subproblem 2 is optimized at $(0, 1, 0)^T = \mathbf{z}_2^2$ with $z(KP_2) = 5$.

$z(KP_2) > z(KP_1) > 0$, so we choose λ_2^2 as the new column. Compute $\mathbf{B}^{-1}\mathbf{a}_2^2$ and append the column to the revised simplex tableau and pivot:

z	11	8	16	-1	-5	29	-5
λ_1^2	-1	0	-1	1	1	0	1
λ_1^1	1	0	1	0	-1	1	-1
λ_2^5	1	1	1	-1	-1	1	0
λ_2^1	0	0	-1	0	1	0	1
λ_2^3	-1	-1	0	1	1	0	0

The updated tableau is

z	6	8	11	4	0	29	0
λ_2^2	-1	0	-1	1	1	0	1
λ_1^1	0	0	0	1	0	1	0
λ_2^5	1	1	1	-1	-1	1	0
λ_2^1	1	0	0	-1	0	0	0
λ_2^3	-1	-1	0	0	1	1	0

Subproblem 1:

$$\text{Max} \quad z(KP_1) = (10-6)z_{11} + (7-8)z_{21} + (5-11)z_{31} + 4$$
$$\text{s.t.} \quad 9z_{11} + 6z_{21} + 3z_{31} \leq 11$$
$$z_{11}, z_{21}, z_{31} \in \{0, 1\}$$

Subproblem 1 is optimized with $z(KP_1) = 0$.

Subproblem 2:

$$\text{Max} \quad z(KP_2) = (6-6)z_{12} + (8-8)z_{22} + (11-11)z_{32} + 0$$
$$\text{s.t.} \quad 5z_{12} + 7z_{22} + 9z_{32} \leq 18$$
$$z_{12}, z_{22}, z_{32} \in \{0, 1\}$$

Subproblem 2 is optimized with $z(KP_2) = 0$.

$z(\text{KP}_2) = z(\text{KP}_1) = 0$, so the optimal solution to the original problem is found, which is $z_1 = (1, 0, 0)$ and $z_2 = (0, 1, 1)$.

In the previous example, we were lucky enough not to obtain any fractional solution to the restricted master problem, so branching was not needed. However, in practice, when the number of columns in the master problem is large, encountering fractional solutions is very common. Therefore, we use the following example to explain the branching scheme for the generalized assignment problem.

13.4.1   GAP Branching Scheme

Suppose at some iteration the solution to the restricted master problem includes $\lambda_1^1 = 1/3$, $\lambda_1^2 = 2/3$. In terms of the original variables, this is equivalent to $z_{11} = 1/3$, $z_{12} = 2/3$, which is infeasible. Hence, we need to branch on either z_{11} or z_{12}. Let us choose z_{11} arbitrarily. The two children problems will be created by setting $z_{11} = 0$ and $z_{11} = 1$. It is easy to see that $z_{11} = 0$ implies that λ_1^1 and λ_1^4 are ruled out of the master problem, while $z_{11} = 1$ implies that λ_1^2 and λ_1^3 are ruled out. So at each child node, the master problem is reduced by two columns. Furthermore, in subproblem 1 at each branch, z_{11} is already fixed to either 0 or 1, and hence subproblem 1 is also reduced by one column. Interested readers may try other numerical examples to practice this branching scheme.

Standard branching on the λ variable creates a problem along a branch where a variable has been set to zero. Recall that $\mathbf{z}_j^k = (z_{1j}^k, z_{2j}^k, \ldots, z_{mj}^k)$ represents a particular solution to the jth knapsack problem. Thus, setting $\lambda_j^k = 0$ implies that the solution $\mathbf{z}_j^k$ is excluded. However, it is quite likely that the next time the knapsack problem for the same jth machine is solved and the optimal solution is also the same one represented by $\mathbf{z}_j^k$. In this case, it would be required to find the next second best solution. At the l-level of the branch-and-bound tree, we may need to find the lth best solution, which is very hard.

Fortunately, this difficulty can be overcome by applying a simple branching rule. Instead of branching on the λ's in the restricted master problem, we use a branching rule that corresponds to branching on the original variables y_{ij}. When $y_{ij} = 1$, all existing columns in the master problem that do not assign task i to machine j are deleted and task i is permanently assigned to machine j (i.e., variable z_{ij} is fixed to 1 in the jth knapsack). Conversely, when $y_{ij} = 0$, all existing columns in the master problem that assign task i to machine j are deleted and task i cannot be assigned to machine j (i.e., variable z_{ij} is removed from the jth knapsack problem). In either case, each of the knapsack problems contains one fewer variable after the branching has been done.

Note that the branching scheme discussed here is specific to the GAP. This is typical of branch-and-price algorithms. Each problem requires its own problem-specific branching scheme.

13.4.2   Tailing-Off Effect of Column Generation

One of the difficulties encountered in applying the column generation is the so-called *tailing-off effect of column generation*. This effect manifests itself as: after a large number of generated columns, the improvement in the objective value becomes very small. Clearly, it is ineffective to continue column generation until an optimum is found. Therefore, the alternative is to terminate the column generation prematurely. The objective value to the current restricted master problem gives a lower bound on the final LP value.

Lagrangian duality can be used to obtain an upper bound on the final LP value. To illustrate this, consider the alternative GAP formulation (13.16)–(13.19). An associated Lagrangian relaxation can be obtained by dualizing the assignment constraints:

$$\max \quad \sum_{j=1}^{n}\sum_{k=1}^{K_j}\left(\sum_{i=1}^{m}p_{ij}z_{ij}^{k}\right)\lambda_{j}^{k} + \sum_{i=1}^{m}u_{i}\left(1-\sum_{j=1}^{n}\sum_{k=1}^{K_j}z_{ij}^{k}\lambda_{j}^{k}\right)$$

$$\text{subject to} \quad \sum_{k=1}^{K_j}\lambda_{j}^{k}=1 \qquad\qquad j=1,2,\ldots,n$$

$$\lambda_{j}^{k}=0 \text{ or } 1 \qquad\qquad j=1,2,\ldots,n;\ k=1,2,\ldots,K_j$$

which provides an upper bound on the optimal value of the LP for any dual vector $\mathbf{u}=(u_1, u_2, \ldots, u_m)^{\mathrm{T}}$.

The reader can verify that the objective function can be rewritten as follows:

$$\sum_{i=1}^{m}u_{i} + \sum_{j=1}^{n}\max\sum_{k=1}^{K_j}\left[\sum_{i=1}^{m}(p_{ij}-u_i)z_{ij}^{k}\right]\lambda_{j}^{k}$$

$$= \sum_{i=1}^{m}u_{i} + \sum_{j=1}^{n}z(KP_j)$$

This shows that after solving a given pricing problem within the branch-and-price algorithm, all the information needed to compute an upper bound of the final LP solution is available. The difference between these two bounds is called a *duality gap*. The width of this gap may be used as a stopping rule to terminate the column generation process.

13.4.3 Treatment of Identical Machines

Should the machines be identical, there is a modification of the above methodology available. Because the machines are identical, the variables λ_{j}^{k} can be aggregated into a single variable

$$\lambda^{k} = \sum_{j}\lambda_{j}^{k}$$

and the convexity constraints can be combined into a single new constraint

$$\sum_{k=1}^{K}\lambda^{k}=n$$

where n is an integer. The master problem simplifies to

$$\text{Max} \quad \sum_{k=1}^{K} p_i z_i^k \lambda^k$$

$$\text{subject to} \quad \sum_{k=1}^{K} z_i^k \lambda^k = 1$$

$$\sum_{k=1}^{K} \lambda^k = n$$

where coefficients z_i^k must satisfy the constraint

$$\sum_{i=1}^{m} w_i z_i^k \leq d$$

Furthermore, this problem has only one subproblem

$$(KP) \quad \max z(KP) \quad = \sum_{i=1}^{m} (p_i - u_i) z_i$$

$$\text{subject to} \quad \sum_{i=1}^{m} w_i z_i \leq d$$

$$z_i = 0 \text{ or } 1 \text{ for all } i$$

To check if there exists a column with positive reduced cost, we calculate the value

$$z(KP) - v$$

where $u_i \, (i = 1, 2, \ldots, m)$ and v are the optimum dual prices from the solution to the LP relaxation of the restricted master problem, as usual. Note that this special GAP problem has the structure of a 0–1 cutting stock problem.

With identical machines, many solutions will differ only by the names of the machines; that is, by swapping the assignments of two machines we get two solutions that are the same but have different values for the variables. This property will cause performance problems when branching on the original variables; that is, when a fractional solution is excluded at some node of the tree, it pops up again with different variable values somewhere else in the tree. Another property—the large number of alternate optima dispersed throughout the tree—excludes pruning by bounds as a viable pruning strategy.

To resolve these performance problems, a special branching scheme exists that works directly on the master problem but is focused on the pairs of tasks: consider rows of the master with respect to tasks r and s. First divide the solution space into a pair of sets in which r and s appear together; tasks r and s can be combined into one task when solving the knapsack. The other branch occurs when the solution space is divided into another pairs of sets in which r and s must appear separately, in which case a constraint $z_r + z_s \le 1$ is added to the knapsack. Note that the structure of the two subproblems differs depending on the branch. Specifically, branch 1 is

$$\sum_{i=1}^{m} w_i z_i \le d \quad \text{and} \quad z_r + z_s \le 1$$

and branch 2 is

$$\sum_{i=1}^{m} w_i z_i \le d \quad \text{and} \quad z_r = z_s$$

13.4.4 Branch-and-Price Algorithm

The application of branch-and-cut algorithm on the branch-and-bound tree is summarized below.

Step 1 (Solving restricted master problem). Find a feasible solution to the LP relaxation of the restricted master problem. If the solution is integral, go to step 3. Otherwise, go to step 2.

Step 2 (Branching). If λ_j^k is a noninteger, then by $y_{ij} = \sum_{k=1}^{K_j} z_{ij}^k \lambda_j^k$ we know that the corresponding y_{ij} is also a noninteger. Branch on y_{ij} by setting it to 0 and 1, respectively. At each branch, y_{ij} is also fixed in the subproblem i. Go to step 3.

Step 3 (Solving Subproblems). Pass the dual solutions obtained from the solution of the restricted master problem to the subproblem(s). Solve each subproblem. Go to step 4.

Step 4 (Pricing). If none of the subproblems has a positive reduced cost, stop. Current solution to the restricted master problem is optimal to the original IP problem. Otherwise, choose one subproblem with a positive reduced cost. The optimal assignment associated with it corresponds to the new column that can be generated. Go to step 1.

13.5 OTHER APPLICATION AREAS

Branch-and-price has been applied to many other MIP problems. Several examples are described below.

- Airline crew scheduling (Vance, Barnhart, Johnson, and Nemhauser)
- Vehicle routing with time windows (Desrochers, Desrosiers, and Solomon)
- Machine scheduling (Bard and Rojanasoonthon)
- Origin–destination integer multicommodity flow problems (Barnhart, Hane, and Vance)

Airline crew scheduling deals with finding a minimum cost assignment of flight crews to a given flight schedule, while satisfying restrictions dictated by collective bargaining agreements and the Federal Aviation Administration. Traditionally, the problem has been formulated as a set partitioning problem. An alternative formulation allows the use of a branch-and-price algorithm.

Vehicle routing problems with time windows is to find the minimum number of vehicles required to visit all customers subject to capacity constraints and time windows defined by the earliest and the latest times when the customer will permit the start of service. The LP relaxation of the set partitioning formulation is solved by branch-and-price algorithm.

Machine scheduling is to schedule n jobs on m nonhomogeneous parallel machines with multiple time windows and job priorities. The objective is to maximize the weighted number of jobs scheduled, where a job in a higher priority class has an infinite more weight than a job in a lower priority class. A branch-and-price algorithm is used to solve the problem.

The origin–destination integer multicommodity flow problem is a constrained version of the linear multicommodity flow problem in which flow of a commodity may use only one path from origin to destination. A new branching rule is devised that allows columns to be generated efficiently as well as allow cuts or cover inequalities to be generated at each node of the branch-and-bound tree. The use of column generation and cut generation is called *branch-and-price-and-cut*.

13.6   NOTES

Sections 13.1 and 13.3

The term branch-and-price first appeared in Savelsbergh (1997) for solving the generalized assignment problem, even though the application of column generation using Dantzig–Wolfe decomposition principle appeared at least 5 years earlier in Desrochers et al. (1992) for solving a generalized vehicle routing problem.

Section 13.2

The decomposition principle originated by Dantzig and Wolfe (1960).

Section 13.4

Applications of branch-and-price are airline crew scheduling (Vance et al, 1997), vehicle routing with time windows (Desrochers et al., 1992, Bard et al., 2002), and machine scheduling (Bard and Rojanasoonthon, 2006). For application of branch-and-price-and-cut, see integer multicommodity flow problems (Barnhart et al., 2000).

13.7 EXERCISES

13.1 Solve the following LP problem using Dantzig–Wolfe decomposition.

$$\begin{aligned}
\text{Maximize} \quad & -3x_1 + x_2 - x_3 + x_4 \\
\text{subject to} \quad & 2x_1 - x_2 - x_3 + x_4 \le 8 \\
& -2x_1 + 2x_2 + 2x_3 + 3x_4 \le 10 \\
& -x_1 + x_2 - 3x_3 + x_4 \le 3 \\
& \mathbf{x} \ge \mathbf{0}
\end{aligned}$$

13.2 Solve the following problem using Dantzig–Wolfe decomposition:

$$\begin{aligned}
\text{Minimize} \quad & 4x_1 + 2x_2 + x_3 - 2x_4 \\
\text{subject to} \quad & x_1 - x_2 + 2x_3 + x_4 \le 11 \\
& x_2 + x_3 + x_4 \ge 17 \\
& 3x_1 - x_3 - 3x_4 \ge 5 \\
& \mathbf{x} \ge \mathbf{0} \text{ and integer}
\end{aligned}$$

13.3 Compare the Dantzig–Wolfe decomposition applied to the GAP problem with its generic form in Section 13.2. Is there any difference? If there is difference, is it simply an equivalent expression or is it completely different in concept?

13.4 Solve the following GAP problem with branch-and-price using the data given below: $m = 3$, $n = 2$,

$$\mathbf{P} = \begin{pmatrix} 10 & 8 \\ 13 & 14 \\ 9 & 10 \end{pmatrix} \qquad \mathbf{W} = \begin{pmatrix} 10 & 9 \\ 14 & 11 \\ 8 & 13 \end{pmatrix} \qquad \mathbf{d} = \begin{pmatrix} 23 \\ 27 \end{pmatrix}$$

14

SOLUTION VIA HEURISTICS, RELAXATIONS, AND PARTITIONING

14.1 INTRODUCTION

This chapter introduces a variety of primal heuristic algorithms that can be used to obtain a good solution or an approximate solution for an integer program or a combinatorial optimization problem (COP). Both classical and artificial intelligence (AI) heuristic algorithms are provided. The traveling salesman problem (TSP) and other combinatorial optimization problems are used for the purpose of illustration. This chapter also (a) describes various relaxation methods for solving integer programming (IP) problems, (b) lists examples of IP model types to which the Lagrangian relaxation approach is applied, (c) derives the associated Lagrangian dual problems for both linear and integer programs, (d) provides three efficient methods for solving the Lagrangian dual, and (e) develops a decomposition algorithm for integer programming.

14.2 OVERALL SOLUTION STRATEGIES

Linear programming problems have been shown to be much easier to solve than integer programming problems. However, the simplex algorithm and the capability it

Applied Integer Programming: Modeling and Solution, By Der-San Chen, Robert G. Batson, and Yu Dang
Copyright © 2010 John Wiley & Sons, Inc.

provides to efficiently solve a sequence of linear programs is basic to solving integer programs, be they pure, mixed, or binary. If the IP has special structure, the solution to the LP relaxation is sometimes the solution to the IP (see Chapter 10). Otherwise, the solution strategy recommended involves selecting from the following:

- Preprocessing
- Branch-and-bound (B&B)or its descendents
  - branch-and-cut (B&C)
  - branch-and-price
- Heuristics to develop
  - Good, approximate solutions
  - Tighter lower bounds
- Relaxations to develop tighter upper bounds

Furthermore, preprocessing, heuristics, and relaxations can be used at each node in branch-and-cut. This exemplifies the strategy of using general-purpose algorithms to control the overall MIP solution process and special-purpose approaches to improve their overall effectiveness. The user of MIP solver software generally has the option of selecting from these approaches, that is, creating a "customized" approach.

14.2.1 Better Formulation by Preprocessing

Preprocessing was discussed in detail in Chapter 4. Recall that a better formulation of an MIP is one that is easier to solve. It is widely accepted that almost any MIP formulation can be improved by preprocessing—this is why modern MIP solvers include a set of rules (see Section 4.2) bundled together into the so-called preprocessor or presolver, automatically applied on behalf of the user. In Section 4.3, we introduced basic preprocessing techniques for tightening bounds, fixing variables, and identifying redundant constraints and infeasibility in general integer programming. In Section 4.4, the reader can find techniques for the same functions, but specially designed for pure 0–1 integer programs. The key idea of preprocessing is to reformulate the problem statement created by the modeler in such a way that the difference in objective function optimal values for the linear programming relaxation and the respective integer program is as small as possible.

In general, preprocessing introduces tighter constraints that dominate existing constraints, which are removed from the reformulated model. So, the problem size generally is improved. Reformulation is completely independent of the solution of the linear programming relaxation of the original model. Results of preprocessing MIP models are reported to reduce problem size by a factor of 5 and runtime by a factor of 10; hence, preprocessing is valuable at the start of any attempt to solve an MIP model. If you are given this option, use it.

14.2.2 LP-Based Branch-and-Bound Framework

LP-based branch-and-bound remains central to state-of-the-art MIP solvers. The application of branch-and-bound to general integer programs was presented in Section 11.1. Recall that branch-and-bound can be viewed as a divide-and-conquer approach to solving IP problems, in which a branching process for dividing and a bounding process for conquering are applied. In the enumeration that keeps up with progress toward optimality in this implicit enumeration approach, pathways (branches) that cannot lead to a better solution than the best already identified are systematically pruned (fathomed). At points where branching does occur, two linear programs are generally solved and the resulting information is used to guide the so-called "intelligent" search for the IP optimal solution.

Versions of the classic branch-and-bound algorithms specialized to binary integer programs and mixed integer programs are well known and may be founded in Hillier and Lieberman (2005). In all these versions, branch-and-bound is LP based; that is, it depends on solution of an intelligently chosen sequence of linear programs to approach the optimal integer solution. Because upper and lower bounds are generated to aid with fathoming and testing for optimality, the algorithm provides built-in measure of solution quality if the user wants to terminate the search before optimality is reached. This is often necessary if branch-and-bound is used on large-scale MIPs.

Heuristic and relaxation for systematically tightening lower and upper bounds, respectively, are in fact important solution strategies in integer programming. These general strategies will be introduced briefly in Sections 14.2.3 and 14.2.4 and in detail in Sections 14.3 and 14.4.

14.2.3 Heuristics for Tightening Lower Bounds

Tightening bounds on variables was presented in Section 4.3. Here, we are concerned with tightening bounds on the optimal value of an MIP. Prior to or during application of a solution algorithm to solve the MIP, if a tighter lower bound on the optimal value is known, searches can be limited to more promising portion of the relaxed LP feasible region. In general, bounds obtained from LP relaxation may be too weak to guide the search for the MIP optimum. Also, the heuristics presented in Section 14.3 often find good solutions to the MIP—that is, solutions that are within a few percent of optimal. Such solutions can be accepted as "good enough" or can become the starting point for an algorithm (e.g., branch-and-bound), significantly reducing the algorithm's number of iterations to converge.

Three heuristic approaches that have been successfully applied to MIPs are presented in Section 14.3:

- Local search
- Tabu search
- Genetic algorithms

Some challenging MIPs can only be solved by heuristics, and these solutions are of course approximate (each resulting in a lower bound on the optimal solution of the original, or primal, minimizing MIP). Another weakness of heuristics is that they provide no upper bound on the MIP optimal value, so the user will not know how close to optimum he may be. Upper bounds require dual problem solution, as discussed next.

14.2.4 Relaxations for Tightening Upper Bounds

Section 14.4 presents three "relaxation" approaches to obtaining a lower bound on the optimal value of a primal, minimizing MIP:

- Linear programming relaxation
- Combinatorial relaxation
- Lagrangian relaxation

The first two create a revised (more extensive) feasible region, but leave the objective function as is. The third approach substitutes another minimizing objective function for **cx**, one that is the same or smaller value on the fixed feasible region.

14.2.5 Strong Cuts for Tightening Solution Polyhedron

The concept of cuts, additional constraints that cut off (reduce the extent of) the relaxed LP solution space while leaving the MIP solution space unchanged, was introduced in Chapter 12. Cuts are generated based on model data and are adjoined to the (current) model to cut off a relaxation solution $\mathbf{x}^*$ in the solution polyhedron $\mathbf{P}$. Stronger cuts produce smaller solution polyhedrons, which still contain the MIP feasible region. Three cut-generating techniques were presented in Section 12.3: rounding, disjunction, and lifting. MIP solvers today give the user many options for cutting planes, both general and structure dependent. In Section 12.3, the reader will find general cuts from sets involving pure integer variables and sets involving mixed integer variables. More specialized cuts generated from 0–1 knapsack sets and sets containing 0–1 coefficients and 0–1 variables are also presented along with cuts from sets with special structure.

The branch-and-cut approach that first appeared in the mid-1980s was a breakthrough that generalized the branch-and-bound method. It built upon the branch-and-bound framework with additional cuts generated and imposed on each node in the branch-and-bound tree, prior to pruning and branching. B&B applies two simple bound cuts at each node and takes advantage of fast reoptimization of the LP at each node. B&C activity at each node may include generating stronger cuts, problem preprocessing, or even application of a primal heuristic. So, many cuts may be applied at each node; the trade-off is that a tighter bound is generated at the node, prior to pruning and branching. Branch-and-cut options are a standard feature of commercial MIP solvers today.

14.3 PRIMAL SOLUTION VIA HEURISTICS

A practicing engineer or operations analyst would say that a heuristic is a simple procedure (or algorithm) that is meant to provide a good but not necessarily optimal solution to a particular difficult problem easily and quickly. In MIP, a *heuristic* is an approximate algorithm designed to quickly find good solutions, but the solution may not be optimal. The word "heuristic" invokes the concept of purposeful search, because the word derives from the Greek "heuriskein" that means "to discover." Shapiro (2001) distinguishes between problem-specific heuristics that use "rules of thumb" to arrive at good feasible solutions to MIPs and general-purpose methods for intelligently searching the space of feasible solutions. He suggests that the latter "may be *combined* with problem-specific heuristics to improve their effectiveness." A term that is sometimes used when one heuristic controls another at a lower level of activity is *metaheuristic*.

Local search heuristics start with a given feasible solution and by limited changes (often called interchange) in one or a few coordinates, attempt to improve the objective function value while retaining feasibility. Hence, this sort of heuristic applies a rule to select an element from a set. For example, the traveling salesman problem presents an obvious simple heuristic: go to the next closest city not yet visited. Starting at home base and systematically moving from one city to the nearest unvisited city until all cities have been visited, and then returning home generates a feasible tour with no subtours. The route prescribed may not be optimal, in fact may be far from optimal. Two more sophisticated local search methods, tabu search and simulated annealing, will be discussed in the context of solving MIP problems in sections to follow. Finally, the general-purpose heuristic method known as genetic algorithms will be described.

There are many reasons to include a section on heuristics in any applied integer programming text:

1. Good heuristics are available for many integer and combinatorial optimization problems due to their structure.
2. Solving real-world MIPs by the approaches of earlier chapters can be too slow—solutions are needed in seconds or minutes, not hours or even days.
3. The MIP formulation is too difficult for branch-and-bound, branch-and-cut, and other LP-based approaches.

Tempering these practical considerations is required to assure quality (near optimal) solutions. Unlike the earlier methods to solve MIPs, heuristic search can become trapped at a local optimum. The improved local search heuristics, tabu search and simulated annealing, have features that enable the search to escape from a local optimum. Genetic algorithms build new, improved solutions from pairs of solutions, mimicking the genetics of natural selection. These primal heuristics have another weakness. While providing lower bound on the MIP minimum optimal value, which improve as the heuristic discovers better solutions, they provide no upper bounds to enable quantification of how close the heuristic solution is to optimal. This requires dual problem solution, to be discussed in Section 14.4. Nemhauser and Wolsey (1988) observed that "often, primal and dual heuristic solutions can be found in pairs. The complimentary slackness conditions are one way of pairing heuristic solutions. The dual solution provides an upper bound on the deviation from optimal of the primal solution."

14.3.1 Local Search Approaches

Local search heuristics for MIP problems were included in Chapter 9 of the famous text by Garfinkel and Nemhauser (1972). The same year Woolsey (1972) observed that "many of those who actually solve (MIP) problems turn to heuristic methods to get good starting solutions, followed by some kind of branch-and-bound solution to take every possible advantage of problem structure." Papers published in the 1980s by Zanakis and Evans (1981), Haessler (1983), and Hillier (1983) all discussed the proper situation and role of local search heuristics. A comprehensive reference is Walser (2008). Generally, there are three roles for local search heuristics in MIP:

1. Locate a feasible solution as starting place for an MIP algorithm, because finding simply a feasible solution for an MIP can be difficult in practice.
2. Local search methods themselves can benefit from a feasible starting solution.
3. Local search methods can be integrated into general search method (e.g., genetic algorithms) or MIP methods (e.g., branch-and-cut).

The following is an example of two local search heuristics.

Example 14.1 (Example 2.5 Wolsey (1998))   Consider the following six city symmetric traveling salesman problem (STSP) where city 1 is home base and the matrix $\mathbf{C}$ consists of distances (costs) for every city pair.

$$\mathbf{C} = \begin{pmatrix} - & 9 & 2 & 8 & 12 & 11 \\ & - & 7 & 19 & 10 & 32 \\ & & - & 29 & 18 & 6 \\ & & & - & 24 & 3 \\ & & & & - & 19 \\ & & & & & - \end{pmatrix}$$

Wolsey applies a "greedy algorithm" that builds up the traveling salesman tour segment-by-segment, always choosing the least costly segment not creating a subtour if added to the current set of segments: The sequence is (1, 3) with cost 2; (4, 6) with cost 3; (3, 6) with cost 6; (1, 2) with cost 9; (2, 5) with cost 10; and (4, 5) with cost 24. The greedy tour is (1, 3, 6, 4, 5, 2, 1) with total cost 54. The nearest-neighbor heuristic applied to the same matrix $\mathbf{C}$ yield the following: (1, 3) with cost 2; (3, 6) with cost 6; (6, 4) with cost 3; (4, 5) with cost 24; (5, 2) with cost 10; and (2, 1) with cost 9. The nearest-neighbor tour is (1, 3, 6, 4, 5, 2, 1) with total cost 54, the same as the greedy tour.

Examples of specific MIP models for which local search algorithms have been developed are as follows:

- Real-time scheduling of jobs on mixed model assembly lines (Bolat et al., 1994)
- Trim-loss problem in slitting rolls of paper (Ramirez-Beltran and Aguilar-Ruggiero, 1997)
- Capacitated production scheduling (Walser et al., 1998)
- Job shop scheduling with earliness and tardiness costs (Danna et al., 2003)
- Vehicle routing with time windows (Danna, 2004).

To motivate an example application of generic algorithms to appear in Section 14.3.2, let us describe an application of a local search algorithm (*nearest neighbors*) that appears in Shapiro (2001)[1]. The specific problem here is the local delivery problem: m customers require delivery from a depot each day; each customer has an integer demand for the product on a given day; and an unlimited number of delivery trucks are available at the depot, each with identical capacity (10, in this case). The cost of sending a truck on a particular route involves a fixed cost and a variable cost depending on the cost associated with traveling each leg of the truck's assigned route. The supplier wants to meet the daily delivery demand, yet minimize the number of trucks used (fixed cost); and within each route, minimize the route cost (an embedded traveling salesman problem). The nearest unserved neighbor is used to created the routes, of course staying within the capacity constraint of the truck and Shapiro's side constraint that no unserved customers more than 60 miles distant from the currently included customers on a route will be considered.

Two versions of the heuristic are used; one, where the current unserved customer *closest* to the depot is used to start each new route; and the other where the current unserved customer *farthest* from the depot is used to start each new route. In the example (5.1) in Shapiro (2001), the depot is labeled 0 and customer numbered 1–13. Besides a 13×13 matrix of distances, the demand at each customer (an integer between 2 and 9) is provided. The result of the first version of the heuristic was as follows:

Route	Customers	Cost ($)
1	0–11–10–0	182
2	0–5–6–9–0	174
3	0–12–7–0	169
4	0–2–1–0	197
5	0–4–3–0	214
6	0–8–0	170
7	0–13–0	200
Total cost of routing solution 1		1306

[1] From Shapiro, *Modeling the Supply Chain*, 1E Copyright 2001 South-Western, a part of Centage Learning, Inc. Reproduced with permission, www.centage.com/permissions.

Note that in route 2, the shortest sequence (TSP solution) is 0–9–6–5–0 for a route cost of 74, or a total trip cost of 174. For routes with one or two customers, the order of listing yields the minimum distance (TSP) order of delivery.

The result of the second version of the heuristic was as follows:

Route	Customers	Cost ($)
8	0–13–3–5–0	241
9	0–1–2–6–0	252
10	0–10–7–0	192
6	0–8–0	170
11	0–4–9–0	199
12	0–12–0	138
13	0–11–0	132
Total cost of routing solution 2		1324

Note in routing solution 1, the larger cost routes are assigned later, whereas in routing solution 2, the largest cost routes appear earlier. Also, the solutions are quite different except for the repetition of the route 0–8–0. In the section on generic algorithms, we will show how to select routes from these two preliminary solutions (parents) to create a new solution (offspring) with 11% better performance than routing solution 1. But for now, routing solution 1 is the best solution.

14.3.2 Artificial Intelligence Approaches

Artificial intelligence is a broad and intensely important field of study, encompassing many approaches (expert systems, neural networks, heuristics, etc.) with an ever-expanding list of applications for each approach. The discussion here is limited to three heuristics that have proven themselves valuable for solving integer and combinatorial programming problems:

- Tabu search
- Simulated annealing
- Genetic algorithms

Genetic algorithms are attributed to Holland (1975), though the most popular reference to date was written by Holland's student Goldberg (1989). Simulated annealing was conceived by Kirkpatrick et al. (1983) and independently by Cerny (1985). However, the article by Glover (1986) suggested strongly that artificial intelligence heuristics were appropriate for many of the difficult problems being encountered while solving large MIPs and COPs with special structure. In this same article, Glover is credited with creating (and coining the name) tabu search. Glover promoted a "blend of heuristics and algorithms" and stated that "effective strategies for combinatorial problems can require methods that formal theories are unable to justify," that is, are not guaranteed to converge.

Glover (1986) noted that "perhaps the most conspicuous limitation of a heuristic method for problem solving involving discrete alternatives is the ability to become trapped at a local optimum." He suggested four classes of heuristics to transcend the problem of local optimality. Tabu search was one class, and simulated annealing was introduced as a new entrant in the "controlled randomization" class, which was already established with "random restart" and "random shakeup" approaches. From the AI point of view, Glover states that "tabu search deviates to an extent from what might be expected of intelligent human behavior" and that simulated annealing, based on physical behavior of molten metal as it cools, resembles human behavior only in that "a human may take non-goal-directed moves with greater probability at greater distances from a perceived destination."

Generic algorithms are considered general-purpose or global-search heuristics. Because they mimic both natural (genetic) adaptations and adaptations observed in human behavior, they are properly called an artificial intelligence approach. The processes found in genetic algorithms, such as crossover, selection, and mutation, were adapted from evolutionary biology.

$$\text{Let } F = \left\{ (x, y) \mid \sum_{j=1}^{n} a_{ij} x_j + \sum_{k=1}^{p} g_{ik} y_k \leq b_i \right\}$$

where $i = 1, \ldots, m; x_j \geq 0, j = 1, \ldots, n;$ and y_k integer, $k = 1, \ldots, p$

that is, F is the feasible region of an MIP, as defined in Chapter 2. In the application of heuristics, a solution $S \in F$ that is currently the best solution is called the *incumbent*. Local search heuristics, including tabu search and simulated annealing, define a *neighborhood* $Q(S)$ of solutions close to S within F. How $Q(S)$ is formed is problem specific. For instance, in the local routing problem, a neighborhood of a given (incomplete) route might be to add any yet unassigned customer to the route, subject to constraints that define F in that problem, or take away one of the customers assigned to the route already. Wolsey (1998) illustrates neighborhood formation rules for two COPs: uncapacitated facility location and the graph equipartition problem.

The term "tabu" in *tabu search* derives from a short-term memory feature of the heuristic that prevents revisiting solutions that have been visited in the recent past. A list of fixed or randomly varying size of recently visited solutions is maintained, called the *tabu list*. Another approach is to prohibit visits to solutions that have certain attributes.

Wolsey (1998) provided a description of a basic tabu search algorithm, which we reproduce here as Figure 14.1, with permission.

Pedroso (2006) has described how to apply tabu search to the solution of a bounded version of the generic MIP defined in Chapter 2:

$$\text{Min } z = \mathbf{cx} + \mathbf{dy}$$
subject to $(\mathbf{x}, \mathbf{y}) \in F$ (defined above)
and for integer $y_k, \ l_k \leq y_k \leq u_k, \quad k = 1, \ldots, p$

1.   Initialize an empty tabu list.
2.   Get an initial solution S.
3.   While the stopping criterion is not satisfied,
        3.1. Choose a subset $Q'(S) \subseteq Q(S)$ of nontabu solutions.
        3.2. Let $S' = \arg \min\{f(T):T \in Q(S)\}$.
        3.3. Replace S by S' and update the tabu list.
4.   On termination, the best solution found is the heuristic solution.

    The parameters specific to tabu search are as follows:
(i)   The choice of subset $Q'(S)$. Here, if $Q(S)$ is small, one takes the whole
        neighborhood, while if $Q(S)$ is large, $Q'(S)$ can be a fixed number of
        neighbors of S, chosen randomly or by some heuristic rule.
(ii)   The tabu list consists of a small number t of most recent solutions or
        modifications. If $t = 1$ or 2, it is not surprising that cycling is still
        common. The magic value $t = 7$ is often cited as a good choice.
(iii) The stopping rule is often just a fixed number of iterations, or a certain
        number of iterations without any improvement of the goal value of the
        best solution found.

FIGURE 14.1   Tabu search algorithm (From Wolsey, *Integer Programming*. Copyright 1998
John Wiley & Sons, Inc. Reprinted with permission of John Wiley & Sons).

Pedroso's strategy involves fixing the integer variables **y** by tabu search heuristic
and then obtaining the corresponding optimal objective value z and continuous
variables **x** by solving a linear programming problem via the simplex algorithm.
A critical measure at the end of this LP solution is the sum **v** of the constraint
violations, if any. Pedroso calls a solution S^i better than S^j if the extent of constraint
violation $v_i < v_j$, or if $v_i = v_j$ and $z_i < z_j$. As Pedroso explains,

- The initial solution is obtained by rounding the LP relaxation optimal values for
  integer variables to obtain a feasible solution.
- Tabu search starts operating on this solution by making changes exclusively
  in the integer variables, after which the continuous variables are recomputed
  through LP optimal solution.
- Modifications of the solutions are made using a simple neighborhood structure:
  incrementing or decrementing one unit to the value of an integer variable's value
  in the incumbent solution; hence, **y**$'$ is a neighbor solution to **y** if $y_k' = y_k + 1$ or
  $y_k' = y_k - 1$, for one index k, and $y_j' = y_j$ for all indices $j \neq k$.
- Moves are tabu if they involve a variable that has been changed recently.

Pedroso states that "this is tabu search based on short-term memory, as described in
Glover (1989). As suggested in Glover (1990), we complement this simple tabu search
with intensification and diversification." See the notes for more on tabu search.

As stated earlier, *simulated annealing* includes a probabilistic "controlled randomization" feature that leads to evaluation of random neighbors of the incumbent, and due to an embedded looping process (the Metropolis algorithm popularized by Hastings in 1970), moves to solutions with higher objective function value with nonzero probability. In the Metropolis algorithm, an initial solution **S** (state of material), an objective function f (internal energy), and initial temperature T are known. A neighbor **S′** is chosen randomly, and the energy levels of the incumbent $f(\mathbf{S})$ and the neighbor $f(\mathbf{S'})$ are compared. If the energy level is lower, keep **S′** with probability 1; if the energy level is higher, keep **S′** with probability $e^{f(\mathbf{S'})-f(\mathbf{S})/T}$. The temperature is gradually reduced by a factor r, $0 < r < 1$, known as the cooling ratio. This probability of accepting **S′** over **S** when $f(\mathbf{S'}) > f(\mathbf{S})$ decreases as temperature reduces (metal cools). A stopping criterion is when the metal is "frozen," that is, when $0 < T < 1$. The probabilistic feature in the looping process periodically moves the incumbent away from local optima, but the repeated application of r forces the proportion of nonimproving interchanges to decrease over time.

Simulated annealing got its unusual name because it mimics the metallurgical process known as annealing, a technique involving controlled cooling of a molten metal to increase the size of its crystals. The heating causes atoms to wander randomly through states of higher energy; gradually lowering the temperature enables perfect crystals to form and allows probabilistic changes in state, the goal is to bring the system from its initial state to a state with the minimum possible energy. According to Kirkpatrick et al. (1983), simulated annealing offered the dual benefits of

- ability to escape local minima at nonzero temperature, and
- divide and conquer outcomes, where gross features of the final state appear at high temperature while finer details appear at lower temperatures.

Wolsey (1998) provided a description of the simulated annealing heuristic, which we have reproduced as Figure 14.2, with permission.

Simulated annealing has been used to solve a variety of COPs as shown below and is reputed to be very successful:

- Graph partitioning problem in Johnson et al. (1989)
- Graph coloring problem in Johnson et al. (1991)
- Traveling salesman problem in Kirkpatrick et al. (1983)
- Quadratic assignment problem in Wilhelm and Ward (1987)

Genetic algorithms employ a *probabilistic search* approach. Unlike tabu search and simulated annealing, genetic algorithms maintain a pool (or population) of candidate solutions, and this population evolves based not only on *mutation* (like SA) but also on *parent selection* of certain pairs from the population, *combination* (or crossover) to create one or two *offspring*, and finally on *population selection* based on a fitness criterion—a new population is selected by replacing members of

1. Get an initial solution S.

2. Get an initial temperature T and a reduction factor r with $0 < r < 1$.

3. While not yet frozen, do the following:

    3.1 Perform the following loop L times:

        3.1.1 Pick a random neighbor S' of S.

        3.1.2 Let $\Delta = f(S') - f(S)$.

        3.1.3 If $\Delta \le 0$, set $S = S'$.

        3.1.4 If $\Delta > 0$, set $S = S'$ with probability $e^{-\Delta/T}$.

    3.2 Set $T \leftarrow rT$. (Reduce the temperature.)

4. Return the best solution found.

Note that as specified above, the larger the Δ is, the lesser is the chance of making a move to a solution worse by Δ. Also, as the temperature decreases, the chances of making a move to a worse solution decrease.

Exactly as for local exchange heuristics, one has to define

(i)   A solution
(ii) The neighbors of a solution
(iii) The cost of a solution
(iv) How to determine an initial solution

The other parameters specific to simulated annealing are then

(v)   The initial temperature T
(vi)   The cooling ratio r
(vii) The loop length L
(viii) The definition of "frozen," or the stopping criterion

FIGURE 14.2   Simulated annealing algorithm (From Wolsey, *Integer Programming*. Copyright 1998 John Wiley & Sons, Inc. Reprinted with permission of John Wiley & Sons).

the original population by an identical member of offspring. Thus, the population evolves from one generation to the next. This process from evolutionary biology can be modeled in a continuous generational genetic algorithm pseudocode in Figure 14.3.

To illustrate the application of a genetic algorithm to a COP, consider the two solutions obtained for the local routing problem introduced in Section 14.3.1 on local search. The following example[2] is taken from Shapiro (2001), who calls genetic algorithms "perhaps the most popular class of heuristics for analyzing combinatorial optimization problems arising in supply chain management." Shapiro provides an excellent explanation of genetic algorithms in terms of chromosomes. He notes that

---

[2] From Shapiro, *Modeling the Supply Chain*, 1E Copyright 2001 South-Western, a part of Centage Learning, Inc. Reproduced with permission, www.centage.com/permissions.

---

1. Choose initial population.
2. Evaluate the fitness of each individual in the initial population.
3. Repeat until termination (time limit, or sufficient fitness achieved by a population member, or plateau reached):
    (i)   Select best-ranking individuals to reproduce
    (ii)   Create offspring by crossover and/or mutation
    (iii)   Evaluate the individual fitness of the offspring
    (iv)   Replace worst-ranked part of population with offspring

---

FIGURE 14.3   Genetic algorithm pseudocode.

"both crossovers and mutations are invoked probabilistically. . . this allows variations in offsprings. The likelihood that a chromosome is included in an offspring depends on its fitness value, which is defined relative to the objective of the optimization." Chromosomes in our example are the various routes included in each solution (1 and 2) given earlier. The underlying legs in the routes, for example, "3–13" in route 8 "0–3–3–5–0" are variously called schemata, building blocks, or genes.

In Figure 14.4, from Shapiro (2001), crossover is applied to the two solutions of the local routing problem given earlier. Recall that there were a total of 13 routes between the 2 (parent) solutions. At the top of the figure, the routes are reordered by their fitness, measured in cost per tons delivered. A probability of selection P(select) has been attached to each route, and the probabilities decrease with fitness. The form of the probability rule was $P(x) = Ke^{-\lambda x}$ for $x \leq 15$, where $K = 1.953$ and $\lambda = 0.04463$ were chosen so that $P(15) = 1.0$, $P(x)$ decreases with increasing x, and $P(20) = 0.80$. Hence, any route approaching fitness value 15 should have a very high probability of being selected in a crossover solution, and routes with fitness value 20 should have a fairly high (0.8) probability of being selected.

A crossover solution is shown in the middle of the figure. A route was selected for potential inclusion of its P(select) was greater than a randomly chosen number r, $0 < r < 1$. But, the route is not included in the crossover solution if it visits a customer already covered by an earlier route selected for inclusion in its crossover solution. Of course, different streams of random numbers could be used to generate many crossover solutions besides this one. At the end of the GA application, all but one customer (customer 6) are in the crossover solution. Customer 6, with 2 units of demand, could have been added to routes 6, 4, or 11. Shapiro says, "this new route can be viewed as a mutation of the original chromosome route 11." The resulting crossover solution is approximately 11% better than the lowest cost parent (feasible solution 1) with a cost $1306.

Applications of genetic algorithms to COPs include the following:

- Manufacturing cell design (Joines et al., 1996)
- Traveling salesman problem (Chatterjee et al., 1996) and (Katayama et al., 2000)
- Generalized assignment problem (Wolsey, 1998).

	A	B	C	D	E	F	G
1			**Routes Ordered by Fitness Value**				
2	Selection				Delivered	Cost/	
3	order		Route	Cost	tons	delivery ton	P(select)
4	1	Route 3	0–12–7–0	169	10	16.90	0.92
5	2	Route 1	0–11–10–0	182	10	18.20	0.87
6	3	Route 13	0–11–0	132	7	18.86	0.84
7	4	Route 6	0–8–0	170	8	21.25	0.76
8	5	Route 10	0–10–7–0	192	9	21.33	0.75
9	6	Route 5	0–4–3–0	214	9	23.78	0.68
10	7	Route 8	0–13–3–5–0	241	10	24.10	0.67
11	8	Route 4	0–2–1–0	197	8	24.63	0.65
12	9	Route 2	0–5–6–9–0	174	7	24.86	0.64
13	10	Route 9	0–1–2–6–0	252	10	25.20	0.63
14	11	Route 11	0–4–9–0	199	7	28.43	0.55
15	12	Route 12	0–12–0	138	4	34.50	0.42
16	13	Route 7	0–13–0	200	3	66.67	0.10
17							
18							
19			**Crossover Analysis**				
20	Selection		Random			Overlap	Include
21	order		probability	Route	Cost	earlier route	route?
22	1	Route 3	0.69	0–12–7–0	169	No	Yes
23	2	Route 1	0.56	0–11–10–0	182	No	Yes
24	3	Route 13	0.30	0–11–0	132	Yes	No
25	4	Route 6	0.32	0–8–0	170	No	Yes
26	5	Route 10	0.66	0–10–7–0	192	Yes	No
27	6	Route 5	0.79	0–4–3–0	x	x	No
28	7	Route 8	0.55	0–13–3–5–0	241	No	Yes
29	8	Route 4	0.24	0–2–1–0	197	No	Yes
30	9	Route 2	0.80	0–5–6–9–0	x	x	No
31	10	Route 9	0.35	0–1–2–6–0	252	Yes	No
32	11	Route 11	0.10	0–4–9–0	199	No	Yes
33	12	Route 12	0.98	0–12–0	x	x	No
34	13	Route 7	0.92	0–13–0	x	x	No
35							
36					x = Rejected by probability test		
37							
38							
39		**New Feasible**					
40		**Solution**					
41				Route 3	0–12–7–0	169	
42				Route 1	0–11–10–0	182	
43				Route 6	0–8–0	170	
44				Route 8	0–13–3–5–0	241	
45				Route 4	0–2–1–0	197	
46			Augmented route 11*		0–4–6–9–0	207	
47					Total cost	1166	
48							
49				*Customer 6 was added to route 11 (0–4–9–0)			

FIGURE 14.4 Genetic algorithm applied to local delivery problem (From Shapiro, *Modeling the Supply Chain*, 1E Copyright 2001 South-Western, a part of Centage Learning, Inc. Reproduced with permission, www.centage.com/permissions)

14.4   DUAL SOLUTION VIA RELAXATION

Section 14.3 was devoted to finding a first feasible solution to an MIP and then finding improved feasible solutions through heuristics. No guarantee of optimality came with the heuristic, but each improved solution gave a tighter upper bound on a minimizing objective function for the MIP. In this section, we consider the dual problem of finding lower bounds for a minimizing problem. Because the duality theory of integer programming has been extremely developed for *pure* integer programs, we shall assume the problem at hand is IP, as follows:

$$(\text{IP}) \quad z_{\text{IP}} = \min\{\mathbf{cx} : \mathbf{x} \in \mathbf{F}\}, \mathbf{F} = \{\mathbf{x} \in \mathbf{Z}_+^n : \mathbf{Ax} \geq \mathbf{b}\}$$

where $\mathbf{c}$ is $n \times 1$, $\mathbf{A}$ is $m \times n$, and $\mathbf{b}$ is $m \times 1$ in dimension.

If a suitably tight lower bound on z_{IP} can be found, then in combination with the primal upper bounds discussed in Section 14.3, one can develop criteria for stopping any algorithmic approach to find z_{IP} once the current objective function value falls within known bounds. This section presents three "relaxation" approaches to obtaining a lower bound on z_{IP}. The first two, linear programming relaxation and combinatorial relaxation, create a revised (more extensive) feasible region but leave the objective function as is. The third approach, Lagrangian relaxation, substitutes another minimizing objective function for $\mathbf{cx}$, one that has the same or smaller value on the fixed feasible region. Another approach to finding a lower bound on z_{IP} is based on duality, and will be presented in Section 14.5.

Definition 14.1   A *relaxation* of an IP is any minimization problem

$$(\text{RP}) \quad z_{\text{R}} = \min\{z_{\text{R}}(\mathbf{x}) : \mathbf{x} \in \mathbf{F}_{\text{R}}\}$$

with the properties

$$(\text{R1}) \quad \mathbf{F} \subseteq \mathbf{F}_{\text{R}}$$
$$(\text{R2}) \quad z_{\text{R}}(\mathbf{x}) \leq \mathbf{cx}, \text{ for all } \mathbf{x} \in \mathbf{F}$$

Proposition 14.1   If RP is infeasible, so is IP. If IP is feasible, then $z_{\text{R}} \leq z_{\text{IP}}$.

In Sections 14.4.1 and 14.4.2, one or more constraints will be dropped from $\mathbf{F}$, and $z_{\text{R}}(\mathbf{x}) = \mathbf{cx}$.

14.4.1   Linear Programming Relaxation

The *linear programming relaxation* of IP is given by

$$(\text{LP}) \quad z_{\text{LP}} = \min\{\mathbf{cx} : \mathbf{x} \in \mathbf{L}\}, \quad \text{where} \quad \mathbf{L} = \{\mathbf{x} \in \mathbf{R}_+^n : \mathbf{Ax} \geq \mathbf{b}\}$$

An optimal solution $\mathbf{x}^*$ to z_{LP} is optimal for IP if $\mathbf{x}^*$ has all integer values. Another condition for $\mathbf{x}^*$ to be optimal for IP is if $\mathbf{cx}^* = z_{\text{IP}}$, which of course is unknown.

So, with the right set of conditions on the upper bound, $\mathbf{x}^*$, and $z_{LP} = \mathbf{cx}^*$, the LP relaxation may provide insight into the optimal vector and value of z_{IP}. Also, recall from Section 4.1 that many combinatorial optimization problems have the property that their LP relaxation has feasible region $\mathbf{L}$ equal to the convex hull of the basic feasible integer solutions:

- Assignment
- Transportation
- Transshipment
- Maximum flow
- Linear minimum cost flow

These are the so-called "easy integer programs" of Chapter 10, and solving their LP relaxation provides the integer optimum as well.

Consider the following example to see how helpful the LP relaxation can be.

Example 14.2   Suppose the integer program is

$$
\begin{aligned}
\text{Min} \quad z &= 5x_1 + 4x_2 \\
3x_1 + 2x_2 &\geq 5 \\
2x_1 + 3x_2 &\geq 7 \\
x_1, x_2 &\geq 0 \text{ and integer}
\end{aligned}
$$

Some obvious feasible points with their z-value are shown below:

$$
\begin{aligned}
(2, 1) \quad & z = 14 \\
(1, 2) \quad & z = 13 \\
(3, 1) \quad & z = 19 \\
(0, 3) \quad & z = 12 \\
(4, 0) \quad & z = 20
\end{aligned}
$$

So, an upper bound on z^* is 12. Now, the solution to the LP relaxation of the integer program is $\mathbf{x}^* = (0.2, 2.2)$ with $z^* = 9.8$. At this point, we know the optimal value of the IP must be either 10, 11, or 12. In fact, the point $(0, 3)$ turns out to be the optimal integer solution with optimal value 12.

Another obvious property of the linear programming relaxation, in fact any relaxation of IP, is that if the relaxation is infeasible, the original IP is infeasible.

14.4.2   Combinatorial Relaxation

As the name implies, sometimes when we remove one or more constraints from IP, we create an instance of a combinatorial optimization problem. If that COP turns out to be easy to solve, then a lower bound on z_{IP} can be generated rapidly. A five-city (asymmetric) traveling salesman problem is used to illustrate the opportunity.

Example 14.3   Consider a matrix of distances for the TSP as follows:

$$\begin{pmatrix} \infty & 11 & 3 & 6 & 9 \\ 5 & \infty & 5 & 4 & 2 \\ 4 & 9 & \infty & 7 & 8 \\ 7 & 1 & 3 & \infty & 4 \\ 3 & 2 & 6 & 5 & \infty \end{pmatrix}$$

Removing the constraints that no subtours are permitted, the relaxation is well known to be the assignment problem. The optimal assignments and their distances are as follows:

$1 \to 3$	3
$3 \to 1$	4
$2 \to 5$	2
$5 \to 4$	5
$4 \to 2$	1
	15

Two subtours $(1 - 3 - 1)$ and $(2 - 5 - 4 - 2)$ arise, and the optimal value 15 is a lower bound on the optimal length of a five-city tour. Note this would also have been the optimal value of the LP relaxation of the TSP. To get an upper bound, arbitrarily choosing the tour $1 - 2 - 3 - 4 - 5 - 1$ yields a length of $11 + 5 + 7 + 4 + 3 = 30$. A much tighter bound is obtained using the nearest-neighbor heuristic, which yields $1 - 3 - 4 - 2 - 5 - 1$ with a length of $3 + 7 + 1 + 2 + 3 = 16$. So, using the COP relaxation and a simple heuristic, we have obtained the optimal value of this TSP to be either 15 or 16.

Wolsey (1998) provides examples including the TSP, the symmetric TSP, the quadratic 0–1 problem, and the knapsack problem. We use a simple knapsack problem to illustrate the power of bounding solutions.

Example 14.4 (Integer Knapsack Problem)

$$z = \max 42x_1 + 26x_2 + 35x_3 + 71x_4 + 53x_5$$
$$14.4x_1 + 10x_2 + 12.1x_3 + 25x_4 + 20x_5 \le 69.9$$
$$x_1, x_2, x_3, x_4, x_5 \ge 0 \text{ and integer}$$

The COP relaxation of this problem is obtained by rounding each coefficient in the constraint down to the largest integer less than or equal to the coefficient given:

$$z = \max 42x_1 + 26x_2 + 35x_3 + 71x_4 + 53x_5$$
$$14x_1 + 10x_2 + 12x_3 + 25x_4 + 20x_5 \le 69$$
$$x_1, x_2, x_3, x_4, x_5 = 0 \text{ and integer}$$

The optimal LP relaxation of COP has $\mathbf{x}^* = (4.93, 0, 0, 0, 0)$ and $z^* = 207$. So, 207 is an upper bound on z_{IP}. Some obvious feasible vectors for COP and their objective function values are as follows:

$$
\begin{array}{ll}
(4,0,0,0,0) & z = 168 \\
(0,6,0,0,0) & z = 156 \\
(0,0,5,0,0) & z = 175 \\
(0,0,0,2,0) & z = 143 \\
(0,0,0,0,3) & z = 159 \\
(3,0,0,1,0) & z = 197 \\
(1,5,0,0,0) & z = 172 \\
(0,0,4,0,1) & z = 193 \\
(0,0,0,1,2) & z = 177
\end{array}
$$

So, the optimal value of IP is somewhere between the lower bound 197 and upper bound 207. The optimal IP solution is $(4, 0, 1, 0, 0)$ with $z^* = 206$; the COP relaxation provided a tight bound in this case.

14.4.3   Lagrangian Relaxation

The (IP) above can be rewritten as

$$(IP) \quad z_{IP} = \min\{\mathbf{cx} : \mathbf{x} \in \mathbf{F}\}$$

where $\mathbf{F} = \{\mathbf{x} \in \mathbf{Z}^n_+ : \mathbf{A}_1\mathbf{x} \geq \mathbf{b}_1, \mathbf{A}_2\mathbf{x} = \mathbf{b}_2\}$

$$\mathbf{F} = \{\mathbf{x} \in \mathbf{Z}^n_+ : \mathbf{A}_1\mathbf{x} \geq \mathbf{b}_1, \mathbf{A}_2\mathbf{x} = \mathbf{b}_2\}$$

and $\mathbf{A}_1$ is $m_1 \times n$, $\mathbf{A}_2$ is $m_2 \times n$, $\mathbf{b}$ is $m_1 \times 1$, and $\mathbf{b}$ is $m_2 \times 1$.

Note the original m constraints have been partitioned into two sets, with $m_1 + m_2 = m$. It is traditional to think of one set, say $\mathbf{A}_1\mathbf{x} \geq \mathbf{b}_1$, as "complicated" constraints and the other, $\mathbf{A}_2\mathbf{x} \geq \mathbf{b}_2$, as simple or "easy" constraints. For example, $\mathbf{A}_2\mathbf{x} \geq \mathbf{b}_2$ might correspond to the constraint set of a COP, or they might just be an expression of the lower bounds on $x_1, x_2, \ldots, x_n$. Instead of merely dropping the complicated constraints from the problem as a relaxation, they can actually be assigned a multiplier $\lambda \in \mathbf{R}^n_+$ to form the *Lagrangian relaxation* of IP with respect to $\mathbf{A}_1\mathbf{x} \geq \mathbf{b}_1$:

$$LR(\lambda) \quad z_{LR}(\lambda) = \min\{\mathbf{cx} + \lambda(\mathbf{b} - \mathbf{A}_1\mathbf{x}) : \mathbf{A}_2\mathbf{x} \geq \mathbf{b}_2, \quad \mathbf{x} \in \mathbf{Z}^n_+\}$$

$LR(\lambda)$ does not contain the complicated constraints and $\lambda \geq 0$ forces $\mathbf{b} - \mathbf{A}_1\mathbf{x} \leq 0$ (as desired) to minimize the overall (penalized) objective function.

Proposition 14.2 $z_{LR}(\lambda)$, as defined above for $LR(\lambda)$, has the property $z_{IP} \geq z_{LR}(\lambda)$, for all $\lambda \geq 0$.

Nemhauser and Wolsey (1988), in their Example 6.1, provide an extended explanation of the bounds and geometry resulting from applying Lagrangian relaxation to a maximizing integer program with two variables and five constraints. Four of the five constraints are treated as "nice" and one as "complicating," so λ is a scalar (which enhances the geometric explanation).

Example 14.5   The Lagrangian relaxation of the IP given in Section 14.4.1 will be developed using the first constraint $3x_1 + 2x_2 \geq 5$ as the complicating constraint. The multiplier λ will, therefore, be a scalar in the formulation:

$$
\begin{aligned}
\text{Min} \quad & z = 5x_1 + 4x_2 + \lambda(5 - 3x_1 - 2x_2) \\
& 2x_1 + 3x_2 \geq 7 \\
& x_1, x_2 = 0 \text{ and integer}
\end{aligned}
$$

which can be rewritten as

$$
\begin{aligned}
\text{Min} \quad & z = (5 - 3\lambda)x_1 + (4 - 2\lambda)x_2 + 5\lambda \\
& 2x_1 + 3x_2 = 7 \\
& x_1, x_2 = 0 \text{ and integer}
\end{aligned}
$$

The solution of this problem would provide a lower bound on the original IP.

Lagrangian relaxation has been applied in integer programming to such difficult problems as

- Airline scheduling (Yan and Lin, 1997)
- Generalized assignment problem (Nauss, 2006)
- Ship scheduling (Rana and Vickson, 1988).

14.5   LAGRANGIAN DUAL

As stated in Section 14.4, relaxation and duality are the two ways of finding lower bounds on z_{IP} in

$$
(\text{IP}) \quad z_{\text{IP}} = \min\{\mathbf{cx} : \mathbf{x} \in \mathbf{F}\}, \mathbf{F} = \{\mathbf{x} \in \mathbf{Z}_+^n : \mathbf{Ax} \geq \mathbf{b}\}
$$

where $\mathbf{c}$ is $n \times 1$, $\mathbf{A}$ is $m \times n$, and $\mathbf{b}$ is $m \times 1$ in dimension.

There is a theory of Lagrangian duality for both linear programming and integer programming; Nemhauser and Wolsey (1988) is a comprehensive reference upon which this section is based. As we shall see, any dual feasible solution provides a lower bound on z for the primal. Note that in the relaxation methods of Section 14.4, the relaxation must be solved to optimality to determine a lower bound.

14.5.1   Lagrangian Dual in LP

Consider LP as defined earlier, with the constraints partitioned into $A_1x \geq b_1$ (complicated) and $A_2x \geq b_2$ (easy) constraint sets. Now

$$\text{(LP)}\quad z_{LP} = \min\{cx : x \in L\}$$

where $L = \{x: A_1x \geq b_1, A_2x \geq b_2, x \in R^n_+ \}$.

The Lagrangian dual of LP is

$$\text{(LDLP)}\quad w_{LDLP} = \max w(u), \text{ where}$$
$$w(u) = ub_1 + \text{minimum}(c - uA_1)x, x \in L'$$
$$u(1 \times m_1) \geq 0 \text{ and } L' = \{x \in R^n_+ : A_2x \geq b_2\}$$

Proposition 14.3 (Weak Duality)   If x_0 is feasible for LP and u_0 is feasible for LDLP, then $cx_0 \geq w(u_0)$.

Proposition 14.4   If L' is nonempty and bounded and LP has a finite optimal solution, then $z_{LP} = w_{LDLP}$.

14.5.2   Lagrangian Dual in IP

Definition 14.2   The two problems

$$\text{(IP)}\quad z = \min\{cx : x \in F\}, \quad \text{and}$$
$$\text{(D)}\quad w = \max\{w(u) : u \in U\}$$

form a *weak dual pair* of $w(u) \leq cx$ for all $x \in F$ and all $u \in U$. When $z = w$, they are said to form a *strong dual pair*.

Proposition 14.5   The integer program IP and the dual of its linear programming relaxation LP, as given below, form a weak dual pair:

$$\text{(DLP)}\quad w_{LDLP} = \max\{ub : u \in U\}, \quad \text{where} \quad U = \{u \in R^m_+ : uA \leq c\}$$

More generally, if a problem is dual to any relaxation of IP, it is a weak dual to IP.

Example 14.6   Consider the integer knapsack problem in Section 14.4.2. The LP relaxation of this problem has dual LP:

$$w = \min 69.9\,u$$
$$14.4u \geq 42$$
$$10u \geq 26$$
$$12.1u \geq 35$$
$$25u \geq 71$$
$$20u \geq 53$$
$$u \geq 0$$

$u^* = 2.92$ and $w^* = 203.88$, so 203.88 is a lower bound on z_{IP}.

The COP relaxation of the IP has dual LP:

$$w = \min 69\, u$$
$$14u \geq 42$$
$$10u \geq 26$$
$$12u \geq 35$$
$$25u \geq 71$$
$$20u \geq 53$$
$$u \geq 0$$

Here, $u^* = 3$ and $w^* = 207$, so a better lower bound for z_{IP} is 207. *Note*: $u = 3$ could have been quickly determined to solve all five constraints (hence, was feasible), so its objective function value 207 is known to be a lower bound of z_{IP} without solving the dual problem to optimality.

Proposition 14.6 Suppose IP and D are a weak dual pair. If $\mathbf{x}^* \in \mathbf{F}$ and $\mathbf{u}^* \in \mathbf{U}$ can be found such that $w(\mathbf{u}^*) = \mathbf{c}\mathbf{x}^*$, then $\mathbf{u}^*$ is optimal for D and $\mathbf{x}^*$ is optimal for IP.

14.5.3 Properties of the Lagrangian Dual

For any $\mathbf{u} \in \mathbf{R}_+^{m_1}$, define the following integer programming problem

$$\text{LRIP}(\mathbf{u}) \quad w(\mathbf{u}) = \mathbf{u}\mathbf{b}_1 + \text{minimum } (\mathbf{c} - \mathbf{u}\mathbf{A}_1)\mathbf{x}$$

where $\mathbf{u} \geq \mathbf{0}$ and $\mathbf{X} = \{\mathbf{x} \in \mathbf{Z}_+^n : \mathbf{A}_2\mathbf{x} = \mathbf{b}_2\}$

The vector $\mathbf{u}$ is called the *dual variable (Lagrange multiplier)* associated with the constraint $\mathbf{A}_1\mathbf{x} \geq \mathbf{b}_1$, just as in the LP case. Note the function $w(\mathbf{u})$ could alternatively be written as

$$w(\mathbf{u}) = \min_{x \in X} \mathbf{c}\mathbf{x} + \mathbf{u}(\mathbf{b}_1 - \mathbf{A}_1\mathbf{x})$$

and $w(\mathbf{u}) \leq z_{IP} = \min\{\mathbf{c}\mathbf{x} : \mathbf{x} \in \mathbf{F}\}$, for all $\mathbf{u} \geq \mathbf{0}$.

Definition 14.3 The problem LRIP($\mathbf{u}$) is called a *Lagrangian relaxation* of IP with parameter $\mathbf{u}$.

To find the largest lower bound over all possible values of u, we need to solve the *Lagrangian dual problem*.

$$\text{LDIP} \quad w_{\text{LDIP}} = \max\{w(\mathbf{u}) : \mathbf{u} \in R_+^{m_1}\}$$

Proposition 14.7 The problem LPIP($\mathbf{u}$) is a relaxation of IP, for all $\mathbf{u} \geq \mathbf{0}$.

Proposition 14.8 Given a specific $\mathbf{u} \in \mathbf{R}_+^{m_1}$, if the following two conditions are satisfied, then the optimal vector $\mathbf{x}^*(\mathbf{u})$ of LRIP($\mathbf{u}$) is optimal for IP:

1. $\mathbf{A}_1 \mathbf{x}^*(\mathbf{u}) \geq \mathbf{b}_1$
2. $(\mathbf{A}_1 \mathbf{x}^*(\mathbf{u}))_i = \mathbf{b}_{1i}$ whenever $\mathbf{u}_i > 0$

Wolsey (1998) provides application of this proposition to the uncapacitated facility location problem and the symmetric traveling salesman problem.

Proposition 14.9 $w_{\mathrm{LDIP}} = \text{minimum}\{\mathbf{cx}: \mathbf{A}_1\mathbf{x} \geq \mathbf{b}_1, \mathbf{x} \in \text{Conv}(\mathbf{X})\}$
This proposition states that the (primal) linear program given is dual to the IP Lagrangian dual problem LDIP.

14.6   PRIMAL–DUAL SOLUTION VIA BENDERS' PARTITIONING

In Sections 14.4 and 14.5, we started with a pure IP and described various solution approaches based on relaxations and duality. In each case, a set of "complicating constraints" was assumed. In this section, while retaining the minimizing objective function, we return to the MIP formulation used in most of the chapters of this book:

$$
\begin{aligned}
(\text{MIP}) \quad z_{\mathrm{MIP}} &= \min \mathbf{c}_1\mathbf{x}_1 + \mathbf{c}_2\mathbf{y}_2 \\
\text{subject to} \quad & \mathbf{A}_1\mathbf{x} + \mathbf{A}_2\mathbf{y} \geq \mathbf{b}_1 \\
& \mathbf{Dy} \geq \mathbf{b}_2 \\
& \mathbf{x} \geq \mathbf{0}, \mathbf{y} = \mathbf{0} \text{ and integer}
\end{aligned}
$$

where $\mathbf{A}_1$ is $m \times n$, $\mathbf{A}_2$ is $m \times p$, $\mathbf{D}$ is $m' \times p$, $\mathbf{x}$ is $n \times 1$, $\mathbf{y}$ is $p \times 1$, $\mathbf{b}_1$ is $m \times 1$, $\mathbf{b}_2$ is $m' \times 1$, $\mathbf{c}_1$ is $1 \times n$, and $\mathbf{c}_2$ is $1 \times p$.

Benders' (Benders, 1962) partitioning approach applies to programming problems that involve groupings of either different types of variables (e.g., continuous and integer) or different types of constraints (e.g., linear and nonlinear). We are interested here in explaining how Benders' general theory of decomposition can be applied to MIPs. The integer variable $\mathbf{y}$ can be viewed as a "complicating variable"; with $\mathbf{y}$ fixed, MIP becomes a linear program that can be readily solved. Setting $\mathbf{y}$ to a specific integer value, say

$$
\mathbf{y} = \mathbf{y}' = \begin{bmatrix} y_1' \\ y_2' \\ \vdots \\ y_p' \end{bmatrix}
$$

the dual of the remaining LP is

$$
\begin{aligned}
(\text{DLP1}) \quad & \max \mathbf{u}(\mathbf{b}_1 - \mathbf{A}_2\mathbf{y}') \\
& \text{subject to} \quad \mathbf{uA}_1 \leq \mathbf{c}_1 \\
& \qquad\qquad\quad \mathbf{u} \geq \mathbf{0}
\end{aligned}
$$

Next, adjoin a constraint $\mathbf{uE} \leq M$ to DLP1, where $\mathbf{E}$ is an $m \times 1$ vector of 1s and M is an appropriately large positive integer. This minor adjustment to DLP1 (hence to MIP) results in a revised dual problem

$$(\text{DLP2}) \quad \max \ \mathbf{u}(\mathbf{b}_1 - \mathbf{A}_2 \mathbf{y}')$$
$$\text{subject to} \quad \mathbf{uA}_1 \leq \mathbf{c}_1$$
$$\mathbf{uE} \leq M$$
$$\mathbf{u} \geq \mathbf{0}$$

Properties of DLP1 and DLP2 are as follows:

- If DLP1 is bounded, then DLP1 and DLP2 are equivalent LPs ($\mathbf{uE} \leq M$ is redundant, for M large enough).
- DLP1 and DLP2 are functions of $\mathbf{y}$.
- If for a given value of $\mathbf{y}$, DLP1 is unbounded, then DLP2 will be bounded and the constraint $\mathbf{uE} \leq M$ will be binding.
- Thus, if we solve DLP2 and $\mathbf{uE} \leq M$ is binding (slack variable $= 0$), then we conclude DLP1 is unbounded.
- The feasible regions of both problems are independent of $\mathbf{y}$, so regardless of $\mathbf{y}$, the optimal solution of DLP2 is a vertex of the feasible region they share.
- The dual variable $\mathbf{u}$ is $1 \times m$, so all these vertices are in $\mathbf{R}_+^m$; say $\{u^1, u^2, \ldots, u^T\}$.

Benders (1962) derived the following pure integer program that is equivalent to MIP:

$$(\text{IP}) \quad \min z$$
$$\text{subject to} \quad z \geq \mathbf{c}_2 \mathbf{y} + u^i(\mathbf{b}_1 - \mathbf{A}_2 \mathbf{y}) \quad i = 1, \ldots, T$$
$$\mathbf{Dy} \geq \mathbf{b}_2$$
$$\mathbf{y} \geq \mathbf{0} \text{ and integer}$$

This problem has T constraints (one for each of the vertices of DLP2) in addition to the m' constraints expressed in $\mathbf{Dy} \geq \mathbf{b}_2$.

The number of vertices (T) can be very large, so instead of solving IP with all T constraints, IP is relaxed to start with only one constraint. One successively generates vertex-related constraints (cuts) for the pure IP by alternately solving DLP2 (to obtain a particular u^i) and the relaxed pure problem. This is why the algorithm is referred to as a primal–dual algorithm.

Figure 14.5 presents the basic version of *Benders' partitioning algorithm* for MIPs, adapted from McDaniel and Devine (1977); these researchers made modifications to create the Figure 14.4 version to gain convergence in fewer iterations. Garfinkel and Nemhauser (1972) presented a more general version of Benders' decomposition for MIPs; many versions of Benders' decomposition for MIPs with special structure have appeared since the 1970s.

Each successive solution z^t of the relaxation of IP results in a tighter lower bound on z_{MIP}, that is, the sequence $\{z^t\}$ is monotonically increasing. At iteration t, the best

Step 0 (Initialization). Set $t = 1$, $B_{\mathbf{u}} = +\infty$ and select some ε (convergence criterion). Select some $\mathbf{u}^1$ that is feasible for DLP2.

Step 1 (Iteration t). Solve the relaxed pure integer program

(IP(t))   min z subject to

$$z \geq \mathbf{c}_2 \mathbf{y} + \mathbf{u}^i (\mathbf{b}_1 - \mathbf{A}_2 \mathbf{y}) \qquad i = 1, \dots, t$$

$$\mathbf{Dy} \geq \mathbf{b}_2$$

$$\mathbf{y} \geq \mathbf{0} \text{ and integer}$$

Let z' and $\mathbf{y}'$ be the solution. If z is unbounded from below, take $\mathbf{y}'$ to be some value that gives z' some arbitrarily large negative value.

Step 2. Generate the most violated constraint of IP by solving the linear program

(DLP2(t))   max: $u_0 = \mathbf{u}(\mathbf{b}_1 - \mathbf{A}_2 \mathbf{y}')$ subject to

$$\mathbf{uA}_1 \leq \mathbf{c}_1$$

$$\mathbf{uE} \leq M$$

$$\mathbf{u} \geq \mathbf{0}$$

Let the solution of this LP be optimal value u_0^{t+1} at $\mathbf{u}^{t+1}$.

Step 3. Check convergence criterion. Set $B_{\mathbf{u}} = \min\{B_{\mathbf{u}}, u_0^{t+1} + \mathbf{c}_1 \mathbf{y}'\}$. If $z' > B_{\mathbf{u}} - \varepsilon$, stop; the optimal solution has been reached. Otherwise, add the constraint $z \geq \mathbf{c}_2 \mathbf{y} + \mathbf{u}^{t+1} (\mathbf{b}_1 - \mathbf{A}_2 \mathbf{y})$ to IP(t). Set $t = t + 1$. Return to step 1.

FIGURE 14.5   Benders' algorithm for MIP.[3]

upper bound on z_{MIP} is given by

$$B_{\mathbf{u}} = \min\{u_0^{i+1} + \mathbf{c}_1 \mathbf{y}^i\}, \quad i = 1, \dots, t$$

[3] Reprinted by permission of authors (see Bibliography). Copyright (1977), the Institute for Operations Research and Management Sciences, 7240 Parkway Drive, Suite 300, Hanover, MD 21076.

14.7   NOTES

Section 14.3

A comprehensive reference on local search heuristics is Walser (2008).

There is a huge literature on tabu search. For instance, for MPS files and test results comparing Pedroso's tabu search approach with branch-and-bound, see the entire chapter in Pedroso (2006). A problem-specific application of tabu search appears in Rolland et al. (1997). Glover and Laguna (1997) include a chapter on tabu search in integer programming, and another on tabu search applications—many of which are MIPs. A recent edited collection by Rego and Alidaee (2005) contains chapters on advances for solving classical problems, in addition to the chapter by Pedroso mentioned above.

Nemhauser and Wolsey (1988) state that "the efficiency of simulated annealing depends on its neighborhood structure. For some combinatorial optimization problems, such as the traveling salesman problem, simulated annealing has found much better solutions than those obtained by a random-start interchange algorithm."

The standard reference on genetic algorithms remains (Goldberg, 1989). Nieminen (2001) reports the development of a genetic algorithm customized to find the (first) feasible solution of an MIP to start the branch-and-bound algorithm. His strategy is quite similar to Pedroso's from tabu search in that he fixed the integer values in the MIP and solves the remaining problem as an LP. If the solution is infeasible, its fitness is the sum of the infeasibilities; if the solution is feasible, the fitness is the optimal value of the LP. This value is then used to determine a new generation of genomes (individuals) made up of vectors of genes (integer values for y_i that satisfy the integer constraints). Hua and Huang (2006) report a variable grouping-based genetic (VGGA) algorithm for large-scale integer programs. The MIP's LP relaxation is solved first, then the integer variables are grouped and a standard genetic algorithm is applied to the subproblem of each group. VGGA uses variable grouping to reduce the dimensionality of the genetic algorithm's search space.

Section 14.5

The presentation here is derived from Nemhauser and Wolsey (1988).

Section 14.6

Benders' decomposition is discussed in Nemhauser and Wolsey (1988, Section II.3.7). An application of Benders' decomposition to ship scheduling appears in Scott (1995). An alternative decomposition approach (the Dantzig–Wolfe approach) was presented as the basis for branch-and-price in our Chapter 13.

14.8   EXERCISES

14.1 Apply the "greedy" and "nearest (unserved)-neighbor" heuristics to the following five-city symmetric traveling salesman problem:

$$
C = \begin{pmatrix}
- & 6 & 2 & 1 & 11 \\
6 & - & 9 & 22 & 10 \\
2 & 9 & - & 30 & 17 \\
1 & 22 & 30 & - & 25 \\
11 & 10 & 17 & 25 & -
\end{pmatrix}
$$

Are the routes the same, as in Example 14.1?

14.2 (Shapiro, 2001)[4]. The depot described in Section 14.3.1 is faced with their new demands for local delivery:

Customer:	1	2	3	4	5	6	7	8	9	10	11	12	13
Tons:	5	4	2	7	3	5	4	4	6	5	3	2	6

(a) Apply the heuristic described in Section 14.3.1 to determine a feasible routing solution for the new demands; all other factors of the vehicle routing problem remain the same. In particular, apply the heuristic twice by changing the rule for selecting the first customer in a route. First, use the heuristic to select the first customer on each route to be the unserved customer that is nearest to the depot. Second, use the heuristic to select the first customer on each route to be the unserved customer that is farthest from the depot.

(b) Apply the genetic programming algorithm outlined in Figure 14.3 to perform crossover operations on the two feasible routing solutions found in part (a) using the probability function $P(x) = 1.953e^{-0.04463x}$, where x is the cost per delivered ton of a route. Use random numbers $1 \le r \le 99$ from a calculator or a random number table and compare your offspring with those of your classmates. If any customers are not visited in your crossover solution, apply a heuristic as in Figure 14.3.

14.3 (Bazaraa et al., 1990)[5]. Consider the problem: Minimize $x_1 + 2x_2$ subject to $3x_1 + x_2 \ge 6$, $-x_1 + x_2 \le 2$, $x_1 + x_2 \le 8$, and $x_1, x_2 \ge 0$. Let $X = \{\mathbf{x}: -x_1 \ x_2 \le 2, x_1 + x_2 \le 8, x_1, x_2 \ge 0\}$.

(a) Formulate the Lagrangian dual problem.

(b) Show that $f(w) = 6w + \text{minimum}\{0, 4 - 2w, 13 - 14w, 8 - 24w\}$. (*Hint*: Examine the second term in $f(w)$ and enumerate the extreme points of X graphically.)

(c) Plot $f(w)$ for each value of w.

(d) From part (c) locate the optimal solution to the Lagrangian dual problem.

(e) From part (d) find the optimal solution to the primal problem.

---

[4] From Shapiro, *Modeling the Supply Chain*, 1E Copyright 2001 South-Western, a part of Centage Learning, Inc. Reproduced with permission, www.centage.com/permissions.
[5] From Bazaraa, Jarvis, Sherali, *Linear Programming and Network Flows*, 2nd ed. Copyright 1990 John Wiley & Sons, Inc. Reprinted with permission of John Wiley & Sons, Inc.

14.4   Consider Benders' reformulation IP of the MIP given in Section 14.6. Now, consider if the MIP had been the maximization problem

$$\text{Max } \mathbf{c}_1\mathbf{x} + \mathbf{c}_2\mathbf{y}$$
$$\text{subject to} \quad \mathbf{A}_1\mathbf{x} + \mathbf{A}_2\mathbf{y} \le \mathbf{b}_1$$
$$\mathbf{D}\mathbf{y} \le \mathbf{b}_2$$
$$\mathbf{x} \ge \mathbf{0}, \mathbf{y} \ge \mathbf{0} \text{ and integer}$$

(a) Write out Benders' reformulation of this MIP.

(b) Write out explicitly the Benders' reformulation of the mixed integer program.

$$\text{Max} 14x_1 + 10x_2 + 4y_1 + 2y_2 + 6y_3$$
$$35x_1 + 24x_2 + 9y_1 + 4y_2 + 14y_3 \le 80$$
$$-2x_1 + 4x_2 - y_1 - 2y_2 + 3y_3 \le 10$$
$$\mathbf{x} \ge \mathbf{0}, \mathbf{y} \ge \mathbf{0} \text{ and integer}$$

15

SOLUTIONS WITH COMMERCIAL SOFTWARE

This final chapter (a) provides some practical considerations when algorithms are implemented as software, (b) describes the key components and features of a typical software system to model and solve integer programming problems, and (c) introduces three commonly used modeling languages (AMPL®, LINGO®, MPL®) and solvers in more depth than earlier chapters. AMPL is from AMPL Optimization LLC, LINGO® is from LINDO Systems, Inc., and MPL is from Maximal Software, Inc.

The purpose of this chapter is to introduce the reader to components of software systems one might encounter, or be asked to implement, working as an operations research analyst or programming specialist. Such implementations may require repetitive solution of a model with different input data and often involve embedding of the model and/or solver in an application (e.g., an inventory control system). There are numerous options available; this chapter should help prepare the reader for such responsibilities. It also enables the reader to understand that most of the topics of earlier chapters, such as preprocessing, branch-and-cut, and primal heuristics, have been implemented in commercial software, some as default settings and others as options the user can control.

Of necessity, in describing the software components and their role in the system, specific examples make the explanation more meaningful. It is not the authors' intention to recommend any particular modeling language or solver for linear and integer programming. Nor is it our intention to be comprehensive in the descriptions we do provide; that is the job of the developers of the many commercial modeling languages and solvers currently available and well indexed in the most recent survey

Applied Integer Programming: Modeling and Solution, By Der-San Chen, Robert G. Batson, and Yu Dang
Copyright © 2010 John Wiley & Sons, Inc.

in a long series published by INFORMS (Fourer, 2007). Each software product in turn has a Web site where the reader can typically find detailed descriptions of the product, tutorials, free trial downloads, bundling options for modeling languages with solvers, contacts for licenses/pricing, and other information.

15.1 INTRODUCTION

In practice, after a linear program (LP) or mixed integer program (MIP) is formulated, some computer software package (e.g., the CPLEX® solver) is typically used to solve the problem. CPLEX is from ILOG®, an IBM® company. Hence, an input mechanism is needed to translate the mathematical/algebraic description of the problem into a format that the software recognizes. Such input mechanisms are often referred to as modeling tools. Fourer et al. (2003) call this the problem of translation from "modeler's form" to "algorithm's form," the latter referring to the simplex algorithm and the simplex-based branch-and-bound and branch-and-cut methods, found in commercial solvers. Common modeling tools for LP or MIP problems were developed chronologically, and they fall into three categories: (1) MPS format files, (2) LP-format files, and (3) algebraic modeling languages.

The MPS format (Murtagh, 1981) was originally developed at IBM in the early 1960s and is widely used in both academia and industry. An MPS format file is a column-oriented (i.e., input fields must fit within prespecified columns) text file in which there are sections specifying various components of an LP or MIP problem. The MPS format is a legacy from the mainframe era and is not as flexible as an LP-format or algebraic modeling language. It is very difficult, if not entirely impossible, to manually write an MPS file for a large-scale LP or MIP problem; one would always resort to a software tool to generate an MPS file for his or her problem. "Even though it is lengthy and rather cryptic to the human eye, the MPS format became a standard for specifying and exchanging mathematical programming problems, and it is still supported by modern commercial mathematical programming systems" (Atamturk and Savelsbergh, 2005).

There are several variants of LP-formats, for example, CPLEX LP-format and the LINDO equivalent, and each LP-format provider has documentation describing components of an LP or MIP problem. The primary difference between the MPS format and an LP-format is that whereas in the MPS format only the objective function coefficients, constraint coefficients, and right-hand side elements are specified, in an LP-format the objective function and constraints are explicitly written in algebraic forms. Hence, an LP-format is more readable than the MPS format. To write a large-scale problem in an LP-format, one still needs assistance of a software tool.

In linear and integer programming, the major conversion challenge has been producing a compact representation of the constraint matrix. In the early 1980s, the programs written and commercialized to handle this task were called matrix generators, but these proved to be difficult to debug and maintain. Algebraic modeling languages evolved in the mid-to-late 1980s as an alternative to matrix generators, enabling the direct linkage from "modeler's form" to "algorithms (solver) form."

These algebraic modeling languages provide "computer readable equivalents" of notation used in the algebraic expression of LP and MIP models. An algebraic modeling language overcomes the deficiencies of an LP-format by introducing sets, symbolic constants, indexed variables, indexed constraints, aggregate operators such as summations, and other logical and flow control expressions. Furthermore, most modeling languages separate the model and the data. Such separation is also important to the maintainability of the models (e.g., model documentation, ease of reuse). As described in Chapter 1, much more powerful solvers were being developed in parallel over the past two decades. In fact, one solver developer has reported a 2360-fold speedup due to software improvements and an additional 800-fold speedup due to advances in hardware, 1988–2002 (Atamturk and Savelsbergh, 2005).

15.2   TYPICAL IP SOFTWARE COMPONENTS

In this section, we briefly describe the typical software components that make up a software system to solve integer programming problems. The intent is to simply familiarize the reader with these components. However, solution methods from previous chapters that form the basis of certain components are identified and much more extensive software references are provided. Also, we attempted to include illustration of the implementation of these components in the descriptions of leading commercial software that follows. The discussion will progress from the solution "engine" or solver, back through modeling languages and option control, to the user who is typically sitting at a PC as the input/output device.

15.2.1   Solvers

An LP or MIP model actually describes an infinite number of possible problems. An "instance" of the problem is when specific data are assigned to the model, and an optimal solution is sought. A solver is a software program that accepts the instance as input, applies one or more of the solution techniques we described in Chapters 9–14, and returns information about the optimal solution. For example, in solving an LP, the information might be that the problem is unbounded, or it might be a unique optimal vector with its optimal value, along with the dual values of the constraints. In solving a MIP, the information would be related to the branch-and-bound method, the basis of all state-of-the-art MIP solvers: for instance, how many nodes were explored, how close is the "best solution found" to optimal, and so on. Of course, any MIP solver is dependent on the speed of the LP solver that it repeatedly invokes; another feature that determined MIP solver performance is the so-called "branching control" to separate and select subproblems in the enumeration scheme.

In addition, "today's MIP codes have become increasingly complex with the incorporation of sophisticated algorithmic components . . . the behavior of the branch-and-bound algorithm can be altered significantly by changing the parameter settings that control these components. Through extensive experimentation, integer program-ming software vendors have determined default settings that work well in most instances encountered in practice" (Atamturk and Savelsbergh, 2005). In particular, branch-and-cut (Chapter 12), branch-and-price (Chapter 13), and primal heuristics (Section 14.3) have been applied by default in MIP solvers. As described in Chapters 12 and 14, cutting plane generators, using a variety of available cuts, are used to tighten the upper bound on a maximizing objective; heuristics may be used to find feasible MIP solutions and to tighten lower bounds. For example, Rothberg (2003) reported that CPLEX using the following cuts we described in Chapter 12: knapsack cover cuts, clique cuts, flow cover cuts, GUB cuts, Gomory mixed integer cuts, mixed integer rounding (MIR) cuts, flow path cuts, and disjunctive cuts. Yet another feature of solvers is their implementation of preprocessing (Chapter 4), in what is called a presolver, discussed next.

15.2.2   Presolvers

Presolve is a feature found in both solvers and algebraic modeling languages; virtually all commercial software products of these two classes included a presolver. In Chapter 4, we systematically detailed the standard techniques of preprocessing the problem instance prior to solvers attempt to find an optimal solution. Techniques for tightening bounds on variables and preprocessing a pure binary integer program (BIP) were given, along with examples of their effectiveness in creating a "better formulation," meaning a formulation that is easier to solve. All these techniques guarantee that the integer optimal solution of the original problem has not been eliminated from the feasible region of the reformulation. The presolver is automatically invoked and attempts to fix or eliminate variables, tighten bounds on variables, and tighten constraints by modifying coefficients or reformulating. Special preprocessing techniques are applied in the case of BIP. If problem infeasibility is detected by the presolver, the user is notified and the solver is not invoked.

MIP solvers use a variety of preprocessing techniques at the root node of the branch-and-bound tree; that is, before the first LP relaxation is solved. Preprocessing is also applied at each subsequent call to the LP solver as branch-and-bound proceeds. All this activity goes on behind the scenes and is not reported to the user. Solvers return solution information to the user through the algebraic modeling language in terms of the original model formulation. However, user options concerning preprocessing are provided in most solvers. We discuss this briefly in Section 15.2.4. For example, Rothberg (2003) reported that CPLEX's presolver used the following preprocessing approaches described in Chapter 4: tightening bounds by rounding a fractional bound, rounding by division with GCD, variable fixing, inactive constraints, and redundant constraints. To review AMPL's presolver, see (Fourer et al., 2003, Section 14.1).

15.2.3   Modeling Languages

In practice, an analyst or operations research team (Batson, 1987) would decide that the appropriate mathematical formulation of real problem is an MIP and proceed to use the conventions and notation found in this book. As the initial data to describe this model instance are located in various databases, the user quickly realizes that the problem is too large to write out completely and that the data are quite extensive and will probably change many times as subsequent problem instances are explored. At one time, the analyst would have had to develop an MPS description of the problem. An algebraic modeling language is software designed to efficiently

- Formulate the model
- Access the data and move each datum into proper location as a model parameter
- Communicate the problem instance to a solver
- Communicate solver results back to the user
- Document the model that was solved
- Handle subsequent modifications to the model and/or updates to the data.

In summary, a modeling language provides efficient documentation, two-way communication between user and solver, and reformulation—helping the analyst to manage the model and the associated data. We shall discuss the communication aspects in more detail in Section 15.2.5. Because of the close relationship between the modeling language and the solver in use, they are often acquired as bundled products from the respective sources. Solver providers may "offer integrated systems that provide a modeling environment specifically for their own solver" (Fourer, 2007). Features of three popular modeling languages (AMPL, LINGO®, and MPL) and how they are used with a solver are discussed in subsequent sections in this chapter.

15.2.4   User's Options/Intervention

Assuming the user is working in a given modeling language, he/she may have more than one solver to choose from on the organization's network. So, the most basic option arises when an analyst indicates which solver is to be invoked. More generally, all modeling languages and solvers provide users with options.

In a modeling language, a command interface is provided that will respond user commands expressed in text or through a graphical user interface. Standard commands related to solutions and sensitivity information desired are conveyed in this manner, along with options the user wants to enable or disable. For instance, the LINGO® solver enables the user to control the preprocessing operation by changing the value of the parameter *prelevel* to turn on or off presolver options such as simple presolve, variable fixing, coefficient reduction, elimination of variables, and elimination of constraints. Or, the user may "turn off" the modeling language's presolver and pass responsibility for presolve to the preprocessing routine of the chosen solver.

User options in the solver enable the user to change certain parameter settings from their default values, either because the user has the skill to identify and communicate special features in the structure of the problem instance, or because the default settings of the solver did not give acceptable results on the first attempt at problem solution. These options may be accessible through the graphical user interface of the modeling software, once the solver is called. A large portion of the paper by Atamturk and

Savelsbergh (2005) is devoted to user control of the following options in three state-of-the-art solvers:

- *Node selection* in branch-and-bound, expressed by the focus given to decrease the global upper bound or to increase the global lower bound. Typically, four or five search options are provided to the user and he/she can choose only one, or a hybrid.

- *Branching* in branch-and-bound: Choosing integer variable whose value at a currently active node is noninteger, for the next branching into two new nodes. Simple branching was described in Section 11.1, based on rounding the noninteger value up or down; a number of other advanced options, such as strong branching, branching based on pseudo-costs, or pseudo-reduced costs, are often available as alternatives.

- *Cutting planes* to provide improved LP bounds: There are two categories of cuts implemented in MIP solvers, general cuts and strong special cuts such as knapsack and fixed-charge flow, all discussed in Chapter 12. Solvers provide users with parameters that control which cuts are enabled and how aggressive the solver should be in "looking for cuts."

- *Preprocessing* prior to solving the relaxed LP at a node: All the preprocessing techniques described in Chapter 4 are typically available in commercial solvers and may be turned on or off by the user.

- *Primal heuristics* attempt to find feasible solutions to an MIP because such solutions automatically provide a lower bound at that node in the branch-and-bound tree. A good lower bound enables the tree to be pruned, and the search reduced. Heuristics of the three types described in Section 14.3 may be options, as well as heuristics developed by the solver provider.

Overriding the default settings in modeling languages and solvers can make the integer program at hand more tractable, but the user must develop skills and insights over time in order to choose wisely from among available options and learn how to carefully monitor the solution process and duration to decide when to revert to default on some options. Such skills and insights can only be developed with experience.

15.2.5 Data and Application Interfaces

Modeling languages must be capable of reading data in standard spreadsheet and database formats. In fact, a main purpose of the modeling language is to retrieve data from such structured data sources, on command, and generate a matrix that the solver can use. Virtually all modeling languages and solvers can also handle model instances expressed in simple text formats, especially the MPS and LP-formats discussed at the start of this section. For the LINGO® modeling language, we provide a detailed listing of database management systems and programming languages that interface with LINGO® in Section 15.4. Similar capabilities to those described are provided in any commercial modeling language.

There are other web-based interfaces that provide users access to the organization's modeling languages and solvers over network connections. Application program interfaces (APIs) have been developed in such languages as C++ and Java for calling each of the modeling languages and solvers mentioned in this chapter. These APIs enable modeling languages and/or solvers to be embedded in customized applications and provide the programmer with solution query methods and routines to access information about the results of applying an optimization method to a problem object. For instance, for MIP problem objects, the solver LINDO's API provides access to values of variables and constraint slacks; methods and routines are provided to retrieve other information about the optimization process, for example, the number of nodes searched, the objective value of the best remaining node, and so on. All such commercial solvers provide API libraries, and these enable the application program to interface directly with an embedded solver.

Embedding of solvers in applications has a longer history than embedding of models developed in modeling systems. Fourer (2007) has noted that it is now possible to embed an entire modeling system, or a particular model, or an instance of a model.

15.3 THE AMPL MODELING LANGUAGE

The AMPL® algebraic modeling language (Fourer et al., 2003) attains a very high level of readability in that a model written in AMPL resembles the algebraic notation in which one would formulate or describe an LP or MIP problem. An important feature is that AMPL facilitates separation of a model structure and its data. Fourer et al. (2003) state that "the separation of model and data is the key to describing more complex linear programs in a concise and understandable fashion." This also enables one to run the same model with different input data and then compare or analyze the results. AMPL has many other features that assist a user in efficiently building an LP or MIP model. The text by Fourer et al. serves as a user's guide and reference manual. In Section 15.3.1 we will describe the basic components of the AMPL modeling language. Then we will introduce several useful modeling techniques of AMPL through examples in Section 15.3.2.

15.3.1 Components of the AMPL Modeling Language

For every LP or MIP problem, the following are the essential parts of the problem description:

- Decision variables
- Objective function
- Constraints
- Variable bounds, which can be in the forms of nonnegativity constraints, unrestricted variables (i.e., variables whose upper and lower bounds are ∞ and $-\infty$), or variables with finite upper or lower bounds
- Integrality requirements on decision variables

The above are specified by the following basic components in AMPL:

- Sets, used to index symbolic constants and variables
- Parameters, that is, symbolic constants as input data
- Variables
- Objective to be minimized or maximized
- Constraints

15.3.2   An AMPL Example: the Diet Problem

We will use a diet problem in Winston (1994) to illustrate these basic components of AMPL. For the purpose of illustration, we have modified the original data so that the optimal solution to the LP problem takes fractional values:

$$
\begin{aligned}
\text{Minimize} \quad & z = 50x_1 + 20x_2 + 30x_3 + 80x_4 \\
\text{subject to} \quad & 400x_1 + 200x_2 + 150x_3 + 500x_4 \geq 500 \\
& 3.2x_1 + 2.5x_2 \geq 6 \\
& 2x_1 + 2x_2 + 4x_3 + 4x_4 \geq 10 \\
& 1.8x_1 + 4.5x_2 + x_3 + 5.6x_4 \geq 8 \\
& x_1, x_2, x_3, x_4 \geq 0
\end{aligned}
$$

Setting

$$
c = \begin{bmatrix} c_1 \\ c_2 \\ c_3 \\ c_4 \end{bmatrix} = \begin{bmatrix} 50 \\ 20 \\ 30 \\ 80 \end{bmatrix}, \quad
x = \begin{bmatrix} x_1 \\ x_2 \\ x_3 \\ x_4 \end{bmatrix}, \quad
b = \begin{bmatrix} b_1 \\ b_2 \\ b_3 \\ b_4 \end{bmatrix} = \begin{bmatrix} 500 \\ 6 \\ 10 \\ 8 \end{bmatrix}, \quad \text{and}
$$

$$
A = \begin{bmatrix}
a_{11} & a_{12} & a_{13} & a_{14} \\
a_{21} & a_{22} & a_{23} & a_{24} \\
a_{31} & a_{32} & a_{33} & a_{34} \\
a_{41} & a_{42} & a_{43} & a_{44}
\end{bmatrix} = \begin{bmatrix}
400 & 200 & 150 & 500 \\
3.2 & 2.5 & 0 & 0 \\
2 & 2 & 4 & 4 \\
1.8 & 4.5 & 1 & 5.6
\end{bmatrix}
$$

we have the following equivalent model:

$$
[\text{LP}] \quad \text{minimize} \quad z = \sum_{j=1}^{4} c_j x_j
$$

$$
\text{subject to} \quad \sum_{j=1}^{4} a_{ij} x_j \geq b_i \quad (i = 1, \ldots, 4)
$$

$$
x_j \geq 0 \qquad (j = 1, \ldots, 4)
$$

To index constants or input data c_j, a_{ij}, b_i, $i = 1, \ldots, 4$, $j = 1, \ldots, 4$, and decision variables x_j, $j = 1, \ldots, 4$, we need to define sets $I = \{1, 2, 3, 4\}$ and $J = \{1, 2, 3, 4\}$. This is done by the AMPL statements

```
Set I := 1..4;
Set J := 1..4;
```

With *I* and *J* defined as above, we can declare the constants by the `param` statements

```
param c {J};
param a {I, J};
param b {I};
```

The decision variables and nonnegativity constraints are specified by

```
var x {J} >= 0;
```

Now we can set up the objective function and constraints by the following statements:

```
minimize z: sum {j in J} c[j]*x[j];
subject to con {i in I}: sum {j in J} a[i, j]*x[j] >= b[i];
```

Up to this point, we have set up the complete model structure for the diet problem. Before we can solve the problem, we need to input data into matrix *A* and vectors *b* and *c*. This is accomplished by the following section in the AMPL model, beginning with a `data` statement:

```
data;

param c :=
  1  50
  2  20
  3  30
  4  80;

param a:
        1     2     3     4 :=
  1   400   200   150   500
  2   3.2   2.5     0     0
  3     2     2     4     4
  4   1.8   4.5     1   5.6;

param b :=
  1  500
  2    6
  3   10
  4    8;
```

Putting all the above together, we have the complete AMPL model shown in Figure 15.1.

```
set I := 1..4;
set J := 1..4;
param c {J};
param a {I, J};
param b {I};
var x {J} >= 0;

minimize z: sum {j in J} c[j]*x[j];
subject to con {i in I}: sum {j in J} a[i, j]*x[j] >= b[i];

data;

param c :=
  1   50
  2   20
  3   30
  4   80;

param a:
      1    2    3    4 :=
1  400  200  150  500
2  3.2  2.5    0    0
3    2    2    4    4
4  1.8  4.5    1  5.6;

param b :=
  1  500
  2    6
  3   10
  4    8;
```

FIGURE 15.1   The AMPL model of the diet problem.

After we save the AMPL model to a file named, for example, diet.mod, we can load the model by issuing the following command at an AMPL prompt:

ampl: **model** diet.mod;

We can also examine the model we have entered by an expand command:

```
ampl: expand;
minimize z:
  50*x[1] + 20*x[2] + 30*x[3] + 80*x[4];

subject to con[1]:
  400*x[1] + 200*x[2] + 150*x[3] + 500*x[4] >= 500;

subject to con[2]:
  3.2*x[1] + 2.5*x[2] >= 6;

subject to con[3]:
  2*x[1] + 2*x[2] + 4*x[3] + 4*x[4] >= 10;
```

```
subject to con[4]:
  1.8*x[1] + 4.5*x[2] + x[3] + 5.6*x[4] >= 8;
```

Suppose we would allow only integral values for the decision variables x in the diet problem. Hence, we have the following MIP problem:

$$[\text{MIP}] \quad \text{minimize} \quad z = \sum_{j=1}^{4} c_j x_j$$

$$\text{subject to} \quad \sum_{j=1}^{4} a_{ij} x_j \geq b_i, \qquad i = 1, \ldots, 4$$

$$x_j \geq 0 \text{ and integer}, \quad j = 1, \ldots, 4$$

To model this problem in AMPL, all we need to do is to replace the `var` statements in Figure 15.1 by the following:

```
var x {J} integer >= 0;
```

and we have created the MILP model in Figure 15.2.

```
set I := 1..4;
set J := 1..4;
param c {J};
param a {I, J};
param b {I};
var x {J} integer >= 0;

minimize z: sum {j in J} c[j]*x[j];
subject to con {i in I}: sum {j in J} a[i, j]*x[j] >= b[i];

data;

param c :=
  1   50
  2   20
  3   30
  4   80;

param a:
     1    2    3    4  :=
1  400  200  150  500
2  3.2  2.5    0    0
3    2    2    4    4
4  1.8  4.5    1  5.6;

param b :=
  1  500
  2    6
  3   10
  4    8;
```

FIGURE 15.2   The MILP model of the diet problem.

Now we have covered the basics in building an LP or MILP model in AMPL. In Section 15.3.3, we will describe some enhanced modeling techniques.

15.3.3   Enhanced AMPL Modeling Techniques

In this section, we will illustrate the following modeling enhancement techniques with AMPL:

1. Separation of the model structure and input data
2. Adding or deleting a constraint
3. Relaxing integrality constraints on some variables

It is often the case that after we have built a model, we would like to run the same model with different sets of input data. The AMPL modeling language enables us to do that through separating the model structure from the input data. Take the diet problem in Figure 15.1 for example. We can slightly modify the statements in the file diet.mod and split them into two files, for example, diet1.mod and diet1a.dat shown in Figures 15.3 and 15.4.

Note the difference between the model in diet1.mod and diet1a.dat combined and that in diet.mod is that the sizes of sets I and J are determined in the data file after the following two statements are executed:

```
param M:= 4;
param N:= 4;
```

Other than the above, the statements in the two models are identical. To load the model structure and input data in AMPL, issue the following commands:

```
ampl: model diet1.mod;
ampl: data diet1a.dat;
```

As we have already mentioned, the advantage of separating the model structure from the input data is that we can feed the same model with different input data. Suppose we have a similar LP problem with five constraints and six variables instead of four constraints and four variables as specified in diet1a.dat. We specify the objective coefficients, constraint coefficients, and right-hand side in a file named diet1b.dat as indicated in Figure 15.5.

```
param M;
param N;
set I := 1..M;
set J := 1..N;
param c {J};
param a {I, J};
param b {I};
var x {J} >= 0;

minimize z: sum {j in J} c[j]*x[j];
subject to con {i in I}: sum {j in J} a[i, j]*x[j] >= b[i];
```

FIGURE 15.3   The model structure of the diet problem—file diet1.mod.

```
data;

param M := 4;
param N := 4;

param c :=
  1    50
  2    20
  3    30
  4    80;

param a:
      1    2    3    4  :=
1 400  200  150  500
2 3.2  2.5    0    0
3   2    2    4    4
4 1.8  4.5    1  5.6;

param b :=
  1 500
  2   6
  3  10
  4   8;
```

FIGURE 15.4   The input data of the diet problem—file `diet1a.dat`.

```
data;

param M := 5;
param N := 6;

param c :=
  1    50
  2    20
  3    30
  4    80
  5    75
  6    90;

param a:
     1    2    3    4    5    6  :=
1 400  200  150  500  100  300
2 3.2  2.5    0    0    0    0
3   2    2    4    4    1    2
4 1.8  4.5    1  5.6  2.2  0.8
5   0   -3   -4    0    0    0;

param b :=
  1 500
  2   6
  3  10
  4   8
  5 -12;
```

FIGURE 15.5   The input data of the five constraint and six variable problem.

We can issue the following AMPL commands to load the new problem:

```
ampl: model diet1.mod;
ampl: data diet1b.dat;
```

There are various cases in which we need to add or delete a constraint, or even relax integer requirements on certain variables. For example, we would like to compare different production plans or service strategies in our study. Suppose we need to add the following constraint to the model in diet.mod:

$$3x_2 + 4x_3 \leq 3$$

We can issue the following in AMPL after we load diet.mod:

```
ampl: subject to new_con: 3*x[2] + 4*x[3] <= 3;
```

To verify that the new constraint is indeed added, we can use the expand command:

```
ampl: expand;
minimize z:
  50*x[1] + 20*x[2] + 30*x[3] + 80*x[4];

subject to con[1]:
  400*x[1] + 200*x[2] + 150*x[3] + 500*x[4] >= 500;

subject to con[2]:
  3.2*x[1] + 2.5*x[2] >= 6;

subject to con[3]:
  2*x[1] + 2*x[2] + 4*x[3] + 4*x[4] >= 10;

subject to con[4]:
  1.8*x[1] + 4.5*x[2] + x[3] + 5.6*x[4] >= 8;

subject to new_con:
  3*x[2] + 4*x[3] < = 3;
```

To delete the constraint new_con, we simply issue the following command:

```
ampl: delete new_con;
```

And we can verify the deletion by the expand command. Note that we cannot delete an individual constraint in a set of indexed constraints. For example, we cannot delete constraint con[3] only, although we can delete the entire set of constraints by the command

```
ampl: delete con;
```

If we want to temporarily drop a constraint but restore it later on, we use the `drop` and `restore` command instead:

```
ampl: drop con[3];
ampl: ...
ampl: restore con[3];
```

As we have seen, we can drop an individual constraint in a set of indexed constraints.

Finally, we show how to relax integer requirements on variables. Take the MILP problem in Figure 15.2 for example. If we would like to relax the integrality on all the variables, we can change option `relax_integrality` from the default value of zero to a nonzero value:

```
ampl: option relax_integrality 1;
```

To restore the integrality requirements, set option `relax_integrality` back to zero.

If we would like to relax the integer requirement only on a certain variable, we can do so by setting the `.relax` suffix of the variable to 1. For example,

```
ampl: let x[2].relax:= 1;
```

relaxes integrality on variable `x[2]` only.

15.3.4   AMPL Compatible MIP Solvers

A number of linear and MIP solvers have been identified in Fourer and Gay (2006) as supported by AMPL, or see the AMPL Web site www.ampl.com. Detailed descriptions of each of these solvers, and information on their respective providers and Web sites, can be found in Fourer (2007). The CPLEX solution of a diet problem and other standard LP/MIPs modeled in AMPL can be found in Fourer et al. (2003).

15.4   LINGO® MODELING LANGUAGE

LINGO® is a Fortran-based optimization tool designed by LINDO Systems, Inc., first offered in 1988. According to LINDO Systems, LINGO® was their "first product to include a full featured modeling language." In 1993, LINGO® added the first nonlinear solver for PCs that can support general and binary integer restrictions. In 1994, LINGO® was included in Winston (1994), making itself the first IP software to appear in a textbook. The release of Windows version LINGO® came into the market in 1995. LINDO Systems Inc. also markets a solver called LINDO (first appearing in 1979) and a spreadsheet add-in optimizer What's Best® (since 1985).

According to the statement of LINDO System, Inc., "LINDO Systems products are in use at over half the Fortune 500 companies—including 23 of the top 25." The latest trial versions of all three of these software packages may be downloaded from the Web site www.lindo.com. The LINDO Systems company Web site also provides application papers that cover many business and industry fields and, of course, reference manuals for the software. A comprehensive guide to optimization modeling using LINGO® is Schrage (2003). The user's guide and reference manual is LINDO Systems (2004). Below, we first provide some technical details concerning LINGO®, and then briefly describe its modeling conventions. Note that LINGO® contains built-in documentation.

15.4.1 Prescription of Tolerances

When solving optimization problems, many calculations of multiplication and division are involved. This will surely affect the accuracy of the solution due to the limited precision of the computer. Hence, it is very important to set tolerances for the solver. Tolerance is the $+/-$ range within which a value can be viewed as the target value. The major types of tolerances involved in LP/IP solvers include

- Feasibility Tolerances (for LP Models)

    The feasibility tolerance allows that when the basic variables fall "close" enough to the right-hand side values, the constraints are considered satisfied. This tolerance directly affects solver's decision on whether to accept an optimal basis. When it is very hard to maintain problem feasibility during the optimization procedure, these tolerances can be set to lower values.

    LINGO® controls this tolerance in two positions: the initial feasibility tolerance and the final feasibility tolerance. The default values for these tolerances are, respectively, $3e{-}06$ and $1e{-}07$.

- Integrality Tolerances

    For some IP problems, it is hard to obtain a solution that is exactly integer. In such cases, integrality tolerances are employed so that when the solution is "close enough" to some integer number, it is accepted as an integer solution.

    LINGO® defines two types of integrality tolerances: absolute integrality tolerance and relative integrality tolerance. Assume x_z is an integer value, and x_r is the (real) solution obtained. x_r is accepted as the target solution if $|x_r - x_z| \leq$ absolute integrality tolerance, or if $|x_r - x_z| \leq$ relative integrality tolerance $^* x_z$. Default value for the absolute integrality tolerance and relative integrality tolerance are, respectively, $1e{-}06$ and $8e{-}6$.

- Optimality Tolerances

    The optimality tolerance decides how closely a solution must be to the true (theoretical) optimal solution to be "considered" optimal.

Similar to the integrality tolerances, LINGO® uses two parameters to control optimality tolerances: absolute optimality tolerance and relative optimality tolerance. The *absolute optimality tolerance* is a positive number r with default value of $8e{-}8$. When applied in the branch-and-bound solver, this tolerance forces the solver to always search for integer solutions that result in at least r units of improvement, comparing to the best integer solution found so far. The *relative optimality tolerance* is a positive fraction r with default value of $5e-8$. It is used to force the branch-and-bound solver to search only integer solutions with at least $100^*r\%$ improvement. According to the LINGO® User's Guide (2004), typical values for the relative optimality tolerance would be in the range 0.01–0.05.

15.4.2   Presolver—Automatic Problem Reduction

As introduced in Chapter 4, preprocessing and model reformulation are very important for solving IP problems. LINGO® incorporates this method as an essential feature. Presolver enables users to decide which preprocessing technique(s) to apply before the solver actually starts the IP algorithms.

LINGO® presolver includes preprocessing techniques as well as all the cuts and other IP algorithms like lattice approach. Here, we only list the preprocessing techniques:

- G(general) C(common) D(divider)
- Coefficient Reduction
- Disaggregation

Note that LINGO® allows users to turn off the presolver because in some cases preprocessing might cause excessive running time.

15.4.3   Solvers for Linear/Integer Programming

A unique feature of LINGO® is that all solvers (linear, integer, nonlinear, quadratic, etc.) are integrated and directly linked to its modeling environment. When a model is run, LINGO® will automatically pass the problem to the appropriate solver. Hence, LINGO® is capable of solving a wide variety of optimization problems, including linear programming, integer programming (binary, pure, and mixed), and nonlinear programming problems.

The solvers for LP problems employ three approaches: primal simplex, dual simplex, and barrier (or interior point approach). The third approach listed is not addressed in this text, but see Hillier and Lieberman (2005) for a discussion. LINGO® allows users to select which approach to use for a specific problem. The default is set to allow solvers automatically decide the best method.

LINGO® employs branch-and-cut as the major IP algorithm, which starts by solving the LP relaxation. Types of cuts to be generated can be selected by users. These types include

- Flow Cover
- Gomory
- GUB

- Knapsack cover
- Objective
- Plant location

The new *K*-best MIP solver enables the user to see multiple best solutions to the given MIP problem, where the number *K* can be specified by users.

15.4.4 Interfacing with the User

As this text is published, Microsoft Windows continues to be the most popular platform for software users. Although LINGO® supports both Windows and UNIX platforms, we only discuss interfacing with other applications on the Windows platform.

Interfacing with programming languages: LINGO® uses the Dynamic Link Library to allow users to hook up LINGO® functions with external applications. The most recent version of LINGO® supports access from Visual Basic, Visual C/C++, Delphi, Fortran, C#, .NET, and Visual Java.

Interfacing with databases: LINGO® supports connections with data sources such as spreadsheet, text files, and databases. For small or moderate data volume, LINGO® can link to spreadsheets such as Microsoft Excel® or FoxPro® through object linking and embedding (OLE), which is a built-in function of LINGO®. Solutions can be output to spreadsheet using @OLE, as well.

For large data volume, LINGO® has a built-in connection function named @ODBC that helps link to database management systems (DBMS) that have open database connectivity. Such DBMSs include

- Microsoft Access®
- dBase (DB/2)
- PeopleSoft Oracle®
- Paradox
- SQL Server

The most recent release of LINGO® (version 11.0) incorporated some functions such as @TEXT and @POINTER, which allow users to build links for both importing and exporting data.

15.4.5 LINGO® Modeling Conventions

LINGO® uses *sets* at its fundamental building block. Each member of a set may have one or more *attributes* associated with it, such as in a product mix application, the product may have a profit, a monthly demand, and so on. Selected attributes are the decision variables in the optimization model. Variables are assumed to be nonnegative unless the statement @FREE() is invoked. Variables may be specified to be binary or general integer using statements @BIN or @GIN, respectively.

As with MPL (see Section 15.5), the @SUM looping operator is used to specify an objective function in a compact form. The general form is @SUM (set: expression), for example, @SUM (Product (j): Profit (j) * ProductCount (j)). The @FOR operator is another set looping function used to generate constraints over members of a set: @FOR (set: constraint). Below is an example of how @FOR and @SUM are used to specify a series of resource constraints:

```
@FOR (Machine (i):
    @SUM (Product (j): ProdHours Used (i,j) * ProductCount (j))
        < = ProdHours Limit (i);
    );
```

The above LINGO® statement illustrates two aspects of modeling in LINGO®: (1) The use of a derived set based on two or more simple sets (machine, product) = (i, j) and (2) the scalability of the model—once constructed, data sets representing say a new quarter of demand, resource availability, and so on could be input and run simply by updating the DATA statements, leaving the model structure as it is.

A LINGO® model specification consists of three sections:

1. A SETS section that specifies the sets and their attributes. This describes the problem parameters or data structure.
2. A DATA section that provides the data direction in vectors or matrices, or specifies where it will be accessed (e.g., in certain cells of a spreadsheet).
3. A section that provides the mathematical model, often surprisingly compact.

15.4.6 LINGO® Model for the Diet Problem

Here, we will provide the LINGO® model for the

```
MODEL:
    SETS:
        FOOD/1, 2, 3,4/: X, COST;
        REQUIREMENT/1, 2, 3, 4/: RQMT, MIN;
    ENDSETS:
    DATA:
        COST = 50, 20, 30, 80;
        MIN = 500, 6, 10, 8;
        RQMTPROV =
            400   200   150   500
            3.2   2.5     0     0
              2     2     4     4
            1.8   4.5     1   5.6;
```

```
ENDDATA
MIN = @SUM (COST(J)*X(J))
@FOR (RQMT(I):
        @SUM (X(J): RQMTPROV(I,J)*X(J))
                > = MIN (I);
    );
```

The LINDO solution of a wide variety of LP/MIP problems can be found in Schrage (2000).

15.5   MPL MODELING LANGUAGE

MPL®, which stands for Mathematical Programming Language, is a product of Maximal Software, Inc. MPL is another algebraic modeling language, such as AMPL and LINGO® described earlier. MPL can be used with many proprietary or open source solvers (consult www.maximalsoftware.com for the most recent table or see the last paragraph of this section). Student versions of MPL and CPLEX are available for free by download from this Web site. A leading introductory operations research text (Hillier and Lieberman, 2005) contains examples of solving linear and integer programming problems in MPL/CPLEX files.

On its Web site, Maximal Software states, "The size of problems that corporations are dealing with has increased and the speed of commercial optimization packages (solvers) has risen dramatically to meet that demand. This means that users need more advanced tools to collect and manage the data, formulate the model, and deliver it to the solver. This is where an advanced modeling system, such as MPL, can become very valuable." Some features of MPL that lend it to teaching modeling skills include

- Takes full advantage of the graphical user interface of MS Windows
- An easy-to-learn syntax
- Powerful data management capabilities (for interface with spreadsheets and databases)
- An online tutorial

The seven sessions currently available in the online tutorial include vectors and indexes, separating data from the model; special types of constraints, and various data management issues. A session on mixed integer programming is planned. The interested reader should check for availability.

MPL for Windows is the most popular platform, but there are versions for UNIX environments as well. MPL has operators to facilitate importing files (data or indexes) from spreadsheets, databases, or external files. MPL also can export results back to such locations, and MPL models can be embedded into other Windows applications that Maximal claims "makes MPL ideal for creating end-user applications." Supported databases include

- Access®
- ODBC
- Paradox
- FoxPro
- Dbase for the Windows version
- Oracle® for the Motif (UNIX) version.

The MPL Main Window provides access to the following 11 menus:

1. Main Menu
2. File Menu
3. Edit Menu
4. Search Menu
5. Project Menu
6. Run Menu
7. View Menu
8. Graph Menu
9. Options Menu
10. Window Menu
11. Help Menu

MPL provides extensive compatibility with Excel, for those whose transactional data have been aggregated into what Shapiro (2001) calls "decision databases," which are databases that are suitable for MIP modeling. A recent enhancement of MPL (the OptiMax 2000 Component Library) enables the modeler to provide the end-user with the appearance of a spreadsheet model while still using MPL as the behind-the-scenes modeling language and permitting selection of the appropriate solver. This can be important because of the well-known limitations of the built-in solver in Excel, such as problem size limitations, speed of data importation and computing, the fact that the optimization model itself is hidden, and Excel's solver lacks the advanced indexing techniques available in modeling languages. Furthermore, Maximal states "OptiMax allows MPL models to be seamlessly integrated directly into object-oriented programming languages such as Visual Basic, VBA for Excel and Access, C/C++ , Java, Delphi, as well as many popular web-scripting languages."

15.5.1 MPL Modeling Conventions

Some notable features of building optimization models in MPL are

- MPL can dynamically store models of any size like LINGO®; the only limitation is how much memory is available on the machine

- Variables and constants can be written on both sides of a constraint—the so-called free format input of constraints, which means no conversion to standard form is required of the modeler
- Summation over vector variables, of up to eight dimensions
- Expansion of similarly structured constraints; a single line enables you to express multiple constraints of identical form, such as monthly inventory balance you want repeated for each month in a planning horizon.
- Extensive flexibility when working with subsets of indexes, functions of indexes, and multidimensional index sets.

A typical MPL model would begin with a TITLE declaration, and then an INDEX section. Each entity (product, plant, machine, etc.) is assigned arbitrary labels that in turn are used in data files, using colon, parentheses, and semicolon much as in other modeling languages, for example,

```
plant : = (p1, p2, p3, p4, p5);
```

The next section of an MPL model is DATA in which line by line the model is told where data in the form of vectors or matrices will be found and what attributes will be used to index such data; for example, to find the monthly demand for each product at each plant, one would specify

```
Demand [plant, product, month] : = SPARSEFILE
                                  ("Demand.dat");
```

indicating that demand is being stored in a matrix in sparse format, where only nonzero values (and an identification of their index values) are entered in the data file. MPL can access data that are either specified in vectors or matrices typed with the model, or located in external to the model. The third section of an MPL model is VARIABLES, in which each decision variable is given a short name. Inside brackets, there is an indication of indexes associated with the variable, for example,

```
Inventory [plant, product, month] - > Invt;
```

where the arrow is used to create an abbreviation for variable names of more than four letters. Should one or more variables be restricted to either binary or integer, those variables are specified in separate sections labeled BINARY VARIABLES or INTEGER VARIABLES, respectively. The fourth section of an MPL model declaration is called MACROS, and there summations that will ultimately become part of the objective function are specified, using the SUM operator, for example,

```
TotalRevenue : = SUM (plant, product, month: Price*Sales);
TotalInvtCost : = SUM (plant, product, month:
                InvtCost*Inventory);
```

where just inside the first parentheses are the indexes over which the summation runs, and after the colon is the vector product of a data vector with a variable vector. Next, the actual mathematical model is specified using variables, macros, operators, and other mathematical logic, expressed in sections labeled MODEL (objective function) and SUBJECT TO (nonbound constraints), for example,

```
MODEL
MAX Profit = TotalRevenue - Total Cost;
SUBJECT TO
SUM (product: Inventory) < = InvtCapacity; and other applic-
able constraints.
```

Finally, the modeler sets any upper bounds on variables in a BOUNDS section and may declare variables to be free (real) valued, otherwise MPL assumes they are nonnegative. An END statement used after the model formulation is judged complete. Assuming one wanted to solve the model using CPLEX, you would choose *Solve CPLEX* from the Run Menu mentioned earlier and find the solution using the View Menu.

15.5.2 MPL Model for the Diet Problem

Here, we will provide the MPL model for the simple diet problem described in the AMPL section above. Note that BR symbolizes brownies eaten, IC scoops of ice cream eaten, COLA bottles of soda drunk, and PC pieces of pineapple cheesecake eaten daily:

```
INDEX
        FOOD : = (BR, IC, COLA, PC);
        REQUIREMENT : = (R1, R2, R3, R4);
DATA
        COST [FOOD] : = (50, 20, 30, 80);
        MIN [REQUIREMENT] : = (500, 6, 10, 8);
        RQMTPROV [REQUIRMENT, FOOD] =
                400  200  150  500
                3.2  2.5    0    0
                  2    2    4    4
                1.8  4.5    1  5.6;
VARIABLES
        EAT [FOOD];
MACROS
        TOTALCOST = SUM (FOOD; COST*EAT);
SUBJECT TO
        NUTRITIONVALUE [REQUIREMENT] - > NVAL:
                SUM (FOOD; RQMTPROV*EAT) > = MIN;
END
```

15.5.3 MPL Compatible MIP Solvers

The numerous linear and MIP solvers supported by MPL have been identified in a white paper at www.maximalsoftware.com. Detailed descriptions of each of these solvers, and information on their respective providers and Web sites, can be found in (Fourer, 2007).

REFERENCES

Agarwal, Y., K. Mathur, and H. Salkin (1989), "A Set-Partitioning-Based Exact Algorithm for the Vehicle Routing Problem", *Networks*, Vol. 19, pp. 731–749.

Agnihothri, R. and P. Taylor (1991), "Staffing a Centralized Appoint Scheduling Department in Lourdes Hospital", *Interfaces*, Vol. 21, pp. 1–15.

Ahuja, R., T. Magnanti, and J. Orlin (1993), *Network Flows: Theory, Algorithms, and Applications*, Prentice Hall, Englewood Cliffs, NJ.

Anbil, R., E. Gelman, B. Patty, and R. Tanga (1991), "Recent advances in Crew-Pairing Optimization at American Airline", *Interfaces*, Vol. 21, pp. 62–74.

Andrew, B. and H. Parsons (1993), "Establishing Telephone-Agent Staffing Levels through Economic Optimization", *Interfaces*, Vol. 23, pp. 14–20.

Aneja, Y. and H. Kamoun (1999), Scheduling of Parts and Robot Activities in a Two Machine Robot Cell", *Computer and Operations Research*, Vol. 26, pp. 297–312.

Applegate, D., R. Bixby, V. Chvátal, and W. Cook (2006), *The Traveling Salesman Problem: A Computational Study*, Princeton University Press, Princeton, NJ.

Applegate, D., R. Bixby, V. Chvátal, and W. Cook (2007), *Concorde*, available at www.tsp.gatech.edu.

Arabeyre, T., J. Fearnkey, F. Steiger, and W. Teather (1969), "The Airline Crew Scheduling Problem: A Survey", *Transportation Science*, Vol. 3, pp. 140–163.

Atamturk, A., G. Nemhauser, and M. Savelsbergh (1998), "Conflict Graphs in Integer Programming", Report LEC-98-03, Georgia Institute of Technology, Atlanta, GA.

Atamturk, A. and M. Savelsbergh (2005), "Integer-Programming Software Systems", *Annals of Operations Research*, Vol. 140, No. 1, pp. 67–124.

Ausiello, G., P. Crescenzi, G. Gambosi, V. Kann, A. Marchetti-Spaccamela, and P. Protasi (1999), *Complexity and Approximation*, Springer-Verlag, Berlin, Germany.

Aykin, T. (1996), "Optimal Shift Scheduling with Multiple Break Windows", *Management Science*, Vol. 42, pp. 591–602.

Baker, K. (1974), *Introduction to Sequencing and Scheduling*, John Wiley & Sons, New York, NY.

Baker, B. and M. Fisher (1981), "Computational Results for Very Large Air Crew Scheduling Problems", *Omega*, Vol. 9, pp. 613–618.

Balinski, M. and R. Quandt (1964), "On an Integer Program for a Delivery Problem", *Operations Research*, Vol. 12, pp. 300–304.

Balinski, M. (1965), "Integer Programming: Methods, Uses, Computation", *Management Science*, Vol. 12, pp. 253–313.

Ball, M., L. Bodin, and R. Dial (1983), "A Matching Based Heuristic for Scheduling Mass Transit Crews and Vehicles", *Transportation Science*, Vol. 17, pp. 4–31.

Bard, J., G. Kontoravdis, and G. Yu (2002), "A Branch-and-Cut Procedure for the Vehicle Routing Problem with Time Windows", *Transportation Science*, Vol. 36, pp. 250–269.

Bard, J. and S. Rojanasoonthon (2006), "A Branch-and-Price Algorithm for Parallel Machine Scheduling with Time Windows and Job Priorities", *Naval Research Logistics*, Vol. 53, pp. 24–44.

Barnhart, C., E. Johnson, G. Nemhauser, M. Savelsbergh, and P. Vance (1998), "Branch-and-Price: Column Generation for Solving Huge Integer Programs", *Operations Research*, Vol. 46, pp. 316–329.

Barnhart, C., C. Hane, and P. Vance (2000), "Using Branch-and-Bound and Price-and-Cut to Solve Origin–Destination Integer Multicommodity Flow Problems", *Operations Research*, Vol. 48, pp. 318–326.

Barvinok, A., E. Gimadi, and A. Serdyukov (2002), "The Maximum TSP", in *The Traveling Salesman Problem: A Guided Tour of Combinatorial Optimization*, E. Lawler, K. Lenstra, A. Rinoony Kan, and D. Shmoys (Eds), John Wiley & Sons, pp. 585–607.

Batson, R. (1979), Stability Theory for Mathematical Programming Problems with Unbounded Convex Feasible Regions: A Point-to-Set Submap Approach, Ph.D. Thesis, Department of Mathematics, The University of Alabama, Tuscaloosa, AL.

Batson, R. (1987), "The Modern Role of OR/MS Professionals in Interdisciplinary Teams", *Interfaces*, Vol. 17, No. 3, pp. 83–93.

Bazaraa, M., J. Jarvis, and H. Sherali (2005), *Linear Programming and Network Flows*, 3rd ed., John Wiley & Sons, New York, NY.

Beale, E. (1955), "Cycling in the Dual Simplex Algorithm", *Naval Research Logistics Quarterly*, Vol. 2, pp. 269–276.

Beasley, J. (Ed.) (1996), *Advances in Linear and Integer Programming*, Oxford University Press, Oxford, England.

Bellmore, M. and S. Hong (1974), "Transformation of the Multisalesman Problem to the Standard Traveling Salesman Problem", *Journal of the ACM*, Vol. 21, pp. 500–504.

Benders, J. (1962), "Partitioning Procedure for Solving Mixed Variable Programming Problems", *Numerishe Matematik*, Vol. 4, pp. 238–252.

Bianco, L., S. Ricciardelli, G. Rinaldi, and A. Sassano (1988), "Scheduling Tasks with Sequencing-Dependent Processing Times", *Naval Research Quarterly*, Vol. 35, pp. 177–184.

Bixby, R., M. Fenelon, Z. Gu, E. Rothberg, and R. Wunderling (1999), "MIP: Theory and Practice Closing the Gap", *Proceedings of the 19th IFIP TC7 Conference on System Modeling and Optimization*, Vol. 174, Kluwer B.V., Deventer, The Netherlands, pp. 19–50.

Bixby, R., M. Fenelon, Z. Gu, E. Rothberg, and R. Wunderling (2002), "One Size Fits All? Computational Tradeoffs in a Commercial Mixed Integer Programming Solver", presented at *The Institute for Mathematics and Its Application Workshop*, October 14–19, University of Minnesota, Twin Cities, MN.

Bland, R. (1977), "New Finite Pivoting Rules for the Simplex Method", *Mathematics of Operations Research*, Vol. 2, pp. 103–107.

Bland, R. and D. Shallcross (1989), "Large Traveling Salesman Problems Arising in X-Ray Crystallography: A Preliminary Report on Computation", *Operations Research Letters*, Vol. 8, No. 3, pp. 125–128.

Bolat, A., M. Savsar, and M. Al-Fawzan (1994), "Algorithms for Real-time Scheduling of Jobs on Mixed Model Assembly Lines", *Computer and Operations Research*, Vol. 21, No. 5, pp. 487–498.

Brearley, A., G. Mitra, and H. Williams (1975), "Analysis of Mathematical Programming Problem Prior to Applying the Simplex Algorithm", *Mathematical Programming*, Vol. 8, pp. 54–83.

Brown, G., G. Ellis, G. Graves, and D. Ronen (1987), "Real-Time Wide Area Dispatching of Mobil Tank Trucks", *Interfaces*, Vol. 17, pp. 107–120.

Cebry, M., A. DeSilva, and F. DiLisio (1992), "Management Science in Automating Postal Operations: Facility and Equipment Planning in the United States Postal Service", *Interfaces*, Vol. 22, pp. 110–130.

Cerny, V. (1985), "A Thermodynamical Approach to the Travelling Salesman Problem: An efficient Simulation Algorithm", *Journal of Optimization Theory and Applications*, Vol. 45, pp. 41–51.

Chatterjee, S., C. Carrera, and L. A. Lynch (1996), "Genetic Algorithms and Traveling Salesman Problems", *European Journal of Operational Research*, Vol. 93, pp. 490–510.

Chen, D. (1970), A Group Theoretic Algorithm for Solving Integer Linear Programming Problems, Ph.D. Thesis, Department of Industrial Engineering, State University of New York at Buffalo, Buffalo, NY.

Chen, D., and S. Zionts (1972) "An Exposition of the Group Theoretic Approach to Integer Linear Programming", *Opsearch*, pp. 75–102; also in Chapter 14, *Linear and Integer Programming*, S. Zionts (1974), Prentice-Hall, Englewood Cliffs, NJ.

Chen, D. and S. Zionts (1976), "Comparison of Some Algorithms for Solving the Group Theoretic Integer Programming Problem", *Operations Research*, Vol. 10, pp. 1120–1128.

Clarke, G., and S. Wright (1964), "Scheduling of Vehicles from a Central Depot to a Number of Deliver Points", *Operations Research*, Vol. 12, pp. 568–581.

Cordier, C., H. Marchand, R. Laundy, and L. Wolsey (1999), "*bc-opt*: a branch-and-cut code for mixed integer programs", *Mathematical Programming*, Series A, Vol. 86, pp. 335–353.

Cox, D., M. Burmeister, E. Price, S. Kim, R. Myers (1990), "Radiation Hybrid Mapping: A Somatic Cell Genetic Method for Constructing High Resolution Maps of Mammalian Chromosomes", *Science*, Vol. 250, pp. 245–250.

Croes, G. (1958), "A Method of Solving Travelling Salesman Problems", *Operations Research*, Vol. 6, pp. 791–812.

Crowder, H., E. Johnson, and M. Padberg (1983), "Solving Large-Scale Zero-One Linear Programming Problems", *Operations Research*, Vol. 31, pp. 803–834.

Dakin, R. (1965), "A Tree Search Algorithm for Mixed Integer Programming Problems", *Computer Journal*, Vol. 8, pp. 250–255.

Danna, E., E. Rothberg, and C. LePape (2003), "Integrating Mixed Integer Programming and Local Search: A Case Study on Job-Shop Scheduling Problems", *Proceedings CPAIOR'03*.

Danna, E. (2004), "Integrating Local Search Techniques into Mixed Integer Programming", *40R*, Vol. 2, pp. 321–324.

Dantzig, G. (1963), *Linear Programming and Extensions*, Princeton University Press, Princeton, NJ.

Dantzig, G., D. Fulkerson, and S. Johnson (1954), "Solution of a Large-scale Traveling Salesman Problem", *Operations Research*, Vol. 2, pp. 393–410.

Dantzig, G. and J. Ramser (1960), "The Truck Dispatching Problem", *Management Science*, Vol. 6, pp. 331–344.

Dantzig, G. and P. Wolfe (1960), "Decomposition Principle for Linear Programs", *Operations Research*, Vol. 8, pp. 101–111.

Danusaputro, S., C. Lee, and L. Martin-Vega (1990), "An Efficient Algorithm for Drilling Printed Circuit Boards", *Computers and Industrial Engineering*, Vol. 18, pp. 145–151.

Day, R. (1965), "On Optimal Extracting from a Multiple File Data Storage System: An Application of Integer Programming", *Operations Research*, Vol. 13, pp. 482–494.

Desrochers, M., J. Desrosiers, and M. Solomon (1992), "A New Optimization Algorithm for the Vehicle Problem with Time Window", *Operations Research*, Vol. 40, pp. 342–354.

Dyckhoff, H. (1981), "A New Linear Programming Approach to the Cutting Stock Problem", *Operations Research*, Vol. 29, pp. 1092–1104.

Eaton, D., M. Daskin, D. Simmons, B. Bulloch, and G. Jansma (1985), "Determining Emergency Medical Service Vehicle Deployment in Austin, Texas", *Interfaces*, Vol. 15, pp. 96–108.

Edmonds, J, and E. Johnson (1973), "Matching, Euler Tours and the Chinese Postman", *Mathematical Programming*, Vol. 5, pp. 88–124.

Elshafei, A. (1977), "Hospital Lay-out as a Quadratic Assignment Problem", *Operational Research Quarterly*, Vol. 28, pp. 167–169.

Farkas, J. (1902), "Über die Theorie der Einfachen Ungleichungen", *Journal für die Reine und Angewandte Mathematik*, Vol. 124, pp. 1–27.

Farley, A. (1990), "A Note on Bounding a Class of Linear Programming Problems, Including Cutting Stock Problems", *Operations Research*, Vol. 38 pp. 922–923.

Fortet, R., L'algebre de Boole et ses (1959), "Applications en Recherche Operationnelle", *Cashiers du Centre d'Etudes de Recherche Operationnelle*, 1, pp. 5–36.

Fourer, R. (2007), "Linear Programming Software Survey", *OR/MS Today*, Vol. 34, No. 3, pp. 42–51.

Fourer, R. and D. Gay (2006), "AMPL: New Solver Support in the AMPL Modeling Language," *PowerPoint Presentation at the INFORMS Annual Meeting*, Pittsburgh, PA, November 5–8, www.ampl.com/INFORMS06.pdf.

Fourer, R., D. Gay, and B. Kernighan (2003), *AMPL: A Modeling Language for Mathematical Programming*, 2nd ed., Brooks/Cole—Thompson Learning, Pacific Grove, CA.

Gale, D., H. Kuhn, and A. Tucker (1951), "Linear Programming and the Theory of Games", Chapter 19 in *Activity Analysis of Production and Allocation*, T. C. Koopmans (Ed.), John Wiley & Sons, New York, NY.

Garfinkel, R. (1985), "Motivation and Modeling", in *The Traveling Salesman Problem: A Guide Tour of Combinatorial Optimization*, E. Lawler, K. Lenstra, A. Rinoony Kan, and D. Shmoys (Eds) John Wiley & Sons, Chichester, UK, pp. 17–36.

Garfinkel, R. and G. Nemhauser (1972), *Integer Programming*, John Wiley & Sons, New York, NY.

Geoffrion, A.M. (1974), "Lagrangean Relaxation for Integer Programming", *Mathematical Programming Study*, Vol, 2, pp. 82–114.

Gilmore, P. and R. Gomory (1961), "A Linear Programming Approach to the Cutting Stock Problem", *Operations Research*, Vol. 9, pp. 849–859.

Gilmore, P. and R. Gomory (1963), "A Linear Programming Approach to the Cutting Stock Problem—Part II", *Operations Research*, Vol. 11, pp. 863–888.

Gilmore, P. and R. Gomory (1964), "Sequencing a One State-Variable Machine: A Solvable Case of the Traveling Salesman Problem", *Operations Research*, Vol. 12, pp. 655–679.

Glover, F. and E. Woolsey (1973), "Further Reduction of Zero-One Polynomial Programming Problems to Zero-One Linear Programming Problems", *Operations Research*, Vol. 21, pp. 156–161.

Glover, F. and E. Woolsey (1974), "Converting 0–1 Polynomial Programming Problem to a 0–1 Linear Program", *Operations Research*, Vol. 22, pp. 180–182.

Glover, F. (1986), "Future Paths for Integer Programming and Links to Artificial Intelligence", *Computers and Operations Research*, Vol. 13, pp. 533–549.

Glover, F. (1989), "Tabu Search: Part I", *ORSA Journal on Computing*, Vol. 1, pp. 190–206.

Glover, F. (1990), "Tabu Search: Part II", *ORSA Journal on Computing*, Vol. 2, pp. 4–32.

Glover, F. and M. Laguna (1997), *Tabu Search*, Kluwer Academic Publishers, Boston, MA.

Goldberg, D. (1989), *Genetic Algorithms in Search, Optimization, and Machine Learning*, Addison-Wesley, Reading, MA.

Goldman, A. and A. Tucker (1956), "Theory of Linear Programming", article 4 in *Linear Inequalities and Related Systems*, H. W. Kuhn and W. Tucker (Eds), Princeton University Press, Princeton, NJ.

Gomory, R. (1958, 1963), "An Algorithm for Integer Solutions to Linear Programs", Princeton IBM Mathematical Research Report; also in *Recent Advances in Mathematical Programming* Graves and Wolfe (Eds), McGraw-Hill, New York, NY.

Gomory, R. (1960), *An Algorithm for the Mixed Integer Problem*, RAND Corp., P-1885, Santa Monica, CA.

Gomory, R. (1960, 1963), All Integer Programming Algorithms, IBM Research Center Report RC-189; also in *Industrial Scheduling*, Muth and Thompson (Eds), Prentice-Hall, Englewood Cliffs, NJ.

Gomory, R. (1965), "On the Relation between Integer and Non-Integer Solution to Linear Programs", *Proceedings of National Academy of Science of the United States of America*, Vol. 53, pp. 260–265.

Gorry, A., and J. Shapiro (1972), "An Adaptive Group Theoretic Algorithm for Integer Programming Problems", *Management Science,* Vol. 17, pp. 229–239.

Grotschel, M. and O. Holland (1991), "Solution of Large-Scale Symmetric Traveling Salesman Problem", *Mathematical Programming*, Vol. 51, pp. 141–202.

Grotschel, M., M. Junger, and G. Reinelt (1985), "Facets of the Linear Ordering Polytope", *Mathematical Programming*, Vol. 33, pp. 43–60.

Grotschel, M. and M. Padberg (1985), "Polyhedral Theory", in *The Traveling Salesman Problem: A Guided Tour of Combinatorial Optimization*, E. Lawler, K. Lenstra, A. Rinoony Kan, and D. Shmoys (Eds), John Wiley & Sons, pp. 251–306.

Grünbaum, B. (1967), *Convex Polytopes*, John Wiley & Sons, New York, NY.

Gu, Z., G. Nemhauser, and M. Savelsbergh (1998), "Lifted Cover Inequalities for 0–1 Integer Programs", *INFORMS Journal on Computing*, Vol. 10, pp. 417–426.

Gu, Z., G. Nemhauser, and M. Savelsbergh (1999), "Lifted Flow Covers for Mixed 0–1 Integer Programs", *Mathematical Programming*, Vol. 85, pp. 439–467.

Guéret, C., C. Prins, and M. Sevaus (2002), *Applications of Optimization with Xpress-MP*, Dash Optimization, London, England.

Gutin, G. and A. Punnen (2002), *The Traveling Salesman Problem and Its Variations*, Kluwer Academic Publishers, Boston, MA.

Haessler, R. (1983), "Developing an Industrial-Grade Heuristic Problem-Solving Procedure", *Interfaces*, Vol. 13, No. 3, pp. 62–71.

Hansen, P., B. Jaumard, and V. Mathon (1993), "Converting the 0–1 Polynomial Programming Problem to 1 0–1 Linear Program", *Operations Research*, 22 (1), pp. 180–182.

Hanson, W. and R. Martin (1990), "Optimal Bundle Pricing", *Management Science*, Vol. 36, No. 2, pp. 155–174.

Hastings, W. (1970) "Monte Carlo Sampling Methods Using Markov Chains and Their Applications", *Biometrika*, Vol. 57, pp. 97–109.

Hillier, F. (1983), "Heuristics: A Gambler's Roll", *Interfaces*, Vol. 13, No. 3, pp. 9–12.

Hillier, F. and G. Lieberman (2005), *Introduction to Operations Research*, 8th ed., McGraw-Hill, New York, NY.

Hoffman, K. and M. Padberg (1991), "Improving Representations of Zero-One Linear Programs for Branch-and-Cut", *ORSA Journal of Computing*, Vol. 3, pp. 121–134.

Hoffman, K. and M. Padberg (1993) "Solving Airline Crew Scheduling by Branch-and-Cut", *Management Science*, Vol. 39, pp. 657–682.

Holland, J. (1975), *Adaptation in Natural and Artificial Systems*, University of Michigan Press, Ann Arbor, MI.

Hong, S. and M. Padberg (1977), "A Note on the Symmetric Multiple Traveling Salesman Problem with Fixed Charges", *Operations Research*, Vol. 25, pp. 871–874.

Hu, T. (1968), "On the Asymptotic Integer Algorithm", MRC Report 946, University of Wisconsin, Madison, WI.

Hu, T. (1969), *Integer Programming and Network Flow*, Addison-Wesley, Reading, MA.

Hua, Z. and F. Huang (2006), "A Variable-Grouping based Genetic Algorithm for Large Scale Integer Programming", *Information Sciences*, Vol. 176, pp. 2869–2885.

Ignizio, J. and T. Cavalier (1994), *Linear Programming*, Prentice-Hall, Englewood Cliffs, NJ.

Jeroslow, R., K. Martin, R. Rardin, and J. Wang (1992), "Gainfree Leontief Substitution Flow Problems", *Mathematical Programming*, Vol. 57, pp. 375–414.

Johnson, D., C. Aragon, L. McGeoch, C. Schevon (1989) "Optimizing by Simulated Annealing: An Experimental Evaluation. Part I: Graph Partitioning," *Operations Research*, Vol. 37, pp. 865–892.

Johnson, D., C. Aragon, L. McGeoch, C. Schevon (1991) "Optimization by Simulated Annealing: An Experiment Evaluation. Part II: Graph Coloring and Number Partitioning", *Operations Research*, Vol. 39, pp. 378–395.

Johnson, E., G. Nemhauser, and M. Savelsbergh (2000), "Progress in Linear Programming-Based Algorithms for Integer Programming: An Exposition", *INFORMS Journal on Computing*, Vol. 12, No. 1.

Johnson, E., M. Kostreva, and U. Suhl (1985), "Solving 0–1 Integer Programming Problems Arising from Large Scale Planning Models", *Operations Research*, Vol. 33, pp. 803–819.

Johnson, E. and M. Padberg (1983), "Degree-two Inequalities, Clique Facets, and Bipartite Graphs", *Annals of Discrete Mathematics*, Vol. 16, pp. 169–188.

Joines, J., C. Culbreth, and R. King (1996), "Manufacturing Cell Design: An Integer Programming Model Employing Genetic Algorithms", *IIE Transactions*, Vol. 28, No. 1, pp. 69–85.

Jongens, K. and T. Volgenant (1985), "The Symmetric Clustered Traveling Salesman Problem", *European Journal of Operational Research*, Vol. 19, pp. 68–75.

Karabakel N., A. Gunal, and W. Ritchie (2000), "Supply Chain Analysis at Volkswagen of America," *Interfaces*, Vol. 30, No. 4, pp. 46–55.

Katayama, K., H. Sakamoto, and H. Narihisa (2000), "The Efficiency of Hybrid Mutation Algorithm for the Travelling Salesman Problem", *Mathematical and Computer Modeling*, Vol. 31, pp. 197–203.

Kemeny, J., H. Mirkil, J. Snell, and G. Thompson (1959), *Finite Mathematical Structures*, Prentice-Hall, Englewood Cliffs, NJ.

Kirkpatrick, S., C. Gelatt, Jr., and M. Vecchi (1983), "Optimization by Simulated Annealing", *Science*, Vol. 220, pp. 671–680.

Kontogiorgis, S. and Acharya (1999), "US Airways Automates Its Weekend Fleet Assignment", *Interfaces*, Vol. 3, pp. 52–62.

Koopmans, T. and M. Beckmann (1957), "Assignment Problems and the Location of Economic Activities", *Econometrica*, Vol. 25, pp. 53–76.

Land, A. and A. Doig (1960), "An Automatic Method for Solving Discrete Programming Problems", *Econometrica*, Vol. 28, pp. 497–520.

Lander, E. (2001), "Initial Sequencing and Analysis of the Human genome", *Nature*, Vol. 409, pp. 860–921.

Lasky, J. (1969), Optimal Scheduling of Freight Trucking, M. S. Thesis, Massachusetts Institute of Technology.

Lawler, E., J. Lenstra, A. Rinnooy Kan, and D. Shmoys (Eds) (1985), *The Traveling Salesman Problem: A Guided Tour of Combinatorial Optimization*, John Wiley & Sons, Chichester, UK.

Lenstra, J. and A. Rinnooy Kan (1975), "Some Simple Applications of the Traveling Salesman Problem", *Operational Research Quarterly*, Vol. 26, pp. 717–733.

Li, L., D. Fonseca, and D. Chen (2006), "Earliness-Tardiness Production Planning for Just-in-Time Manufacturing: A Unifying Approach by Goal Programming", *European Journal of Operational Research*, Vol. 175, pp. 508–515.

Lin, S. and B. Kernighan (1973), "An Effective Heuristic Algorithm for the Traveling Salesman Problem", *Operations Research*, Vol. 21, pp. 498–516.

LINDO Systems Inc. (2004), *LINGO: The Modeling Language and Optimizer*, LINDO Systems Inc., Chicago, IL.

Marchand, H., A. Martin, R. Weismantel, L. Wolsey (2002), "Cutting Planes in Integer and Mixed Integer Programming", *Discrete Applied Mathematics*, Vol. 123, pp. 397–446.

Martin, R. (1999), *Large Scale Linear and Integer Optimization: A Unified Approach*, Kluwer Academic Publishers, Boston, MA.

Maximal Software (undated), "Developing Large-Scale Optimization Models with the MPL Modeling System," White Paper, Maximal Software, Inc., www.maximalsoftware.com.

McCloskey, J. and F. Hanssmann (1957), "An Analysis of Stewardess Requirements and Scheduling for a Major Airline", *Naval Research Logistic Quarterly*, Vol. 4, pp. 183–192.

McDaniel, D. and M. Devine (1977), "A Modified Benders' Partitioning Algorithm for Mixed Integer Programming", *Management Science*, Vol. 24, No. 3, pp. 312–319.

Miller, C.E., A.W. Tucker, and R.A. Zemlin, "Integer Programming Formulation of Traveling Salesman Problems", *Journal of the ACM*, Vol, 7, No. 4, pp. 326–329

Miller, H., W. Pierskalla, and G. Rath (1976), "Nurse Scheduling Using Mathematical Programming", *Operations Research*, Vol. 24, pp. 857–870.

Miller, D. and J. Pekny (1991), "Exact Solution of Large Asymmetric Traveling Salesman Problems", *Science*, Vol. 251, pp. 754–761.

Minieka, E. (1978), *Optimization Algorithms for Networks and Graphs*, Marcel Dekker, New York, NY.

Morito, S. (1976), Integer Programming by Group Theory, Ph.D. Thesis, Case Western Reserve University.

Murtagh, B. (1981), *Advanced Linear Programming: Computation and Practice*, McGraw-Hill, New York.

Murty, K. G. (1992), *Network Programming*, Prentice Hall, Englewood Cliffs, NJ.

Nauss, R. (2006), "The Generalized Assignment Problem," Chapter 3 in *Integer Programming: Theory and Practice*, Taylor and Francis, Boca Raton, FL.

Nemhauser, G. and L. Wolsey (1988), *Integer and Combinatorial Optimization*, John Wiley & Sons, New York, NY.

Nieminen, K. (2001), Developing Mixed Integer Programming Methods, Master's Thesis, Systems Analysis Laboratory, Helsinki University of Technology, Helsinki, Finland.

Noon, C. (1988), The Generalized Traveling Salesman Problem, Ph.D. Thesis, Department of Industrial Engineering, University of Tennessee, Knoxville, TN.

Noon, C. and Bean (1993), "An Efficient Transformation of the Generalized Traveling Salesman Problem", *INFOR*, Vol. 31, pp. 39–44.

Orlin, J. (1982), "Minimizing the Number of Vehicles to Meet a Fixed Periodic Schedule: An Application of Periodic Posets", *Operations Research*, Vol. 24, pp. 760–776.

Padberg, M. and M. Grotschel (1985), "Polyhedral Computations", *The Traveling Salesman Problem: A Guided Tour of Combinatorial Optimization*, John Wiley & Sons, New York, NY.

Padberg, M. and G. Rinaldi (1987), "Optimization of a 532-City Symmetric Traveling Salesman Problem by Branch and Cut", *Operations Research Letters*, Vol. 6, No. 1.

Padberg, M. and G. Rinaldi (1991), "A Branch and Cut Algorithm for the Resolution of Large-Scale Symmetric Salesman Problems", *SIAM Review*, Vol. 33, pp. 60–100.

Padberg, M., T. Roy and L. Wolsey (1985), "Valid Linear Inequalities for Fixed Charge Problems", *Operations Research*, Vol. 33, pp. 842–861.

Parker, R. and R. Rardin (1988), *Discrete Optimization*, Academic Press, Orlando, FL.

Pedroso, J. (2006), "Tabu Search for Mixed Integer Programming", Chapter 11 in *Metaheuristic Optimization via memory and Evolution: Tabu Search and Scatter Search*, Springer, New York.

Phillips, D. and A. Garcia-Diaz (1981), *Fundamentals of Network Analysis*, Prentice-Hall, Englewood Cliffs, NJ.

Pinedo, M. (2002), *Scheduling: Theory, Algorithms, and Systems*, 2nd ed., Prentice-Hall, Englewood Cliffs, NJ.

Potvin, J. (1993), "The Traveling Salesman: A Neural Network Perspective", *INFORMS Journal on Computing*, Vol. 5, pp. 328–348.

Potvin, J. (1996), "Genetic Algorithms for the Traveling Salesman Problem", *Annals of Operations Research*, Vol. 63, pp. 339–370.

Ramirez-Beltran, N. and K. Aguilar-Ruggiero (1997), "Application of an Heuristic Procedure to Solve Mixed-Integer Programming Problems", *Computers and Industrial Engineering*, Vol. 33, Nos. 1–2, pp. 43–46.

Ratliff, H. and A. Rosenthal (1981), "Order-Picking in a Rectangular Warehouse: A Solvable Case for the Traveling Salesman Problem", PDRC Report Series No. 81-10, Georgia Institute of Technology, Atlanta, GA.

Ravindran, A., D. Phillips, and J. Solberg (1987), *Operations Research: Principles and Practice*, 2nd ed., John Wiley & Sons, New York, NY.

Rao, M. (1980), "A Note on the Multiple Traveling Salesman Problem", *Operations Research*, Vol. 28, pp. 628–632.

Rana, K. and R. Vickson (1998), "A Model and Solution Algorithm for Optimal Routing of a Time-Chartered Containership", *Transportation Science*, Vol. 22, No. 2, pp. 83–95.

Rayward-Smith, V., I. Osman, C. Reeves, and G. Smith (Eds) (1997), *Modern Heuristic Search Methods*, John Wiley & Sons, New York, NY.

Rego, C. and B. Alidaee (2005), *Metaheuristic Optimization via Memory and Evolution*, Kluwar Academic Publishers, Boston, MA.

Reinelt, G. (1991) "TSPLIB—A Traveling Salesman Library", *ORSA Journal in Computing*, Vol. 3, pp. 376–384.

Reinelt, G. (1994), *The Traveling Salesman: Computational Solutions for TSP Applications*, *Lecture Notes in Computer Science* 840, Springer-Verlag, Berlin, Germany.

Reinelt, G. (1991) "TSPLIB—A Traveling Salesman Library", *ORSA Journal in Computing*, Vol. 3, pp. 376–384.

Reinelt, G. (2007), TSPLIB, www.iwr.uni-heidelberg.de/groups/comopt/software/tsplib95/.

Reiter, S. and G. Sherman (1965), "Discrete Optimizing", *Journal of the Society of Industrial and Applied Mathematics*, Vol. 13, pp. 864–889.

Rockafellar, R. (1970), *Convex Analysis*, Princeton University Press, Princeton, NJ.

Rolland, E., D. A. Schilling, and J. R. Current (1997), "An Efficient Tabu Search Procedure for the P-Median Problem", *European Journal of Operational Research*, Vol. 96, pp. 329–342.

Rothberg, E. (2003), "The CPLEX Library: Presolve and Cutting Planes," www.mpi-inf.mpg.de/conferences/adfocs-03/Slides/Rothberg_4.pdf.

Rothstein, M. (1973), "Hospital Manpower Shift Scheduling by Mathematical Programming", *Health Services Research*, Vol. 8 pp. 60–66.

Salkin, H. and K. Mathur (1989), *Foundations of Integer Programming*, North-Holland, New York, NY.

Salveson, M. (1955), "The Assembly Line Balancing Problem", *Journal of Industrial Engineering*, Vol. 6, pp. 18–25.

Savelsbergh, M. (1994), "Preprocessing and Probing Techniques for Mixed Integer Programming Problems", *ORSA Journal on Computing*, Vol. 6, No. 4, pp. 445–454.

Savelsbergh, M. (1997), "A Branch-and-Price Algorithm for the Generalized Assignment Problem", *Operations Research*, Vol. 45, pp. 831–841.

Schindler, S. and T. Semmel (1993), "Station Staffing at Pan American World Airways", *Interfaces*, Vol. 23, pp. 91–106.

Schrage, L. (2003), *Optimization Modeling with LINGO*, 4th ed., LINDO Systems, Inc., Chicago, IL.

Schriver, A. (1986), *Theory of Linear and Integer Programming*, John Wiley & Sons, New York, NY.

Scott, J. (1995), "A Transportation Model, its Development and Application to a Ship Scheduling Problem", *Asia-Pacific Journal of Operations Research*, Vol. 12, No. 2, pp. 111–120.

Shapiro, J. (1968), "Dynamic Programming Algorithms for the Integer Programming Problem I: The Integer Programming Problem Viewed as a Knapsack Type Problem", *Operations Research*, Vol. 16, pp. 103–121.

Shapiro, J. (2001), *Modeling the Supply Chain*, Duxbury Press, Pacific Grove, CA.

Stigler, G. (1963), "United States V. Loew's, Inc.: A Note on Block Booking", *Supreme Court Review*, Vol. 152.

Taha, H. (1975), *Integer Programming: Theory, Applications, and Computations*, Academic Press, Orlando, FL.

Taha, H. (2007), *Operations Research: An Introduction*, 8th ed., Pearson Prentice-Hall, Upper Saddle River, NJ.

Taylor, P. and S. Huxley (1989), "A Break from Tradition for the San Francisco Police: Patrol Officer Scheduling Using an Optimization-Based Decision Support Tool", *Interfaces*, Vol. 19, pp. 4–24.

Thirez, H. (1968), Airline Crew Scheduling: A Group theoretic Approach, Ph.D. Thesis, Massachusetts Institute of Technology.

Vance, P., C. Barnhart, E. Johnson, and G. Nemhauser (1997), "Airline Crew Scheduling: A New Formulation and Decomposition Algorithm", *Operations Research*, Vol. 45, pp. 188–220.

Vance, P., A. Atamturk, C. Barnhardt, E. Gelman, E. Johnson, A. Krishna, D. Mahidhara, G. Nemhauser, R. Rebello (1997), "A Heuristic Branch-and-Price Approach for the Airline Crew Pairing Problem," Tech Report, TLI/LEC-97-06, Georgia Institute of Technology, Atlanta, GA.

Vanderbeck, F. (2000), "On Dantzig–Wolfe Decomposition in Integer Programming and Ways to Perform Branching in a Branch-and-Price Algorithm", *Operations Research*, Vol. 48, pp. 111–128.

Van Roy, T. and L. Wolsey (1987), "Solving Mixed 0–1 Programs by Automatic Reformulation", *Operations Research*, Vol. 35, pp. 45–57.

Vasko, F., J. Wolfe, and K. Stott (1955), "Optimal Selection of Ingot Sizes via Set Covering", *Operations Research*, Vol. 35, pp. 346–353.

Venter, J. (2001), "The Sequence of the Human Genome", *Science*, Vo. 291, pp. 1304–1351.

Vollman, T., W. Berry, and D. Whybark (1988), *Manufacturing Planning and Control Systems*, 2nd ed., IRWIN, Homewood, IL.

Wagner, H. (1975), *Principles of Operations Research*, 2nd ed., Prentice-Hall, Englewood Cliffs, NJ.

Walser, J., R. Iyer, and N. Venkatsubramanyan (1998), "An Integer Local Search Method with Application To Capacitated Production Planning", *Proceedings of the 14th National Conference on Artificial Intelligence*, AAAI/IAAI, 373–379.

Walser, J. (2008), *Integer Optimization by Local Search: A Domain-Independent Approach*, Springer, New York.

Warner, D. (1976), "Scheduling Nursing Personnel According to Nursing Preference: A Mathematical Programming Approach", *Operations Research*, Vol. 24, no. 5, pp. 760–776.

Watters, L. (1967), "Reduction of Integer Polynomial Programming Problems to Zero-One Linar Programming Problems", *Operations Research*, Vol. 24, no. 5, pp. 1171–1174.

Weismantel, R. (1997) "On the 0/1 Knapsack Polytope", *Mathematical Programming*, Vol. 77, pp. 49–68.

Wilhelm, M. and T. Ward (1987), "Solving Quadratic Assignment Problems by Simulated Annealing", *IIE Transactions*, Vol. 19, No. 1, pp. 107–119.

Williams, H. (1993), *Model Solving in Mathematical Programming*, John Wiley & Sons, Chichester, UK.

Williams, H. (1993), "Logic Applied to Integer Programming and Integer Programming Applied to Logic", *European Journal of Operational Research*, Vol. 81, pp. 605–616.

Williams, H. (1999), *Model Building in Mathematical Programming*, 4th ed., John Wiley & Sons, Chichester, UK.

Williams, H. and S. Brailsford (1999), "Computational Logic and Integer Programming" in *Advances in Linear and Integer Programming*, J. Beasley (Ed.), Oxford University Press, Oxford, UK. pp. 249–281.

Williams, H. and J. Wilson (1998), "Connections Between Integer Linear Programming and Constraint Logic Programming—An Overview and Introduction to the Cluster of Articles", *INFORMS Journal on Computing*, Vol. 10, pp. 261–264.

Winston, W. (1994), *Operations Research: Applications and Algorithms*, 3rd Edition, Duxbury Press, Pacific Grove, CA.

Woolsey, R. (1972), "A Candle to Saint Jude, or Four Real World Applications of Integer Programming", *Interfaces*, Vol. 2, No. 2, pp. 20–27.

Wolsey, L. (1998), *Integer Programming*, John Wiley & Sons, New York, NY.

Wolsey, L. (1989), "Strong Formulations for Mixed Integer Programming: A Survey", *Mathematical Programming*, Vol. 45, pp. 173–191.

Yan, S. and C. Lin (1997), "Airline Scheduling for the Temporary Closure of Airports", *Transportation Science*, Vol. 31, pp. 72–82.

Zanakis, S. and J. Evans (1981), "Heuristic 'Optimization': Why, When, and How to use It", *Interfaces*, Vol. 11, No. 5, pp. 34–91.

Zionts, S. (1968), "On an Algorithm for the solution of Mixed Integer Programming Problems", *Management Science*, Vol. 15, pp. 113–116.

Zionts, S. (1969), "Toward a Unifying Theory of Integer Linear Programming", *Operations Research*, Vol. 17, pp. 404–410.

Zionts, S. (1974), *Linear and Integer Programming*, Prentice-Hall, Englewood Cliffs, NJ.

APPENDIX

ANSWERS TO SELECTED EXERCISES

CHAPTER 1

Problem 1.15

$$\text{Maximize} \quad -3x_1 + 11x_2 - 5x_3 - x_4$$
$$\text{s.t.} \quad x_1 + 5x_2 - 3x_3 + 6x_4 \leq 7$$
$$x_1 - x_2 - x_3 + 2x_4 \leq -3$$
$$x_1, x_2, x_3, x_4 \geq 0$$

Problem 1.16

$$\text{Let} \quad x_1 = x_1^+ - x_1^-$$
$$x_3' = x_3 - (-2) = x_3 + 2 \geq 0(x_3' - 2)$$

Rewrite constraint 2:

$$-x_2 - x_3 - x_4 \leq -13$$

Applied Integer Programming: Modeling and Solution, By Der-San Chen, Robert G. Batson, and Yu Dang
Copyright © 2010 John Wiley & Sons, Inc.

$$\text{LP}: \quad \text{Maximize}: \quad -(x_1^+ - x_1^-) + 5x_2 + 2(x_3' - 2) - 7x_4 - x_5$$
$$\text{s.t.} \quad -x_2 - (x_3' - 2) - x_4 \leq -13$$
$$(x_1^+ - x_1^-) - x_2 + 2x_4 + 2x_5 \leq 4$$
$$x_1^+, x_1^-, x_2, x_3', x_4, x_5 \geq 0$$

$$\text{Maximize}: \quad -x_1^+ + x_1^- + 5x_2 + 2x_3' - 7x_4 - x_5 - 4$$
$$\text{s.t.} \quad -x_2 - x_3' - x_4 \leq -15$$
$$x_1^+ - x_1^- - x_2 + 2x_4 + 2x_5 \leq 4$$
$$x_1^+, x_1^-, x_2, x_3', x_4, x_5 \geq 0$$

Problem 1.17

$$\text{Maximize} \quad 7x_1 + 2x_2 + x_3 - 4x_4$$
$$\text{s.t.} \quad 2x_1 - x_2 + x_3 \leq 10$$
$$x_1 + x_4 = 12$$
$$x_1, x_2, x_4 \geq 0$$
$$x_3 \leq 0$$

Let $x_3 = -x_3' \geq 0$ and rewrite constraint 2 as two $\leq$ inequalities:

$$\text{Maximize} \quad 7x_1 + 2x_2 - x_3' - 4x_4$$
$$\text{s.t.} \quad 2x_1 - x_2 - x_3' \leq 10$$
$$x_1 + x_4 \geq 12$$
$$-x_1 - x_4 \leq -12$$
$$x_1, x_2, x_3', x_4 \geq 0$$

Problem 1.18

$$\text{Minimize} \quad -11x_1 + 13x_2 - 15x_3$$
$$\text{s.t.} \quad x_2 + x_3 = 7$$
$$x_1 - x_3 \leq 3$$
$$x_1 \text{ unrestricted}$$
$$x_2 \geq 5, x_3 \geq 0$$

$$\text{Let} \quad x_1 = x_1^+ - x_1^-$$
$$\text{where} \quad x_1^+ = x_1 \qquad \text{if } x_1 > 0$$
$$= 0 \qquad \text{otherwise}$$
$$x_1^- = -x_1 \qquad \text{if } x_1 < 0$$
$$= 0 \qquad \text{otherwise}$$

Change objective function to maximize.

$$\text{Let} \quad x_2' = x_2 - 5 \geq 0 \quad (x_2 = x_2' + 5)$$

Finally, rewrite constraint 1 as 2 inequalities:

$$\text{Maximize} \quad +11(x_1^+ - x_1^-) - 13(x_2' + 5) + 15x_3$$
$$\text{s.t.} \quad x_2' + x_3 \leq 2$$
$$-x_2' - x_3 \leq -2$$
$$(x_1^+ - x_1^-) - x_3 \leq 3$$
$$x_1^+, x_1^-, x_2', x_3 \geq 0$$

Problem 1.19

$$\text{Let} \quad x_2' = 15 - x_2 \geq 0 \quad (x_2 = 15 - x_2')$$
$$\text{Maximize} \quad x_1 + (15 - x_2') + x_3$$
$$\text{s.t.} \quad -x_1 + (15 - x_2') \geq 8$$
$$x_1 - (15 - x_2') + x_3 \leq 2$$
$$x_1, x_2', x_3 \geq 0$$
$$\text{Maximize} \quad x_1 - x_2' + x_3 + 15$$
$$\text{s.t.} \quad x_1 + x_2' \leq 7$$
$$x_1 + x_2' + x_3 \leq 17$$
$$x_1, x_2', x_3 \geq 0$$

CHAPTER 2

Problem 2.3

Step 1:

Input parameters:	number of beverages n (say 3), number of food items m (say 4), cost of each item c_i, upper bound on daily consumption of each item u_i
Decision variables:	whether or not to select each beverage y_i (binary); $i = 1, \ldots, n$ (3), how much of each beverage and food to consume x_i (continuous ≥ 0), $i = n + m$ (7)
Constraints:	total beverage consumed must equal L, total food consumed must equal W, cannot drink more than two types of beverage, upper bound on amount of each beverage and food item
Objective:	minimize total cost

Step 2:

Item	Amount/Day (oz)	Cost/oz	Daily Limitation (oz)
Water	x_1	c_1	u_1
Tea	x_2	c_2	u_2
Milk	x_3	c_3	u_3
Bread	x_4	c_4	u_4
Rice	x_5	c_5	u_5
Cereal	x_6	c_6	u_6
Apple	x_7	c_7	u_7

Indicator variables for beverages:

$$\text{Water}: \quad y_1 = 1 \quad \text{if water consumed}$$
$$= 0 \quad \text{otherwise}$$

$$\text{Tea}: \quad y_2 = 1 \quad \text{if tea consumed}$$
$$= 0 \quad \text{otherwise}$$

$$\text{Milk}: \quad y_3 = 1 \quad \text{if milk consumed}$$
$$= 0 \quad \text{otherwise}$$

$y_1 + y_2 + y_3 \leq 2$ (cannot consume more than two types of beverages)

$$\text{Minimize}: \quad \sum_{i=1}^{7} c_i x_i$$

$$\text{s.t.} \qquad x_1 + x_2 + x_3 = L$$
$$x_4 + x_5 + x_6 + x_7 = W$$
$$y_1 + y_2 + y_3 \leq 2$$
$$x_1 \leq y_1 u_1$$
$$x_2 \leq y_2 u_2$$
$$x_3 \leq y_3 u_3$$
$$x_i \leq u_i, \quad i = 4, 5, 6, 7$$
$$x_1, x_2, x_3, x_4, x_5, x_6, x_7 \geq 0$$
$$y_1, y_2, y_3 \geq 0 \text{ or } 1$$

Problem 2.4

$$\text{Let} \quad Y_{ij} = 1 \quad \text{if stock } i \text{ is bought in year } j$$
$$= 0 \quad \text{otherwise}$$

$$\text{Max}: \quad 90(Y_{11} + Y_{12} + Y_{13}) + 120(Y_{21} + Y_{22} + Y_{23}) + 100(Y_{31} + Y_{32} + Y_{33})$$
$$+ 80(Y_{41} + Y_{42} + Y_{43}) + 130(Y_{51} + Y_{52} + Y_{53})$$

s.t.
$$10Y_{11} + 15Y_{21} + 12Y_{31} + 9Y_{41} + 13Y_{51} \leq 45$$
$$20Y_{12} + 15Y_{22} + 25Y_{32} + 15Y_{42} + 10Y_{52} \leq 60$$
$$15Y_{13} + 20Y_{23} + 20Y_{33} + 15Y_{43} + 10Y_{53} \leq 50$$
$$Y_{ij} = (0, 1) \quad i = 1, \ldots, 5; j = 1, 2, 3$$

Problem 2.6

Let Y_i = number of nurses of schedule type i, $i = 1, \ldots, 5$.

(a)
$$A = \begin{matrix} 1 & 0 & 1 & 0 & 0 \\ 1 & 0 & 0 & 0 & 0 \\ 0 & 0 & 1 & 1 & 0 \\ 0 & 1 & 0 & 0 & 1 \\ 0 & 1 & 0 & 1 & 1 \\ 0 & 1 & 0 & 1 & 0 \\ 1 & 0 & 1 & 0 & 0 \end{matrix}$$

(b)
$$\text{Min}: \quad 525Y_1 + 470Y_2 + 550Y_3 + 500Y_4 + 425Y_5$$
s.t.
$$Y_1 + Y_3 \geq 20$$
$$Y_1 + Y_5 \geq 25$$
$$Y_3 + Y_4 \geq 26$$
$$Y_2 + Y_5 \geq 26$$
$$Y_2 + Y_4 + Y_5 \geq 30$$
$$Y_2 + Y_4 \geq 30$$
$$Y_1 + Y_3 \geq 35$$
$$Y_1, Y_2, Y_3, Y_4, Y_5 \geq 0 \text{ and integer}$$

Problem 2.7

Let X_i be the number of part nurses hired to work (one day) on day i, $i = 1, 2, s. . ., 7$.

$$\text{Min}: \quad 525Y_1 + 470Y_2 + 550Y_3 + 500Y_4 + 425Y_5 + 150X_1$$
$$+ 150X_2 + 150X_3 + 150X_4 + 150X_5 + 150X_6 + 150X_7$$
$$\text{s.t.} \quad Y_1 + Y_3 + X_1 \geq 20$$
$$Y_1 + Y_5 + X_2 \geq 25$$
$$Y_3 + Y_4 + X_3 \geq 26$$
$$Y_2 + Y_5 + X_4 \geq 26$$
$$Y_2 + Y_4 + Y_5 + X_5 \geq 30$$
$$Y_2 + Y_4 + X_6 \geq 30$$
$$Y_1 + Y_3 + X_7 \geq 35$$
$$Y_1 - 4Y_1' \geq 0$$
$$Y_2 - 4Y_2' \geq 0$$
$$Y_3 - 4Y_3' \geq 0$$
$$Y_4 - 4Y_4' \geq 0$$
$$Y_5 - 4Y_5' \geq 0$$
$$MY_1' + MY_4' - X_1 \geq 0$$
$$MY_1' + MY_5' - X_2 \geq 0$$
$$MY_3' + MY_4' - X_3 \geq 0$$
$$MY_2' + MY_5' - X_4 \geq 0$$

$$MY_2' + MY_4' + MY_5' - X_5 \geq 0$$
$$MY_2' + MY_4' - X_6 \geq 0$$
$$MY_1' + MY_3' - X_7 \geq 0$$
$$Y_1, \dots, Y_5 \geq 0 \text{ and integer}$$
$$X_1, \dots, X_7' \geq 0 \text{ and integer}$$
$$Y_1', \dots, Y_5' \text{ binary}$$

Problem 2.11

$$\text{Let} \quad X_{ij} = \text{amount shipped from DC}_i \text{ to retail partner } j$$
$$i = 1, \dots, 20; j = 1, \dots, 500$$

$$\text{Let} \quad Y_{ij} = 1 \quad \text{if } DC_i \text{ is used to supply partner } j$$
$$= 0 \quad \text{otherwise}$$

$$\text{Min}: \quad \sum_{i=1}^{20}\sum_{j=1}^{500}(C_{ij}*X_{ij} + f_i*Y_{ij})$$

$$\text{s.t.} \quad \sum_{i=1}^{20}X_{ij} = d_j \qquad\qquad j = 1,\ldots,500$$

$$X_{ij} \leq M*Y_{ij} \qquad\qquad i = 1,\ldots,20; j = 1,\ldots,500 \quad (M = \sum_{j=1}^{500}d_j)$$

$$X_{ij} \geq 0 \qquad\qquad i = 1,\ldots,20; j = 1,\ldots,500$$
$$Y_{ij} = (0,1) \qquad\qquad i = 1,\ldots,20; j = 1,\ldots,500$$

Problem 2.13

Let X_{ij}^k be the amount of commodity k (beverage type k) shipped from DC_i to partner j, $k = 1, 2, 3,$ and 4.

Furthermore, assume if any quantity of any commodity is shipped from DC_i to partner j, the fixed cost f_i is incurred (but not repeated for $k = 1, 2, 3, 4$) as in Problem 2.11.

$$\text{Min}: \quad \sum_{k=1}^{4}\sum_{i=1}^{20}\sum_{j=1}^{500}(C_{ij}*X_{ij}^k + f_i*Y_{ij})$$

$$\text{s.t.} \quad \sum_{i=1}^{20}X_{ij}^k = d_j^k \qquad\qquad j = 1,\ldots,500; k = 1,\ldots,4$$

$$\sum_{k=1}^{4}X_{ij}^k \leq M*Y_{ij} \qquad\qquad i = 1,\ldots,20; j = 1,\ldots,500 \quad (M = \sum_{k=1}^{4}\sum_{j=1}^{500}d_j^k)$$

$$X_{ij}^k \geq 0 \qquad\qquad i = 1,\ldots,20; j = 1,\ldots,500; k = 1,\ldots,4$$
$$Y_{ij} = (0,1) \qquad\qquad i = 1,\ldots,20; j = 1,\ldots,500$$

CHAPTER 3

Problem 3.2

$$\text{Given}: \quad A \quad T$$
$$B \quad F$$
$$C \quad F$$
$$D \quad T$$

(1) $C \cap [(A \cup B) \rightarrow D]U \sim A$

                               $F \cap T \cup F$

                               $F \cup F$

                               F

(2) $(A \cap D) \cup [C \leftrightarrow (B \cup D) \cup F]$

                               $T \cup F \leftrightarrow T \cup F$

                               $T \cup F \leftrightarrow T$

                               $T \cup F$

                               T

(3) $D \cup \{A \rightarrow [(C \cap A) \cup B]U \sim D\} \cap (C \cap B)$

                               $T \cup \{T \cap F\}$

                               $T \cup F$

                               T

Problem 3.4

(a)   Want $x_1 = 0$ or $x_2 = 0$ (or both)

                       Let $M = \max\{x_1, x_2\}$

                               $y \varepsilon \{0, 1\}$

                               $x_1 \leq M * y$

                               $x_2 \leq M * (1-y)$

                               $x_1, x_2 \geq 0$

Check:

                    if $y = 1$ then $x_1 = M$ and $x_2 = 0$
                    if $y = 0$ then $x_1 = 0$ and $x_2 = M$

Note: This generalizes to $g_1(\bar{x})^* g_2(\bar{x}) = 0$.

(b)

$$\text{Let } y_1 = 1 \qquad \text{if } 0 < x \le a$$
$$= 0 \qquad \text{otherwise}$$
$$y_2 = 1 \qquad \text{if } a < x \le b$$
$$= 0 \qquad \text{otherwise}$$
$$x_1 = x \qquad \text{if } 0 < x \le a$$
$$= 0 \qquad \text{otherwise}$$
$$x_2 = x \qquad \text{if } a < x \le b$$
$$= 0 \qquad \text{otherwise}$$
$$y_1 + y_2 \le 1$$
$$x_1 \le a y_1$$
$$x_2 \le b y_2$$
$$x_1 \ge 0$$
$$x_2 > a y_2$$
$$y = 3y_1 + 2x_1 - 5y_2 + 3x_2$$
$$y_1, y_2 \varepsilon (0,1); x_1, x_2 \ge 0$$

Problem 3.6

$$\text{Let } \quad Y_i = 1 \qquad \text{if stock } i \text{ is purchased} \quad i = 1, \ldots, 6$$
$$= 0 \qquad \text{otherwise}$$
$$\text{and} \quad C_i = \text{profit of holding stock } i \quad i = 1, \ldots, 6$$
$$\text{and} \quad Z_k = 0, 1 \qquad k = 1, \ldots 2$$

Objective function will be to Max : $\sum_{i=1}^{6} C_i * Y_i$.
Constraints 1 and 2 may be expressed as

$$2 \le Y_1 + Y_2 + Y_3 + Y_4 + Y_5 + Y_6 \le 4$$

Constraint 3: $Y_3 + Y_5 \le 1$.
Constraint 4: Either $Y_1 + Y_2 + Y_3 + Y_4 = 2$ or $Y_3 + Y_4 + Y_5 + Y_6 \ge 2$.
Let M be a large constant. Represent above as

$$\text{Either} \quad Y_1 + Y_2 + Y_3 + Y_4 \le 2 \text{ and} -Y_1 - Y_2 - Y_3 - Y_4 \le -2$$
$$\text{or} \qquad -Y_3 - Y_4 - Y_5 - Y_6 \le -2$$

$$Y_1 + Y_2 + Y_3 + Y_4 - 2 \le M * Z_1$$
$$-Y_1 - Y_2 - Y_3 - Y_4 + 2 \le M * Z_1$$
$$-Y_3 - Y_4 - Y_5 - Y_6 + 2 \le M * (1 - Z_1)$$

Constraint 5: If $Y_4 = 1$, then $Y_1 = 1$ as well, "not $Y_4 = 1$" is "$Y_4 = 0$."

$$
\begin{aligned}
\text{Either} \quad & Y_4 = 0 \text{ or } Y_1 = 1 \text{ (or both)} \\
\text{Either} \quad & Y_4 \leq 0 \text{ or } Y_1 \geq 1 \\
\text{Either} \quad & Y_4 - 0 \leq 0 \text{ or } 1 - Y_1 \leq 0 \\
\text{Either} \quad & Y_4 - 0 \leq M * Z_2 \quad 1 - Y_1 \leq M * (1 - Z_2)
\end{aligned}
$$

Check:

$$
\begin{aligned}
&\text{If } Z_2 = 0, \text{ then } Y_4 = 0 \text{ and } 1 - Y_1 \leq M \text{ (always true)} \\
&\text{If } Z_2 = 1, \text{ then } Y_4 \leq M \text{ and } Y_1 = 1 \\
&\qquad (Y_4 \text{ can be } 0 \text{ or } 1)
\end{aligned}
$$

Note: A much simpler expression is just $Y_4 \leq Y_1$ ($Y_4 - Y_1 \leq 0$).

Final model is

$$
\begin{aligned}
\text{Max}: \quad & C_1 Y_1 + C_2 Y_2 + C_3 Y_3 + \cdots + C_n Y_n \\
\text{s.t.} \quad & Y_1 + Y_2 + Y_3 + Y_4 + Y_5 + Y_6 \leq 4 \\
& -Y_1 - Y_2 - Y_3 - Y_4 - Y_5 - Y_6 \leq -2 \\
& Y_3 + Y_5 \leq 1 \\
& Y_1 + Y_2 + Y_3 + Y_4 \leq 2 + M * Z_1 \\
& -Y_1 - Y_2 - Y_3 - Y_4 \leq -2 + M * Z_1 \\
& -Y_3 - Y_4 - Y_5 - Y_6 \leq 2 + M(1 - Z_1) \\
& Y_4 \leq M * Z_2 \\
& -Y_1 \leq 1 + M(1 - Z_2) \\
& Y_1, \ldots, Y_6, Z_1, Z_2 = (0, 1)
\end{aligned}
$$

Problem 3.10

Let $x_j = $ start time of job $j; j = 1, \ldots, n$

$x_j + p_j = $ end time of job j

Desire $x_j + p_j \leq d_j$ but may not be possible; $j = 1, \ldots, n$

Define tardy time $= t_j = d_j - (x_j + p_j) = d_j - x_j - p_j$

To minimize total tardy time:

$$\text{Min} : \sum_{j=1}^{n} t_j$$

$t_j = d_j - x_j - p_j$

For all $i, j, i = 1, \ldots, n; j = 1, \ldots, n;$ and $i \neq j$

Either $x_i + t_i \leq x_j$ or $x_j + t_j \leq x_i$

that is, $x_i - x_j + t_i \leq My_k$

$x_j - x_i + t_j \leq M(1 - y_k)$

where $y_k \, \varepsilon \, (0, 1); \; k = 1, \ldots, \binom{n}{2}$

To minimize the total number of tardy jobs:

$$\text{Let} \quad y_j = 1 \quad \text{if } x_j + p_j > d_j$$
$$= 0 \quad \text{otherwise}$$

$$\text{Min} : \sum_{j=1}^{n} y_j$$

s.t. if $x_j + p_j \leq d_j$, then $y_j = 0$

which implies : "not $x_j + p_j \leq d_j$ or $y_j = 0$

that is, $-x_j - p_j + d_j \leq M_j z_j$ or $y_j \leq M(1 - z_j)$,

where $M_j = \max\{-x_j - p_j + d_j, y_j\}$ and $z_j \varepsilon \{0, 1\}$

Problem 3.11

Breakpoints are at $a_1 = 0°$, $a_2 = 40°$, $a_3 = 100°$, and $a_4 = 200°$.
Any point on the line segment (t) is
$t = \lambda_1 0 + \lambda_2 40 + \lambda_3 100 + \lambda_4 200$
where $\lambda_1 + \lambda_2 + \lambda_3 + \lambda_4 = 1$ and at most two consecutive λ_i are > 0:

$$\lambda_1 \leq y_1$$
$$\lambda_2 \leq y_1 + y_2$$
$$\lambda_3 \leq y_2 + y_3$$
$$\lambda_4 \leq y_3$$
$$y_i = 0, 1 \text{ for } i = 1, 2, 3$$
$$y_1 + y_2 + y_3 = 1$$

Model

$$f(t) = \lambda_1 f(0) + \lambda_2 f(40) + \lambda_3 f(100) + \lambda_4 f(200)$$

$$= \lambda_1 + 0.08\lambda_2 + 0.02\lambda_3 + 0.32\lambda_4$$

s.t. $\quad \lambda_1 \leq y_1$

$$\lambda_2 \leq y_1 + y_2$$

$$\lambda_3 \leq y_2 + y_3$$

$$\lambda_4 \leq y_3$$

$$\lambda_1 + \lambda_2 + \lambda_3 + \lambda_4 = 1$$

$$y_1 + y2 + y3 = 1$$

$$\lambda_j \geq 0 \quad j = 1, 2, 3, 4$$

$$y_i = 0, 1 \quad i = 1, 2, 3$$

Problem 3.12

Product bundling model with three customer segments $i = 1, 2, 3$: $n_1 = 300, n_2 = 240$, and $n_3 = 600$.

Let x_j = price of bundle $j, \quad j = 1, 2, 3, 4$
and y_{ij} = 1, if customer segment i purchases bundle j
$\quad\quad = 0,$ otherwise
and s_i = consumer surplus achieved by segment i in the "bundle j" chosen

Note: Both y_{ij} and s_i assume customer i will purchase only *one* "bundle" to achieve

$$s_i = \text{Max}_j \{r_{ij} - x_j\}$$

though we all know a customer may order two bundles, for example, HB plus drink, but not combo.

Finally, let $z_{ij} = y_{ij}{}^* x_j, i = 1, 2, 3$ and $j = 1, 2, 3, 4$.

MIP Model

$$\text{Max}:\quad 300(z_{11}+z_{12}+z_{13}+z_{14})+240(z_{21}+z_{22}+z_{23}+z_{24})$$
$$+600(z_{31}+z_{32}+z_{33}+z_{34})$$

s.t. $y_{11}+y_{12}+y_{13}+y_{14}=1$

$y_{21}+y_{22}+y_{23}+y_{24}=1$

$y_{31}+y_{32}+y_{33}+y_{34}=1$

$s_1 \geq r_{11}-x_1$

$s_1 \geq r_{12}-x_2$

$s_1 \geq r_{13}-x_3$

$s_1 \geq r_{14}-x_4$

$s_2 \geq r_{21}-x_1$

$s_2 \geq r_{22}-x_2$

$s_2 \geq r_{23}-x_3$

$s_2 \geq r_{24}-x_4$

$s_3 \geq r_{31}-x_1$

$s_3 \geq r_{32}-x_2$

$s_3 \geq r_{33}-x_2$

$s_3 \geq r_{34}-x_2$

$s_1 = r_{11} * y_{11}-z_{11}+r_{12}*y_{12}-z_{12}+r_{13}*y_{13}-z_{13}+r_{14}*y_{14}-z_{14}$

$s_2 = r_{21}*y_{21}-z_{21}+r_{22}*y_{22}-z_{22}+r_{23}*y_{23}-z_{23}+r_{24}*y_{24}-z_{24}$

$s_3 = r_{31}*y_{31}-z_{31}+r_{32}*y_{32}-z_{32}+r_{33}*y_{33}-z_{33}+r_{34}*y_{34}-z_{34}$

$z_{ij} \leq x_j \quad i=1,2,3; j=1,2,3,4$

$z_{ij} \leq r_{ij}*y_{ij} \quad i=1,2,3; j=1,2,3,4$

$z_{ij} \geq x_j-(1-y_{ij})*M_j \quad i=1,2,3; j=1,2,3,4$

where M_j is an upper bound on $x_j, j=1,2,3,4$

$y_{ij}=(0,1) \quad i=1,2,3; j=1,2,3,4$

$x_j \geq 0 \quad j=1,2,3,4$

$z_{ij} \leq 0 \quad i=1,2,3; j=1,2,3,4$

CHAPTER 4

Problem 4.1

By graphing each feasible region, the student can see that the first one is the better
LP formulation for the problem, since P_1(feasible region of the first formulation)
$\subset P_2$(feasible region of the second graph).

Problem 4.4

(1) $$2y_1 + 7y_2 - 3y_3 + 6y_4 - 9y_5 + y_6 \leq -12$$

(2) $$y_1 - 2y_2 + y_3 + 4y_4 + 2y_5 - 3y_6 \leq 13$$

$$1 \leq y_1 \leq 4$$
$$0 \leq y_2 \leq 7$$
$$4 \leq y_3 \leq 10$$
$$2 \leq y_4 \leq M \quad \text{where } M \text{ is a large constant}$$
$$0 \leq y_5 \leq 2$$
$$0 \leq y_6 \leq M$$
$$y_j \text{ integer } j = 1, \ldots, 6$$

Iteration 1

Constraint (1)	Bound Tightens?
$y_1: u_1 = 1/2^*[-12 - 7(0) - 6(2) - 1(0) + (10) + 9(2)] = 12$	No
$y_2: u_2 = 1/7^*[-12 - 2(1) - 6(2) - 1(0) + 3(10) + 9(2)] = [3.14] = 3$	Yes
$y_3: l_3 = 1/-3^*[-12 - 2(1) - 7(0) - 6(2) - 1(0) + 9(2)] = [2.67] = 3$	No
$y_4: u_4 = 1/6^*[-12 - 2(1) - 7(0) - 1(0) + 3(10) + 9(2)] = [5.66] = 5$	Yes
$y_5: l_5 = 1/-9^*[-12 - 2(1) - 7(0) - 6(2) - 1(0) + 3(10)] = [4/-9] = 0$	No
$y_6: u_6 = 1/1^*[-12 - 2(1) - 7(0) - 6(2) + 3(10) + 9(2)] = 22$	Yes

Updated bounds are $y_2 \leq 3$, $y_4 \leq 5$, and $y_6 \leq 22$.

Constraint (2)	Bound Tightens?
y_1: $u_1 = 1/1^*[13 - 1(4) - 4(2) - 2(0) + 2(3) + 3(22)] = 73$	No
y_2: $l_2 = 1/-2^*[13 - 1(1) - 1(4) - 4(2) - 2(0) + 3(22)] = -33$	No
y_3: $u_3 = 1/1^*[13 - 1(1) - 1(4) - 4(2) - 2(0) + 3(22)] = 76$	No
y_4: $u_4 = 1/4^*[13 - 1(1) - 1(4) - 2(0) + 2(3) + 3(22)] = 20$	No
y_5: $u_5 = 1/2^*[13 - 1(1) - 1(4) - 4(2) + 2(3) + 3(22)] = 36$	No
y_6: $l_6 = 1/-3^*[13 - 1(1) - 1(4) - 4(2) - 2(0) + 2(3)] = -2$	No

Iteration 2

Constraint (1): Exactly as above. No changes beyond those identified in Iteration 1.
Constraint (2): Exactly as above. No bound changes. Stop.

New Model is

$$2y_1 + 7y_2 - 3y_3 + 6y_4 - 9y_5 + y_6 \leq -12$$
$$y_1 - 2y_2 + y_3 + 4y_4 + 2y_5 - 3y_6 \leq 13$$
$$1 \leq y_1 \leq 4$$
$$0 \leq y_2 \leq 3$$
$$4 \leq y_3 \leq 10$$
$$2 \leq y_4 \leq 5$$
$$0 \leq y_5 \leq 2$$
$$0 \leq y_6 \leq 22$$
$$y_j \text{ integer } j = 1, \ldots, 6$$

Problem 4.11

This is a pure 0–1 IP.
Constraint (1): $5x_1 + x_2 + 3x_3 - 2x_4 + x_5 - 3x_6 \leq 9$.
Row bound method:

$$u_1 = 5(1) + 1(1) + 3(1) - 2(0) + 1(1) - 3(0) = 10$$
$$l_1 = 5(0) + 1(0) + 3(0) - 2(1) + 1(0) - 3(1) = -5$$
$$-5 \leq 9 \leq 10$$

so no new conclusion, where $9 = b_1$.

Constraint (2): $2x_1 - 2x_2 + x_3 + x_4 - 2x_5 + x_6 \leq 6$.

$$u_2 = 2(1) - 2(0) + 1 + 1 - 2(0) + 1 = 5$$

Now since $b_2 = 6 > u_2 = 5$, constraint 2 is redundant.
Constraint (3): $x_1 + x_2 - x_3 - x_4 + 2x_5 - x_6 \geq 2$.
Must be rewritten: $-x_1 - x_2 + x_3 + x_4 - 2x_5 + x_6 \leq -2$.

$$u_3 = -1(0) - 1(0) + 1(1) + 1(1) - 2(0) + 1(1) = 3$$
$$l_3 = -1(1) - 1(1) + 1(0) + 1(0) - 2(1) + 1(0) = -4$$

Because $-4 \leq -2 \leq 3$, no new conclusion.
Constraint (4): $2x_1 + x_2 - 2x_3 + 3x_4 - x_5 + x_6 \geq 8$.
Must be rewritten: $-2x_1 - x_2 + 2x_3 - 3x_4 + x_5 - x_6 \leq -8$.

$$U_4 = -2(0) - 1(0) + 2(1) - 3(0) + 1(1) - 1(0) = 3$$
$$L_4 = -2(1) - 1(1) + 2(0) - 3(1) + 1(0) - 1(1) = -7$$

Now $-8 < L_4 = -7$, so constraint is infeasible and must be removed, or entire program is infeasible.
Constraints (1) and (3) remain in the model.

Problem 4.15

To generate knapsack cut, need $a_j > 0$.

$$\text{Let} \quad y_2' = 1 - y_2 (y_2 = 1 - y_2')$$
$$\text{and} \quad y_5' = 1 - y_5 (y_5 = 1 - y_5')$$

Transformed constraint is

$$3y_1 - (1 - y_2') + 2y_3 + 4y_4 - 3(1 - y_5') \leq 5$$
$$3y_1 + y_2' + 2y_3 + 4y_4 + 3y_5' \leq 9$$

$\{y_1, y_4, y_5'\}$ is a knapsack cover because $3*1 + 0 + 0 + 4*1 + 3*1 = 10 > 9$, but $(0, 0, 0, 1, 1)$, $(1, 0, 0, 0, 1)$, and $(1, 0, 0, 1, 0)$ are each feasible.
Hence, $n_g = 3$.

$$\text{So, } y_1 + y_4 + y_5' \leq 2$$
$$\text{that is, } y_1 + y_4 + (1 - y_5) \leq 2$$
$$y_1 + y_4 - y_5 \leq 1$$

Problem 4.17

Let $x_2' = 1000x_2$. Then the model becomes

$$\text{Max}: \quad z = 2x_1 + 3x_2' - x_3$$
$$\text{s.t.} \quad 21x_1 - 5x_2' \leq 13$$
$$-11x_1 + x_3 \leq 9$$
$$x_2' + 4x_3 \geq 17$$
$$x_1, x_2', x_3 \geq 0$$

CHAPTER 5

Problem 5.7

Optimal matching of employees is (1, 6), (2, 7), (3, 8), and (4, 5), with $z^* = 6$.

Problem 5.9

Let x_1 = number of 8ft lengths sold uncut

x_2 = number of 8ft lengths cut into two 4ft lengths

x_3 = number of 14ft lengths cut into 10ft and 4ft lengths

x_4 = number of 14ft lengths cut into 12ft lengths

x_5 = number of 14ft lengths cut into 10ft lengths

x_6 = number of 16ft lengths cut into 12ft and 4ft lengths

Note: It can be shown that other cutting combinations are unprofitable, but if they are included as variables in the model, their optimal value will be zero.
Optimal cutting plan is $x^* = (200, 0, 60, 40, 0, 40)$ with $z^* = \$14{,}800$.

CHAPTER 7

Problem 7.9

$$r(\mathbf{A}) = 2$$
$$r(\mathbf{A} : \boldsymbol{b}) = 2$$

Hence, it is a consistent system with $r = 2 < n = 3$.
Thus, the given system has infinite number of solutions.

Problem 7.10

If $(125/92, 4/23, 91/92) = \bar{\mathbf{x}}^*$, then
$z^* = 2^*(125/92) - 3^*(4/23) + 10^*(91/92) = (250 - 48 + 910)/92$
$= 1112/92 = 12.087$.
First primal constraint: $-3x_1 + x_2 + 9x_3 + x_{s1}$.
$-3^*(125/92) + (4/23) + 9^*(91/92) = 460/92 = 5$. Hence, $x_{s1}* = 0$.
Second primal constraint: $x_1 - 2x_2 + x_3 + x_{s2} = 2$.
$(125/92) - 2^*(4/23) + (91/92) = (125 - 32 + 91)/92 = 184/92 = 2$. Hence, $x_{s2}^* = 0$.
Third primal constraint: $6x_1 + 5x_2 + 2x_3 + x_{s3} = 11$.
$6^*(125/92) + 5^*(4/23) + 2^*(91/92) = (750 + 80 + 182)/92 = 1012/92 = 11$.
Hence, $x_{s3}* = 0$.
By complementary slackness at optimal:

$$z^* = 12.087 = w^*$$
$$x_1^* = 125/92 \quad u_{s1}^* = 0$$
$$x_2^* = 4/23 \quad u_{s2}^* = 0$$
$$x_3^* = 91/92 \quad u_{s3}^* = 0$$
$$x_{s1}^* = 0 \quad u_1^* = ?$$
$$x_{s2}^* = 0 \quad u_2^* = ?$$
$$x_{s3}^* = 0 \quad u_3^* = ?$$

At optimal dual:

$$5u_1 + 2u_2 + 11u_3 = 1112/92$$
$$-3u_1 + u_2 + 6u_3 = 2$$
$$u_1 - 2u_2 + 5u_3 = -3$$
$$9u_1 + u_2 + 2u_3 = 10$$
$$u_1, u_2, u_3 \geq 0$$

Using $\mathbf{A}^T\mathbf{u} = \mathbf{c}$, the student should verify $u^* = 12.1$.

Problem 7.11

Dual is

$$\text{Min}: \quad w = -5u_1 + 17u_2 + 5u_3$$
$$\text{s.t.} \quad -2u_1 + 5u_2 + 2u_3 \geq 11$$
$$u_1 + 4u_2 \geq -13$$
$$4u_1 - u_2 + u_3 \geq 7$$
$$-5u_1 - u_3 \geq 9$$
$$u_1, u_2, u_3 \geq 0$$

Problem 7.12

Dual is

$$\text{Max}: \quad w = 5u_1 + 17u_2 + 5u_3$$
$$\text{s.t.} \quad 2u_1 + 5u_2 - 2u_3 \leq 11$$
$$-u_1 + 4u_2 \leq -13$$
$$4u_1 - u_2 + u_3 \geq 7$$
$$-5u_1 - u_3 = -9$$
$$u_1 \geq 0$$
$$u_2 \leq 0$$
$$u_3 \text{ unrestricted}$$

Problem 7.13

Given:

$$x_1^* = 7$$
$$x_2^* = 10$$
$$x_3^* = x_4^* = 0$$
$$x_5^* = 6$$
$$z^* = 94$$

Primal constraint 1: $x_1 + 2x_2 + 3x_3 + x_4 - 3x_5 + x_{s1} = 9$.
$7 + 2^*10 + 3^*10 + 0 - 3^*6 = 27 - 18 = 9$. Hence, $x_{s1*} = 0$.
Similarly, substituting in primal constraints 2 and 3, $x_{s2*} = 0$ and $x_{s3*} = 0$.
By complementary slackness at optimal:

$$z^* = 94 = w^*$$
$$x_1^* = 7 \qquad u_{s1}^* = 0$$
$$x_2^* = 10 \qquad u_{s2}^* = 0$$
$$x_3^* = 0 \qquad u_{s3}^* = ?$$
$$x_4^* = 0 \qquad u_{s4}^* = ?$$
$$x_5^* = 6 \qquad u_{s5}^* = 0$$
$$x_{s1}^* = 0 \qquad u_1^* = ?$$
$$x_{s2}^* = 0 \qquad u_2^* = ?$$
$$x_{s3}^* = 0 \qquad u_3^* = ?$$

At optimal dual:

$$9u_1 + 10u_2 + 11u_3 = 94$$
$$u_1 + 2u_2 - 3u_3 = 4$$
$$2u_1 - u_2 + 2u_3 = 3$$
$$-3u_1 + u_2 + 2u_3 = 6$$

Using $\mathbf{A}^T\mathbf{u} = \mathbf{c}$, the student should verify $u^* = 94.04 = w^*$.

CHAPTER 8

Problem 8.1

No, because it is not convex.

Problem 8.3

After graphing the problem, we find the direction of min $-2x_1 + x_2$ and hence conclude that the problem is unbounded ($z \to -\infty$).

Problem 8.5

(1)

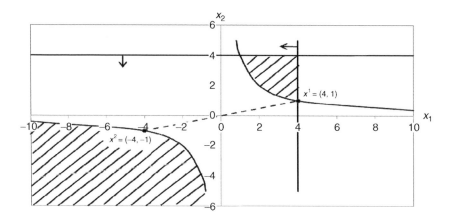

Let $x^1 = (4, 1), x^2 = (-4, -1)$. It can be seen from the graph that all $\alpha x^1 + (1 - \alpha)x^2$ are not in S for each α, $0 \le \alpha \le 1$. Therefore, S is not a convex set.

(2)

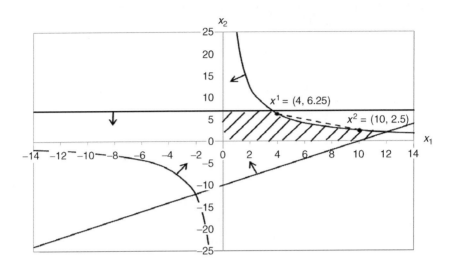

Let $x^1 = (4, 6.25)$, $x^2 = (10, 2.5)$. It can be seen from the graph that all $\alpha x^1 + (1 - \alpha) x^2$ are not in S for each α, $0 \leq \alpha \leq 1$. Therefore, S is not a convex set.

(3)

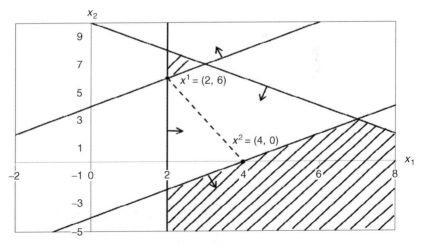

Let $x^1 = (2, 6)$, $x^2 = (4, 0)$. It can be seen from the graph that all $\alpha x^1 + (1 - \alpha)x^2$ is not in S for each α, $0 \leq \alpha \leq 1$. Therefore, S is not a convex set.

Problem 8.12

Extreme points: (0, 0, 0), (0, 0, 1), (0, 1, 0), (1, 0, 1).
There is a single simplex, the entire region.

Problem 8.16

After graphing the problem as shown below, it can be seen that there are three extreme
points; their exact coordinates are as follows:
$2x_1 + x_2 = 7$ and $x_2 = 0$, hence, $x^1 = (3.5, 0)$.
$x_1 - x_2 = 5$ and $x_2 = 0$, hence, $x^2 = (5, 0)$.
$-3x_1 + x_2 = 3$ and $2x_1 + x_2 = 7$, hence, $x^3 = (0.8, 5.4)$.
Now, to find the extreme directions:

$$d^1 + d^2 = 1$$

$$-3d^1 + d^2 \leq 0$$

$$d^1 - d^2 \leq 0$$

$$-2d^1 - d^2 \leq 0$$

$$d^1 \geq 0$$

$$d^2 \geq 0$$

Substituting $d^2 = 1 - d^1$:

$$d^1 \geq \frac{1}{4}$$

$$d^1 \leq \frac{1}{2}$$

$$d^1 \geq -1 \text{ (redundant)}$$

$$d^1 \geq 0 \text{ (redundant)}$$

$$d^1 \leq 1 \text{ (redundant)}$$

So, $1/4 \leq d^1 \leq 1/2$, $1/2 \leq d^2 \leq 3/4$.
Considering $d^1 + d^2 = 1$ and $-3d^1 + d^2 = 0$. Hence, $d^1 = 1/4$ and $d^2 = 3/4$.
Considering $d^1 + d^2 = 1$ and $d^1 - d^2 = 0$. Hence, $d^1 = 1/2$ and $d^2 = 1/2$.

So, extreme points of D (extreme directions) are $d^1 = \begin{bmatrix} 1/4 \\ 3/4 \end{bmatrix}$ and $d^2 = \begin{bmatrix} 1/2 \\ 1/2 \end{bmatrix}$, as shown
in the following figure.

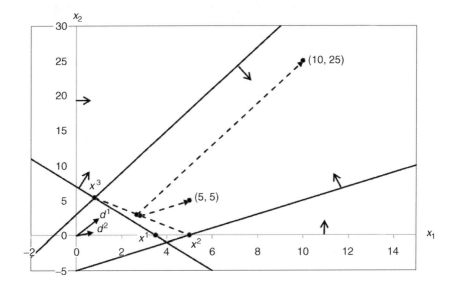

Any point in the polyhedron can be represented by extreme points and extreme directions as

$$x = \sum_{i=1}^{3} \alpha_i x^i + \sum_{j=1}^{2} \beta_j d^j$$

where

$$\sum_{i=1}^{3} \alpha_i = 1, \alpha_i \geq 0 \quad i = 1, 2, 3$$

$$\beta_j \geq 0 \qquad j = 1, 2$$

$$\begin{bmatrix} 10 \\ 25 \end{bmatrix} = \alpha x^2 + (1-\alpha)x^3 + \beta d^1 \text{ (all other } \alpha_i, \beta_j = 0)$$

$$\begin{bmatrix} 10 \\ 25 \end{bmatrix} = \alpha \begin{bmatrix} 5 \\ 0 \end{bmatrix} + (1-\alpha) \begin{bmatrix} 0.8 \\ 5.4 \end{bmatrix} + \beta \begin{bmatrix} 1/4 \\ 3/4 \end{bmatrix}$$

Therefore, $\alpha = 4/9$ and $\beta = 88/3$.

$$\begin{bmatrix} 5 \\ 5 \end{bmatrix} = \alpha x^2 + (1-\alpha)x^3 + \beta d^1 \text{ (all other } \alpha_i, \beta_j = 0)$$

$$\begin{bmatrix} 5 \\ 5 \end{bmatrix} = \alpha \begin{bmatrix} 5 \\ 0 \end{bmatrix} + (1-\alpha) \begin{bmatrix} 0.8 \\ 5.4 \end{bmatrix} + \beta \begin{bmatrix} 1/2 \\ 1/2 \end{bmatrix}$$

Therefore, $\alpha = 23/48$ and $\beta = 35/8$.

Note: The starting point for the ray extending to (10, 25) is (8/3, 3); the starting point for the ray extending to (5, 5) is (2.8125, 2.8125).

Problem 8.18

After graphing the problem, it can be seen that the constraint of $x_1 - 2x_2 \leq 6$ is not necessary in the description of facets, while all others are necessary. Because (6, 0) is a degenerate extreme point and constraint 2 is not on the edge of the facet.

CHAPTER 9

Problem 9.4

In canonical form,

$$\text{Max}: \quad -z + 2x^1 + 3x^2 - 2x^3$$
$$(1) \qquad \text{s.t.} \quad x_1 + x_2 + x_3 + s_1 = 1$$

$$(2) \qquad x_1 - 2x_2 + 2x_3 + s_2 = 2$$
$$x_1, x_2, x_3, s_1, s_2 \geq 0$$

With $n = 5$ variables and two equations, a basic feasible solution will have two basic variables (and typically nonzero) and the other three nonbasic with value zero. Substituting (1/3, 1/3, 1/3) in constraints (1) and (2) yield $s_1 = 0$ and $s_2 = 5/3$. The fact that $s_2 \neq 0$ is an argument that (1/3, 1/3, 1/3) could not be basic.

Problem 9.8

Canonical form:

$$\text{Max}: \quad z - x_1 - 2x_2$$
$$(1) \qquad \text{s.t.} \quad 2x_1 + 5x_2 = 21$$

$$(2) \qquad x_1 - x_2 + s_1 = 10$$
$$x_1, x_2, s_1 \geq 0$$

Adjoining an artificial variable to (1)

$$\text{Max}: \quad z - x_1 - 2x_2$$
$$\text{s.t.} \quad 2x_1 + 5x_2 + x^a = 21$$
$$x_1 - x_2 + s_1 = 10$$
$$x_1, x_2, s_1, x^a \geq 0$$

Phase 1 objective: Max $-z^a + x^a = 0$.

Basic Variable	$-z^a$	x_1	x_2	x^a	s_1	RHS
$-z^a$	1	0	0	1	0	0
x^a	0	2	5	1	0	21
s_1	0	1	-1	0	1	10

The coefficient of x^a in row 0 is nonzero. Row 0−row 1 yields

Basic Variable	$-z^a$	x_1	x_2	x^a	s_1	RHS
$-z^a$	1	-2	-5	0	0	-21
x^a	0	2	5	1	0	21
s_1	0	1	-1	0	1	10

Let x_2 be the entering variable and x^a be the leaving variable.

Basic Variable	$-z^a$	x_1	x_2	x^a	s_1	RHS
$-z^a$	1	0	0	1	0	0
x_2	0	2/5	1	1/5	0	21/5
s_1	0	7/5	0	1/5	1	71/5

Hence, x^a is driven out of basis.
Phase 2 objective: Max $z - x_1 - 2x_2 = 0$.

Basic Variable	z	x_1	x_2	x^a	s_1	RHS
z	1	-1	-2	0	0	0
x_2	0	2/5	1	1/5	0	21/5
s_1	0	7/5	0	1/5	1	71/5

It is not yet in canonical form because the coefficient of x_2 in row 0 is not 0. Let row $0 = $ row 0 + row 1^*2.

Basic Variable	z	x_1	x_2	x^a	s_1	RHS
z	1	$-1/5$	0	2/5	0	42/5
x_2	0	2/5	1	1/5	0	21/5
s_1	0	7/5	0	1/5	1	71/5

Let x_1 be the entering variable, and since $\min\{(21/5)/(2/5), (71/5)/(7/5)\} = 71/7$, so let the leaving variable be s_1.

Basic Variable	z	x_1	x_2	x^a	s_1	RHS
z	1	0	0	3/7	1/7	73/7
x_2	0	0	1	1/7	$-2/7$	1/7
s_1	0	1	0	1/7	5/7	71/7

The optimum solution is $(71/7, 1/7)$ with an objective value of $73/7$. In decimal form, the optimum solution is $(10.14, 0.14)$ with an objective solution of 10.43.

Problem 9.18

Let the price for beef, dog food, bread, bones, and chicken be c_j ($j = 1, \ldots, 5$), respectively. Uno consumes each type of food in the quantity of x_{1j} lb, respectively. Similarly, Dos and Tres consume x_{2j} and x_{3j} for each type of food.

(i)

$$\text{Minimize} \quad z = \sum_{i=1}^{3}\sum_{j=1}^{5} c_j x_{ij} \quad (c_j \text{ is shown in table in the text})$$

s.t. $\quad x_{13} \geq 0.5$

$x_{14} \geq 1.7$

$x_{15} \geq 1.9$

$x_{22} \geq 1.5$

$x_{23} \geq 0.3$

$x_{24} \geq 0.9$

$x_{25} \geq 0.1$

$x_{31} \geq 1.5$

$x_{32} \geq 0.9$

$x_{33} \geq 0.8$

$x_{34} \geq 0.6$

$x_{35} \geq 0.2$

$x_{23} + x_{25} \geq 2.5$

$x_{12} + x_{14} + x_{15} \geq 2.7$

$x_{31} + x_{33} \geq 2.6$

$x_{ij} \geq 0$

(ii) The LINGO model and solution by using sets are shown on the next page. As shown in the solution, the minimum cost is \$16.71 per day. Uno consumes 0 lb beef, 0 lb dog food, 2.4 lb bread, 0.9 lb bones, and 0.1 lb chicken. Tres consumes 1.5 lb beef, 0.9 lb dog food, 1.1 lb bread, 0.6 lb bones, and 0.2 lb chicken.

```
MODEL:
SETS:
DOGTYPE/UNO,DOS,TRES/;
FOODTYPE/BEEF,DOGFOOD,BREAD,BONES,CHICKEN/:C;
LINK(DOGTYPE,FOODTYPE):B,X;
ENDSETS
DATA:
C = 2.5 1 0.8 1.2 1.6;
B = 0 0 0.5 1.7 1.9
    0 1.5 0.3 0.9 0.1
    1.5 0.9 0.8 0.6 0.2;
ENDDATA
MIN = @SUM(FOODTYPE(J):C(J)*@SUM(DOGTYPE(I):X(I,J)));
@FOR(DOGTYPE(I):@FOR(FOODTYPE(J):X(I,J)>=B(I,J)));
X(2,3) + X(2,5) >=2.5;
X(1,2) + X(1,4) + X(1,5) >=2.7;
X(3,1) + X(3,3) >= 2.6;
END
```

```
  Global optimal solution found at iteration:                    12
  Objective value:                                          16.71000
                    Variable          Value      Reduced Cost
                    C( BEEF)        2.500000          0.000000
                 C( DOGFOOD)        1.000000          0.000000
                   C( BREAD)       0.8000000          0.000000
                   C( BONES)        1.200000          0.000000
                 C( CHICKEN)        1.600000          0.000000
               B( UNO, BEEF)        0.000000          0.000000
            B( UNO, DOGFOOD)        0.000000          0.000000
              B( UNO, BREAD)       0.5000000          0.000000
              B( UNO, BONES)        1.700000          0.000000
            B( UNO, CHICKEN)        1.900000          0.000000
               B( DOS, BEEF)        0.000000          0.000000
            B( DOS, DOGFOOD)        1.500000          0.000000
              B( DOS, BREAD)       0.3000000          0.000000
              B( DOS, BONES)       0.9000000          0.000000
            B( DOS, CHICKEN)       0.1000000          0.000000
              B( TRES, BEEF)        1.500000          0.000000
           B( TRES, DOGFOOD)       0.9000000          0.000000
             B( TRES, BREAD)       0.8000000          0.000000
             B( TRES, BONES)       0.6000000          0.000000
```

B(TRES, CHICKEN)	0.2000000	0.000000
X(UNO, BEEF)	0.000000	2.500000
X(UNO, DOGFOOD)	0.000000	1.000000
X(UNO, BREAD)	0.5000000	0.000000
X(UNO, BONES)	1.700000	0.000000
X(UNO, CHICKEN)	1.900000	0.000000
X(DOS, BEEF)	0.000000	2.500000
X(DOS, DOGFOOD)	1.500000	0.000000
X(DOS, BREAD)	2.400000	0.000000
X(DOS, BONES)	0.9000000	0.000000
X(DOS, CHICKEN)	0.1000000	0.000000
X(TRES, BEEF)	1.500000	0.000000
X(TRES, DOGFOOD)	0.9000000	0.000000
X(TRES, BREAD)	1.100000	0.000000
X(TRES, BONES)	0.6000000	0.000000
X(TRES, CHICKEN)	0.2000000	0.000000

Row	Slack or Surplus	Dual Price
1	16.71000	-1.000000
2	0.000000	0.000000
3	0.000000	0.000000
4	0.000000	-0.8000000
5	0.000000	-1.200000
6	0.000000	-1.600000
7	0.000000	0.000000
8	0.000000	-1.000000
9	2.100000	0.000000
10	0.000000	-1.200000
11	0.000000	-0.8000000
12	0.000000	-1.700000
13	0.000000	-1.000000
14	0.3000000	0.000000
15	0.000000	-1.200000
16	0.000000	-1.600000
17	0.000000	-0.8000000
18	0.9000000	0.000000
19	0.000000	-0.8000000

CHAPTER 10

Problem 10.4

(a) This 6×4 matrix does not meet the second sufficient condition of Theorem 10.1, because there are two columns with more than two nonzero elements. So, the theorem cannot be used to prove TU. This is not an interval matrix either. The only way to check TU is to compute the determinant of 209 submatrices. After computation, we can show that each of these has a determinant of $-1, 0,$ or 1. Hence, matrix is TU.

(b) The matrix is not TU because the entire matrix is $-2 \neq 0$, 1, or -1.

(c) The matrix is not TU. It is easy to find a 3*3 submatrix with determinant not equal to -1, 0, and 1, such as

$$\begin{pmatrix} 1 & 0 & 1 \\ -1 & -1 & 0 \\ 0 & -1 & -1 \end{pmatrix} \quad \text{with determinant} = 2$$

Problem 10.6

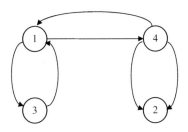

Problem 10.13

```
MODEL:
SETS:
NODES/A,B,C,D,E,F,G,H/:DEMAND;
ARCS(NODES,NODES)/A,B A,C B,D C,D B,E C,G D,E D,G D,F E,H F,H
G,H/:CAPACITY,
FLOW,COST;
ENDSETS
MIN=@SUM(ARCS:COST*FLOW);
@FOR(NODES(I):@SUM(ARCS(I,J):FLOW(I,J))-
@SUM(ARCS(K,I):FLOW(K,I))=DEMAND(I));
@FOR(ARCS:FLOW<=CAPACITY);
DATA:
DEMAND = 25 20 20 5 10 0 -30 -50;
CAPACITY = 30 44 28 19 42 20 26 27 16 23 29 41;
COST = 3 5 3 5 5 4 6 7 2 8 9 7;
ENDDATA
END

Global optimal solution found at iteration:        8
Objective value:                           916.0000
```

Variable	Value	Reduced Cost
FLOW(A, B)	21.00000	0.000000
FLOW(A, C)	4.000000	0.000000

```
FLOW ( B, D)    28.00000    0.000000
FLOW ( C, D)    4.000000    0.000000
FLOW ( B, E)    13.00000    0.000000
FLOW ( C, G)    20.00000    0.000000
FLOW ( D, G)    21.00000    0.000000
FLOW ( D, F)    16.00000    0.000000
FLOW ( E, H)    23.00000    0.000000
FLOW ( F, H)    16.00000    0.000000
FLOW ( G, H)    11.00000    0.000000
```

CHAPTER 11

Problem 11.3

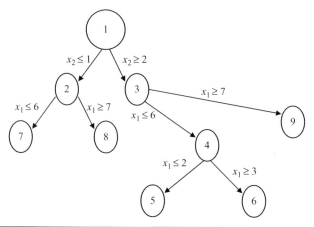

Node	x_1	x_2	x_3	x_4	z
1	7.11	1.56	0	5.44	35.33
2	6.83	1	0	5.17	32.83
3	6.67	2	0	5	34
4	6	2.67	0	4.33	32
5 (Fathom—integer)	6	2	0	5	32
6 (Fathom by lower bound)	5.67	3	0	4	31
7 (Fathom—integer)	6	1	0	6	32

Problem 11.4

- Node 3 is fathomed due to "integer feasibility," and ultimately optimality as well.
- Node 4 is fathomed due to "infeasibility."
- Node 6 is fathomed due to "optimality."
- Node 7 is fathomed due to bound: $37 < \_z = 40$.

Problem 11.11

(b)After obtaining the final tableau, the student should verify

$$\text{Row 1}: \quad x_2 + 0.1x_3 - 0.37x_4 = 0.27$$
$$\text{Row 2}: \quad x_1 + 0.1x_3 - 0.03x_4 = 1.93$$

Hence,

$$\text{Cut 1}: \quad -0.1x_3 - 0.63x_4 - 0.27$$
$$\text{Cut 2}: \quad -0.1x_3 - 0.97x_4 - 0.93$$

Problem 11.12

$$\text{Max} \quad 3y_1 + y_2 + 2y_3 + 3y_4$$
$$\text{s.t.} \quad -y_1 + 3y_2 + y_3 - 2y_4 + s_1 = 17$$
$$7y_1 + 3y_3 + y_4 + s_2 = 23$$
$$y_1 + 2y_2 + s_3 = 11$$
$$y_2 + 3y_4 + s_4 = 13$$
$$-0.118y1 - 0.265s_1 - 0.912s_2 - 0.206s_4 + s_5 = -0.147$$
$$-0.222y_1 - 0.222s_2 - 0.889s_4 - 0.111s_5 + s_7 = -0.778$$
$$-0.065y_1 - 0.258s_1 - 0.839s_2 - 0.226s_7 + s_8 = -0.968$$
$$-0.25y_1 - 0.25s_2 - 0.875s_7 - 0.125s_8 + s_9 = -0.75$$
$$-0.571y1 - 0.571s_2 - 0.286s_8 - 0.714s_9 + s_{10} = -0.714$$
$$-0.99y_1 - 0.99s_2 - 0.99s_9 - 0.01s_{10} + s_{11} = -0.99$$
$$-0.5s_9 - 0.499s_{10} - 0.001s_{11} + s_{12} = -0.5$$
$$y_1, y_2, y_3, y_4 \text{ integer}$$

Initial Solution: The Tableau

Row		Y1	Y2	Y3	Y4	S1	S2	S3	S4	RHS
1	ART	1.471	0.000	0.000	0.000	0.059	0.647	0.000	0.824	26.588
2	Y2	-0.882	1.000	0.000	0.000	0.265	-0.088	0.000	0.206	5.147
3	Y3	2.235	0.000	1.000	0.000	0.029	0.324	0.000	-0.088	6.794
4	S3	2.765	0.000	0.000	0.000	-0.529	0.176	1.000	-0.412	0.706
5	Y4	0.294	0.000	0.000	1.000	-0.088	0.029	0.000	0.265	2.618

Cut 1: Using y_2 as source row in initial tableau, we get

$$-0.118y_1 - 0.265s_1 - 0.912s_2 - 0.206s_4 + s_5 = -0.147$$

Cut 2: Using y_3 as source row in Tableau 2, we get

$$-0.222y_1 - 0.222s_2 - 0.889s_4 - 0.111s_5 + s_7 = -0.778$$

Cut 3: Using y_2 as source row in Tableau 3, we get

$$-0.065y_1 - 0.258s_1 - 0.839s_2 - 0.226s_7 + s_8 = -0.968$$

Cut 4: Using y_3 as source row in Tableau 4, we get

$$-0.25y_1 - 0.25s_2 - 0.875s_7 - 0.125s_8 + s_9 = -0.75$$

Cut 5: Using s_4 as source row in Tableau 5, we get

$$-0.571y_1 - 0.571s_2 - 0.286s_8 - 0.714s_9 + s_{10} = -0.714$$

Cut 6: Using s_1 as source row in Tableau 6, we get

$$-0.99y_1 - 0.99s_2 - 0.99s_9 - 0.01s_{10} + s_{11} = -0.99$$

Cut 7: Using s_7 as source row in Tableau 7, we get

$$-0.5s_9 - 0.499s_{10} - 0.001s_{11} + s_{12} = -0.5$$

Tableau 8 produces the optimal integer solution: $z^* = 25$, $y_1 = 0$, $y_2 = 4$, $y_3 = 6$, and $y_4 = 3$.

Problem 11.14

Max $\quad 3y_1 + y_2 + 2y_3 + 3y_4$

Subject to $\quad -y_1 + 3y_2 + y_3 - 2y_4 + s_1 = 17$

$$7y_1 + 3y_3 + y_4 + s_2 = 23$$

$$y_1 + 2y_2 + s_3 = 11$$

$$y_2 + 3y_4 + s_4 = 13$$

$$-0.118y_1 \quad y_2 - 0.265s_1 - 0.015s_2 - 0.206s_4 + s_5 = -0.147$$

$$-0.235y_1 - 0.029s_1 - 0.324s_2 - 0.339s_4 + s_6 = -0.794$$

$$-0.182y_1 - 0.273s_1 - 0.298s_4 - 0.155s_6 + s_7 = -0.363$$

$$-0.333y_1 - 0.333s_4 - 0.142s_6 - 0.664s_7 + s_8 = -0.666$$

$$y_1, y_3, y_4 \text{ integer} \geq 0 \text{ and } y_2 \geq 0$$

After four cuts, the optimal integer solution is $z^* = 25$, $y_1 = 0$, $y_2 = 4$, $y_3 = 6$, and $y_4 = 3$.

CHAPTER 12

Problem 12.3

Given $S = \{y_\varepsilon(0, 1) : 5y_1 + 2y_2 - 3y_3 - y_4 + 4y_5 \leq 6\}$.

(1) $y_1 = 1$ is invalid because it excludes some feasible points in S, such as the origin, $(0, 1, 1, 1, 1)$ and others.

(2) $y_3 = 0$ is invalid because it excludes some feasible points in S, such as $(1, 1, 1, 0, 0)$.

(3) The inequality $y_1 + y_2 + y_5 \leq 2$ is valid because $C = \{1, 2, 5\}$ is a cover and $y_1 + y_2 + y_5 \leq 2$ is the knapsack cover cut.

(4) $y_3 + y_4 \geq 1$ is invalid because it excludes some feasible points in S, such as the origin or $(1, 0, 0, 0, 0)$.

Problem 12.4

(1) $d = 13$.

Hence, $y_1 + y_2 + y_3 + y_4 \geq 3$.

$(0, 0, 0, 30/13)$ is "cut off" by this integer rounding cut.

(4) $d = 2$.

Hence, $y_1 + 3y_3 + 2y_4 + y_5 \leq 12$.

$(25/2, 0, 0, 0)$ is "cut off" by this integer rounding cut.

Problem 12.5

Optimal tableau of the LP relaxation is as follows:

Basic	y_1	y_2	y_3	s_4	s_5	s_6	RHS
z	0.67	0	0	0.33	0	0.67	10
y_2	0.78	1	0	0.22	0	0.11	3.67
s_5	0.11	0	0	−0.11	1	−0.56	0.67
y_3	0.89	0	1	0.11	0	0.56	6.33

Note: Any row may be used:

Using row 1, the C–G cut is $y_2 \leq 3$.

Using row 2, the C–G cut is $3y_1 + 4y_2 - y_3 \leq 14$.

Using row 3, the C–G cut is $-y_2 + y_3 \leq 6$.

Problem 12.8

$$\text{Max} \quad y_1 + y_2 + y_3 - 2x$$
$$\text{Subject } to: \quad 3.1y_1 + 1.3y_2 + 1.4y_3 - x + s_1 = 19.7$$
$$y_1, y_2, y_3 = 0 \text{ integer}$$

The MIR cut is $1.727y_1 + y_2 + y_3 - 1.182x + 0.727s_1 + s_2 = 15$.
The solution is $z^* = 15$ and $y^* = (0, 13, 2, 0)$. Using the Gomory mixed integer cut, $z^* = 15$, and $y^* = (0, 14, 1, 0)$—an alternate optima.

CHAPTER 13

Problem 13.1

$x^* = (0, 4.5, 0.5, 0)$ with $z^* = 4$.

Problem 13.2

$x^* = (2, 16, 1, 0)$ with $z^* = 41$.

CHAPTER 14

Problem 14.1

Yes, both yield the tour 4–1–3–2–5–4.

Problem 14.2

(a) The heuristic where the first customer of each route is the unserved customer nearest the depot produces as follows:

Route	Customers Visited	Tons Delivered	Cost
1	0–11–2–3–0	9	169
2	0–5–6–12–0	10	197
3	0–9–8– 0	10	180
4	0–7–10–0	9	182
5	0–4–0	7	168
6	0–1–0	5	190
7	0–13–0	6	200
Total			1286

The heuristic where the first customer of each route is the unserved customer farthest from the depot produces as follows:

Route	Customer Served	Tons Delivered	Cost
8	0–13–2–0	10	202
9	0–3–4–0	9	214
10	0–1–11–12–0	10	217
4	0–10–7–0	9	192
3	0–8–9–0	10	180
11	0–6–5–0	8	167
Total			1172

(b) Performing the genetic algorithm's crossover operation will create an off-spring that depends on the random number sequence used by the student. Students should compare their crossover solution with the two parent solutions and the offspring generated by classmates. The second solution above is likely close to optimal.

INDEX

Applied Integer Programming: Modeling and Solution, By Der-San Chen, Robert G. Batson, and Yu Dang
Copyright © 2010 John Wiley & Sons, Inc.